Second Edition

MATERIALS and PROCESS SELECTION for ENGINEERING DESIGN

Second Edition

MATERIALS and PROCESS SELECTION for ENGINEERING DESIGN

Mahmoud M. Farag

CRC Press is an imprint of the
Taylor & Francis Group, an **informa** business

CRC Press
Taylor & Francis Group
6000 Broken Sound Parkway NW, Suite 300
Boca Raton, FL 33487-2742

Printed in the United States of America on acid-free paper
10 9 8 7 6 5 4 3 2

International Standard Book Number-13: 978-1-4200-6308-0 (Softcover)

Library of Congress Cataloging-in-Publication Data

Farag, Mahmoud M., 1937-
Materials & process selection for engineering design / Mahmoud M. Farag. -- 2nd ed.
p. cm.
Includes bibliographical references and index.
ISBN 978-1-4200-6308-0 (hardback : alk. paper)
1. Engineering design. 2. Materials. 3. Manufacturing processes. I. Title.

TA174.F26 2007
620.1'1--dc22 2007031387

Visit the Taylor & Francis Web site at
http://www.taylorandfrancis.com

and the CRC Press Web site at
http://www.crcpress.com

Dedication

To
Penelope, Sherif, Sophie, Eamon,
Hisham, Nadia, and Nadine

Contents

PART I Performance of Materials in Service

PART II Relationship between Design, Materials, and Manufacturing Processes

PART III Selection and Substitution of Materials in Industry

PART IV Appendices

Preface

Introducing a new engineering product or changing an existing model involves making designs, reaching economic decisions, selecting materials, choosing manufacturing processes, and assessing its environmental impact. These activities are interdependent and should not be performed in isolation from one another. This is because the materials and processes used in making the product can have a large influence on its design, cost, and performance in service. For example, making a part from injection-molded plastics instead of pressed sheet metal is expected to involve large changes in design, new production facilities, and widely different economic and environmental impact analysis.

Experience has shown that in most industries it is easier to meet the increasing challenge of producing more economic and yet reliable, esthetically pleasing and environmental friendly product if a holistic decision-making approach of concurrent engineering is adopted in product development. With concurrent engineering, materials and manufacturing processes are considered in the early stages of design and are more precisely defined as the design progresses from the concept to the embodiment and finally the detail stage.

The objective of this book is to illustrate how the activities of design, materials and process selection, and economic and environmental analysis fit together and what sort of trade-offs can be made to arrive at the optimum solution when developing a new product or changing an existing model.

The book starts with an introductory chapter that briefly reviews the stages of product development in industry, recycling of materials, and life cycle costing. The subject matter is then grouped into three parts. Part I consists of three chapters that discuss the performance of materials in service. After a review of different types of mechanical failures and environmental degradation, the materials that are normally selected to resist a given type of failure are discussed. Part II consists of three chapters that deal with the effect of materials and manufacturing processes on design. The elements of industrial and engineering design are first explained, followed by a discussion of the effect of material properties and manufacturing processes on the design of components. Part III consists of four chapters that are devoted to the selection and substitution of materials in industry. After a brief review, the economic and environmental aspects of materials and manufacturing processes, several quantitative and computer-assisted methods of screening, comparing and ranking of alternative solutions, and selecting the optimum solution are presented. The final chapter presents five widely different detailed case studies in materials selection and substitution.

The book is written for junior and senior engineering students who have completed the first course in engineering materials; however, first-year students and practicing engineers will also find the subject matter interesting and useful. To enhance the value of the text as a teaching device, a variety of examples and open-ended case studies are given to explain the subject matter and to illustrate its practical application in engineering. Each chapter starts with an introduction, which includes its goals and objectives, and ends with a summary, review questions, suggestions for students' projects, and selected references for further reading. The International System of Units is used throughout the text, but Imperial units are also given wherever possible. Tables of composition and properties of a wide variety of materials, conversion of units, and a glossary of technical terms are included in the appendices. PowerPoint presentations and solution manuals are available to instructors.

Author

Dr. Mahmoud Farag received a BSc degree in mechanical engineering from Cairo University and an MMet and PhD from Sheffield University, United Kingdom. He is currently professor of engineering at the American University in Cairo (AUC).

Dr. Farag's area of academic interest is in engineering materials and manufacturing and he has published three engineering textbooks, edited one book, and written several chapters for engineering books. He has also authored or coauthored about 100 papers in academic journals and conference proceedings on issues related to the effect of microstructure on the behavior of engineering materials. His current research interests include the behavior of nanostructured materials, with emphasis on NiTi alloys, natural fiber-reinforced plastics, and using quantitative methods in selecting materials for engineering applications. In addition to his academic work, Dr. Farag has a wide range of industrial and consulting experience.

Dr. Farag has more than 30 years of teaching experience and has taught a variety of materials courses at different levels ranging from introductory overviews to sophomore/junior students to advanced topics to master's degree students. He has also taught manufacturing courses with a heavy emphasis on how processing affects the properties of materials. One of Dr. Farag's favorite courses, which he created at AUC and has written textbooks for, is materials selection. This is a capstone course for mechanical/materials engineering senior students that integrates economic analysis with the process of product design and material and process selection.

Dr. Farag was a visiting scientist/scholar at the University of Sheffield, United Kingdom; MIT, Cambridge, Massachusetts and University of Kentucky, Lexington, United States; Aachen Technical University, Germany; and Joint Research Center, Commission of the European Communities, Ispra, Italy. He is a member of the American Society of Mechanical Engineers, the Materials Information Society, ASM International, United States; the Institute of Materials, Minerals and Mining, United Kingdom; and the Egyptian Society for Engineers. He is listed in Marquise *Who's Who in the World, Who's Who in Science and Engineering*, and *Who's Who in Finance and Industry*. Dr. Farag is a recipient of the Egyptian State Award for promotion of science and the First Order of Merit in arts and sciences.

1 Product Design and Development in the Industrial Enterprise

1.1 INTRODUCTION

Product design and development involves interdisciplinary activities with contributions from different segments of an industrial enterprise including design, materials and manufacturing, finance, legal, sales and marketing. This is because in addition to satisfying the technical requirements, a successful product should also be esthetically pleasing, safe to use, economically competitive, and compliant with legal and environmental constraints.

The total development effort depends on the complexity of the product, and project teams can consist of a few people working for a few days or weeks on a simple product like a hand tool to several hundred people working for several months or even years on a complex product like a motorcar or an airplane. The cost of development can range from a few hundred dollars for a simple product to millions of dollars for a complex product.

A product usually starts as a concept that, if feasible, develops into a design and then a finished product. While each engineering product has its own individual character and its own sequence of development events, the following seven phases can be identified in a variety of product design and development projects:

1. Identification of needs, feasibility study, and concept selection
2. System-level design, detail design, and selection of materials and processes
3. Testing and refinement
4. Manufacturing the product
5. Launching the product
6. Selling the product
7. Planning for its retirement

The overall goal of this chapter is to introduce the spectrum of activities that are normally involved in different product development phases. The main objectives are to

1. review the main activities of performing a feasibility study and selecting an optimum concept,
2. discuss the main stages of designing and manufacturing a product,

3. discuss the main activities involved in testing and refining a new product and then launching and selling it,
4. analyze the environmental issues that are involved in making a product and in retiring it, and
5. explain the concepts of life cycle costing and the product life cycle.

Several of these activities will be discussed in more detail later in this book.

1.2 FEASIBILITY STUDY AND CONCEPT SELECTION

A statement describing the function, main features, general shape, and essential features of the product is normally followed by a feasibility study that addresses market, technical, economic, social, environmental, safety, and legal issues.

1.2.1 Market Research

Market research involves a survey to evaluate competing products and their main characteristics in addition to identifying the customer needs. Elements of the market research include the following

1. The range of features and technical advantages and disadvantages of the existing products, the mechanism of their operation, and the materials and processes used in making them.
2. Past and anticipated market growth rate and expected market share by value and volume.
3. The number of companies entering and leaving the market over the past few years, and reasons for those movements.
4. The reasons for any modifications that have been carried out recently and the effect of new technology on the product.
5. Patent or license coverage and what improvements can be introduced over the existing products.
6. Profile of prospective customers (income, age, sex, etc.) and their needs in the area covered by the product under consideration.
7. Ranking of customer needs in the order of their importance.
8. Product price that will secure the intended volume of sales.
9. How long will it take for the competition to produce a competitive product?
10. What is the optimum packaging, distribution, and marketing method?

The preceding information is essential for determining the rate of production, plant capacity, and financial and economic evaluation of the proposed product.

1.2.2 Product Specifications

Product specifications give precise and measurable description of the expected product performance based on the qualitative descriptions of the customer needs.

For example, a specification of "the total weight of the product must be less than 5 kg" can be based on the customer need of a "lightweight product" and the observation that the lightest competing product is 5 kg. Similarly, a specification of "average time to unpack and assemble the product is less than 22 min" can be based on a customer need of "the product is easy to assemble" and the observation that the competing product needs 24 min to unpack and assemble.

1.2.3 Concept Generation, Screening, and Selection

Product specifications are then used to develop different product concepts that satisfy customer needs. Some of the concepts may be generated by the development team as novel solutions, but others may be based on existing solutions or patents. The different concepts are then compared to select the most promising option. The Pugh method is useful as an initial concept-screening tool. In this method, a decision matrix is constructed as shown in Table 1.1. Each of the characteristics of a given concept is compared against a base/reference concept and the result is recorded in the decision matrix as (+) if more favorable, (−) if less favorable, and (0) if the same. Concepts with more (+) than (−) are identified as serious candidates for further consideration.

1.2.4 Economic Analysis

The economic analysis section of the feasibility study normally provides an economic model that estimates the development costs, initial investment that will be needed, manufacturing costs, and income that will probably result for each of the selected concepts. The economic analysis also estimates sources and cost of financing based on the rate of interest and schedule of payment. The model should allow for a "what if" analysis to allow the product development team to assess the sensitivity of the product cost to changes in different elements of cost.

TABLE 1.1
Concept Screening Matrix

Selection Criteria	Reference Concept	Concept A	Concept B	Concept C	Concept D
Criterion 1	0	−	+	+	0
Criterion 2	0	+	+	0	0
Criterion 3	0	+	+	+	−
Criterion 4	0	0	0	+	−
Criterion 5	0	0	−	−	+
Total (+)	0	2	3	3	1
Total (−)	0	1	1	1	2
Total (0)	5	2	1	1	2
Net Score	0	1	2	2	−1
Decision	No	Consider	Consider	Consider	No

1.2.5 Selecting an Optimum Solution

The final stage of the feasibility study identifies an optimum solution. Selection is usually based on economics as well as technical specifications, since the product is expected to satisfy the customer needs at an acceptable price. This process involves trade-offs between a variety of diverse factors, such as

- customer needs,
- physical characteristics of size and weight,
- expected life and reliability under service conditions,
- energy needs,
- maintenance requirements and operating costs,
- availability and cost of materials and manufacturing processes,
- environmental impact,
- quantity of production, and
- expected delivery date.

A quantitative method that can be used in concept selection gives weight to product specifications according to their importance to the function of the product and preference of the customer. The total score of each concept is determined by the weighted sum of the ratings of its characteristics, as shown in Table 1.2.

The optimum solution should be acceptable not only to the consumer of the product, but also to the society in general. If other members of the community object to the product, whether for legal or safety reasons, causing harm to the environment, or merely because of social customs or habit, then it may not be successful. This part of the study requires an understanding of the structure and the needs of the society, and any changes that may occur during the intended lifetime of the product. The following case study uses a hypothetical product—the Greenobile—to illustrate how the issues discussed in concept development and feasibility studies may apply in practice.

TABLE 1.2
Concept Selection

		Concept A		Concept B		Concept C	
Product Specifications/ Selection Criteria	**Weight**	**Rating**	**Weighted Rating**	**Rating**	**Weighted Rating**	**Rating**	**Weighted Rating**
Criterion 1	0.1	2	0.2	4	0.4	4	0.4
Criterion 2	0.2	4	0.8	4	0.8	3	0.6
Criterion 3	0.2	4	0.8	4	0.8	4	0.8
Criterion 4	0.3	3	0.9	3	0.9	5	1.5
Criterion 5	0.2	3	0.6	1	0.2	2	0.4
Total score			3.3		3.1		3.7
Rank			Second		Last		First (optimum)

Note: Rating: 5 = excellent, 4 = very good, 3 = good, 2 = fair, 1 = poor.

Case Study 1.1: Developing the Greenobile

A motorcar company is considering the introduction of an inexpensive, environment-friendly two-passenger (two-seater) model. The idea behind this product is based on the statistics that on about 80% of all trips American cars carry no more than two people, and in a little more than 50% of all trips, the driver is alone. Such cars will be predominantly driven in city traffic, where the average vehicle speed is about 55 km/h (30 mph).

Market Research

Market research is carried out through interviews and discussions with focus groups of 8–12 prospective customers. The questions discussed include the following:

1. Frequency of driving the car, how far is each journey on an average, expected distance traveled per year, and expected life of the car
2. Esthetic qualities: main preferences for body styling and look, number of doors, number of wheels, etc.
3. Level of comfort on a bumpy road
4. Ease of handling and parking
5. Safety issues including stability on the road, especially when turning around sharp corners
6. Expected cost

Specifications

The Greenobile specifications were developed based on the market research and consultations among the company departments involved: research and development, design, materials and manufacturing, legal affairs, marketing, sales, and retailers.

Product description:	A two-seater, inexpensive, and environment-friendly motorcar
Primary use:	Driving in the city to get to work or go for shopping
Main customers:	Middle-class families who may already own a family car
Technical specifications:	Mass, up to 500 kg Maximum speed, 90 km/h Speed maintained on 5% gradient, 60 km/h Acceleration time from 0 to 90 km/h, average 20 s
Cost:	Up to $6000
Safety requirements:	Same as a normal sedan car
Engine emission:	Environmental Protection Agency (EPA) test limits or less

Main features: Seating for two passengers
Small boot/trunk accessible from the back

Main business goals: Production starts in 3 years from the approval of final concept

Units produced in the first year (20,000) to increase by 10% every subsequent year

Concept Generation

The Greenobile development team came up with three concepts that satisfy the product specifications listed earlier.

Concept A is a sedan with a hard roof, four wheels, two seats side by side, internal combustion engine, expected life 5 years, expected weight 500 kg, acceleration from 0 to 90 km/h is 15 s, higher level of comfort, expected cost $6000.

Concept B is a sedan with a hard roof, three wheels (one in front and two in the back), two seats side by side, rechargeable battery-operated engine, expected life 4 years, expected weight 450 kg, acceleration from 0 to 90 km/h is 25 s, medium level of comfort, expected cost $5000.

Concept C is a sedan with a movable roof, three wheels (one in front and two in the back), two seats one behind the other, rechargeable battery-operated engine, expected life 4 years, expected weight 400 kg, acceleration from 0 to 90 km/h is 20 s, lower level of comfort, expected cost $4000.

Concept Selection

Table 1.3 gives the concept selection criteria and their relative importance as weights. The weights were developed on the basis of market research. Table 1.4 gives the concept selection based on the weights and the ratings of each concept. The ratings are based on the specifications.

TABLE 1.3
Concept Selection Criteria and Their Relative Importance for Greenobile

Selection Criteria	Weight
Life	0.10
Esthetic qualities	0.20
Level of comfort	0.10
Ease of handling and parking	0.10
Safety	0.20
Cost	0.30

TABLE 1.4
Concept Selection for Greenobile

Selection Criteria	Weight	Concept A Rating	Concept A Weighted Rating	Concept B Rating	Concept B Weighted Rating	Concept C Rating	Concept C Weighted Rating
Life	0.1	5	0.5	4	0.4	4	0.4
Esthetic qualities	0.2	5	1.0	4	0.8	3	0.6
Level of comfort	0.1	5	0.5	4	0.4	3	0.3
Ease of handling and parking	0.1	4	0.4	5	0.5	3	0.3
Safety	0.2	5	1.0	4	0.8	4	0.8
Cost	0.3	3	0.9	4	1.2	5	1.5
Total score			4.3		4.1		3.9
Rank		First		Second		Third	

Note: Rating: 5 = excellent, 4 = very good, 3 = good, 2 = fair, 1 = poor.

FIGURE 1.1 The Greenobile.

Conclusion

Concept A is the optimum solution and Figure 1.1 shows the model of the car developed using the selected concept. Its longer life, better esthetic qualities and comfort, and higher safety outweigh its higher cost. No legal or environmental objections are expected since it meets the safety requirements and its engine emission meets the EPA test limits.

It is expected, however, that the company would adjust the engine to operate as a hybrid (ethanol and benzene) and then consider replacing the internal combustion engine by a battery-operated engine when future development allows sufficient power to be generated by this type of engine.

1.3 SYSTEM-LEVEL DESIGN

The product specifications of the selected concept are translated to system-level design, which gives a broad description of the product's appearance and architecture in addition to technical specifications and functions. Appearance and architecture are in the realm of industrial design, which is concerned with the esthetic qualities (visual and tactile) in addition to product–human interface, ease of use and maintenance, and safety features. Technical specifications and function are in the realm of engineering design, which is concerned with the level and type of technology on which it is based, performance under service conditions, efficiency, energy consumption, and environmental issues.

A simple product that consists of a few parts can usually be easily drawn schematically to illustrate its appearance and function. More complex products, however, need to be divided into subsystems or subassemblies, each of which performs part of the total function of the product. The product architecture in this case can be schematically represented by blocks representing the different subassemblies and how they interact together to perform the total function of the product. At this stage it may be appropriate to perform a make-or-buy decision to determine whether a subassembly is to be manufactured specially for the product or if there is a ready-made standard alternative. Examples of functional subassemblies in a motorcar include the engine, steering and brake system, electrical system, etc.

Each of the subassemblies that need to be manufactured is then divided into components that can be fitted together to perform the function of that subassembly. The function of each of the components is then identified and its critical performance requirements determined. These requirements are then used to define the material performance requirements. The following example illustrates these concepts.

Case Study 1.2: Planning for a New Model of a Household Refrigerator

Based on market survey, a medium-size manufacturer of household refrigerators is considering the production of a new model based on the established concepts of other models already in production by the company. To plan for the manufacture of the new model, decisions need to be made on which components are to be made in-house and which to be bought from outside vendors.

Figure 1.2 gives a breakdown of the refrigerator into major subassemblies, subassemblies, etc. Some of the subassemblies or parts are obviously bought from specialized outside manufacturers, including the motor-pump unit and the electrical system. Some of the subassemblies that will be manufactured in-house include the refrigerator body and door, as well as the cooler and condenser subassemblies. Some parts of the control system, however, may present an option. The company policy to specialize and concentrate its efforts and skills in one basic line rather than diversify may favor buying the control system from a specialized supplier. However, factors such as quality and reliability of supply may rule in favor of manufacturing some parts in-house.

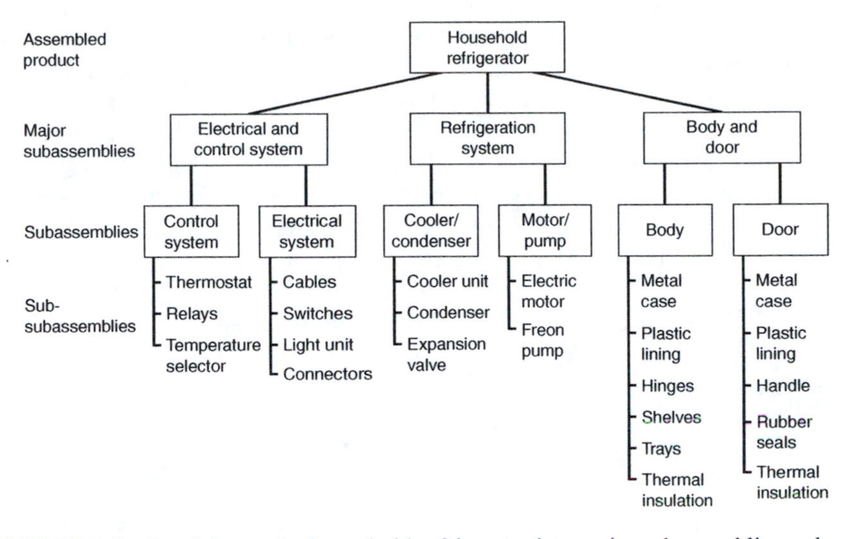

FIGURE 1.2 Breakdown of a household refrigerator into major subassemblies, subassemblies, and sub-subassemblies.

For the subassemblies that will be manufactured in-house, a list of components, detail designs, bill of materials, sequence of manufacturing processes, and estimates of processing and assembly times are normally needed. A master production and purchasing schedule is then prepared to ensure that materials, parts, and subassemblies are available when needed.

1.4 DETAIL DESIGN AND SELECTION OF MATERIALS AND PROCESSES

As progress is made from product specifications to system-level design and then to detail design, the tasks to be accomplished become more narrowly defined. In the detail design stage, the focus is on static and dynamic forces and their effect on the performance of the component under the expected service conditions. This latter task requires thorough knowledge of how materials behave in service and what processes are needed to achieve the final shape of the component. Behavior of materials in service is discussed in Part I and the effect of materials and processes on design is discussed in Part II. A two-step process may be used in developing the final detail design and deciding on the materials and processes: configuration and final detail design.

1.4.1 Configuration (Embodiment) Design

In the configuration, or embodiment, design stage, a qualitative sketch of each part is first developed giving only the order of magnitude of the main dimensions and the main features: wall, bosses, ribs, holes, grooves, etc. The material performance requirements are used to narrow down the field of possible materials and processes to a few promising candidates. In many cases, the different performance requirements that have to be met by a given part present conflicting limitations on the material properties. For example, the material that meets the strength requirements may be difficult to manufacture using the available facilities or the material that resists the corrosive environment may be too expensive. To resolve such problems, compromises or trade-offs have to be made. Quantitative methods of selection can be helpful in ranking the candidate combinations of materials and processes, as discussed in Part III.

At this stage, make-or-buy decisions are made on whether to manufacture the component specially for this product or to use a standard part that is purchased from an external supplier.

1.4.2 Final Detail Design

If more than one combination of material and process prove to be viable, then each of the candidate combinations is used to make a detail design. Each of the detail designs should give complete specification of geometry, tolerance, material treatment, weight, material and manufacturing cost, etc. A final detail design is then selected based on technical performance and economic value.

The design and materials selection for a subassembly that contains several components can be complicated by the fact that a well-matched combination of components needs to be found. It is not sufficient that each individual part is well designed, but the assembled components should function together to achieve the design goals. The issue of successfully matching a group of components should also be addressed when redesigning a component in an existing subassembly. If the material of the new component is too different from the surrounding materials, problems resulting from load redistribution or galvanic corrosion could arise, for example. A detailed account of materials selection and substitution is given in Part III.

1.4.3 Design Reviews

Design reviews represent an important part of each phase of the design process. They provide an opportunity to identify and correct problems before they can seriously affect the successful completion of the design. The design review teams normally have representatives from the materials and manufacturing, quality control, safety, financial, and marketing areas. This ensures that the design is satisfactory not only from the performance point of view but also from the manufacturing, economic, reliability, and marketing points of view.

1.5 TESTING AND REFINEMENT

The testing and refinement phase is normally carried out as part of the R&D function of the company. A first prototype (alpha) is usually built from parts with the same geometry and material as the final product but not necessarily using the same manufacturing processes. Alpha prototypes are tested to ensure that the product works as intended and that it satisfies its main requirements. After modifications, a second prototype (beta) may be built to ensure reliability of the product and to measure its level of performance. Potential customers may also be involved at this stage to incorporate their feedback in making the final product.

1.6 LAUNCHING THE PRODUCT

Launching the product covers the activities of planning and scheduling, manufacturing the product, marketing, and arranging for after-sales services. This stage is best organized on the basis of planning and scheduling schemes, which are drawn to meet the product delivery times, as discussed in this section.

1.6.1 Project Planning and Scheduling

Engineering projects and activities normally have a series of deadlines that are set to meet a final completion or delivery date as part of a contract with penalties for not finishing on time. To avoid delays and in view of the complexity of many of the engineering projects, planning and scheduling should play an important role in project development. The first step in planning is to identify the activities that need to be controlled. The usual way to do that is to start with the entire system and identify the major tasks. These major tasks are then divided into sections, and these in turn are subdivided until all the activities are covered. The following simple example illustrates this process.

Case Study 1.3: Planning for Installation of an Injection Molding Machine

Task

Establish and plan the activities involved in installing and commissioning an injection molding machine for plastics.

Analysis

The activities involved in installing and commissioning the machine can be divided into three major sections:

I. Site preparation
II. Installation
III. Preparation of the machine for production

The foregoing major activities can be divided into the simple activities shown in Table 1.5. The sequence in which the activities should be performed and the time required to complete each activity are also included. Figure 1.3 shows the sequence of the activities on a bar chart or Gantt chart.

TABLE 1.5
Installing and Preparing an Injection Molding Machine for Operation

Major Task	Activity	Description	Immediate Predecessor	Time (h)
I	a	Excavate foundation	—	5
	b	Pour concrete in foundation	a	2
	c	Unpack parts	—	3
II	d	Place machine body on foundation	b, c	2
	e	Level machine body	d	1
	f	Assemble rest of the machine parts	c, e	3
	g	Connect electrical wiring	f	1
	h	Connect cooling water and drainage	f	2
III	i	Install injection molding die	g, h	3
	j	Calibrate temperature controller	i	2
	k	Place plastic pellets in hopper	f	1
	l	Adjust plastic metering device	k	1
	m	Perform experimental runs	j, l	2

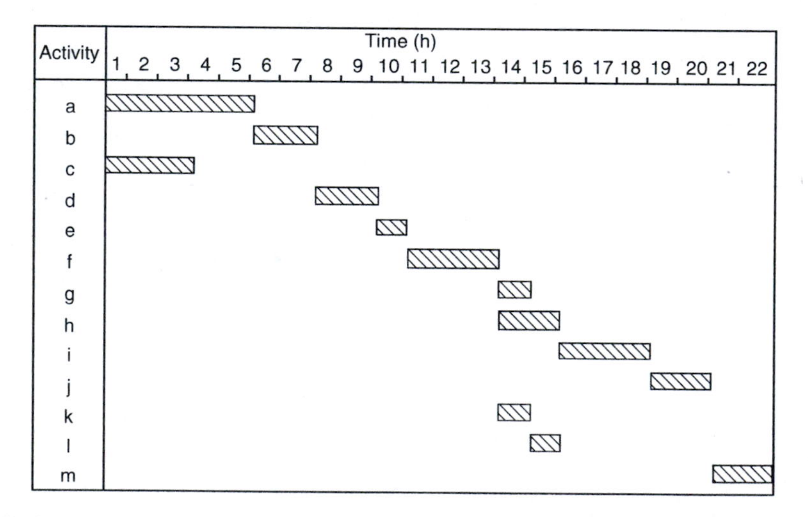

FIGURE 1.3 Bar chart describing the activities of installing and preparing an injection molding machine for operation. See Table 1.5 for a description of the activities.

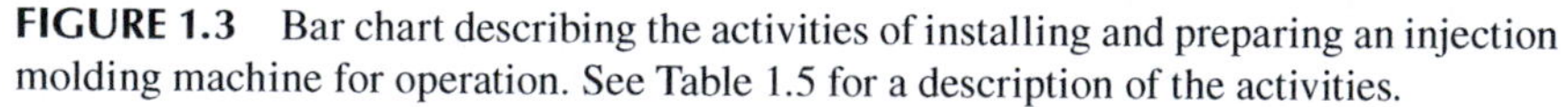

The bar, or Gantt, chart shown in Figure 1.3 is one of the many analytical techniques that have been developed to facilitate the planning and scheduling of a large number of activities that are usually involved in industrial projects. Using network planning models makes it possible to locate the activities that are critical and must be done on time, and the activities that have schedule slack. The critical path method (CPM)

and the program evaluation and review technique (PERT) are widely used network planning models. Some references that give detailed accounts of project planning and scheduling techniques are provided in the bibliography.

1.6.2 Manufacturing

The sequence of manufacturing processes is first established for each part of the product and recorded on a process sheet. The form and condition of the material, as well as the tooling and production machines that will be used are also recorded on the process sheet. An example of a simple process sheet is shown in Figure 1.4. Allocating established standard times and labor costs for each operation, the information in the process sheets is used to estimate the processing time and cost for each part. Further discussion of cost estimation in manufacturing is given in Chapter 8. The information in the process sheets is also used to estimate and order the necessary stock materials; design special tools, jigs, and fixtures; specify the production machines and assembly lines; and plan work schedules and inventory controls.

Before starting large-scale production, a pilot batch is usually made to test the tooling and familiarize the production personnel with the new product and also to identify outstanding problems that could affect the efficiency of production.

1.6.3 Quality Control

Quality control represents an important activity in manufacturing. It could vary from 100% inspection of produced parts to statistical sample inspection, depending

PROCESS SHEET		*Page 1 of 3 pages*
Written by M.M.F. *Date* 1/3/88	*Order no.* 1844 *Date* 1/1/88 *Pcs req'd* 30	*Dwg no.* 12 *Patt. no.* 5
Enters assembly at stage x-23 loader 6	*Part name* 250 mm pulley	
Material condition Gray CI, ASTM A48-74 35, 245 BHN	*Rough weight* 15 kg	*Finish weight* 12 kg

Oper. no.	*Description*	*Set-up time (h)*	*Cycle time (h)*	*Mach. no.*
10	Turn O.D. of body and flanges, face hub speed 200 rpm, feed 0.25 mm per rev tool no. TT-25	0.5	0.5	L-2
20	Turn bore and face other side of hub speed 200 rpm, feed 0.25 mm per rev tool no. TT-25	0.6	0.3	L-2
30	Drill and tap 2 holes, 10 and 12 mm diameter standard metric thread M10 and M12	0.3	0.2	D-1

FIGURE 1.4 Example of a simple process sheet.

on the application and the number of parts produced. In some applications it may be necessary to test subassemblies and assemblies to ensure that the product performs its function according to specifications.

1.6.4 Packaging

Packaging is meant to protect the finished product during its shipping to the consumer. Secondary functions of packaging include advertising and sales appeal, which are important aspects of marketing the product, especially in the case of consumer goods.

1.6.5 Marketing

The marketing personnel should be involved in the various stages of product development to allow them to develop the publicity material that will help in selling the product. In addition to publicity material, installation and maintenance instructions need to be prepared and distributed with the product. Clear installation, operation, and maintenance instructions will make it easier for the user to achieve the optimum performance of the product.

1.6.6 After-Sales Service

Most products require either regular or emergency service during their useful life. The accuracy and speed of delivering the needed service and the availability of spare parts could affect the company's reputation and the sales volume of the product. Feedback from the user is an important factor to be considered in making an improved version of a product and in developing a new product.

1.7 SELLING THE PRODUCT

In a free enterprise, the price of goods and services is ultimately determined by supply and demand. Typical supply and demand curves are illustrated in Figure 1.5. The demand curve shows the relationship between the quantity of a product that customers are willing to buy and the price of the product. The supply curve shows the relationship between the quantity of a product that vendors will offer for sale and the price of the product. The intersection of the two curves determines both the quantity "n" and the price "p" of the product in the free market. The concept of supply and demand is important in engineering economy studies, since proposed ventures frequently involve action that will increase the supply of a product or influence its demand. The effect of such action upon the price at which the product can be sold is an important factor to be considered in evaluating the desirability of the venture.

In some cases, a product is meant to compete directly with the existing products. In such cases, the selling price is already established and the problem is to work backward from it to determine the cost of each of the product elements. This procedure is sometimes called design to cost. An understanding of the elements that make up the cost of a product is, therefore, essential in ensuring that the product will be

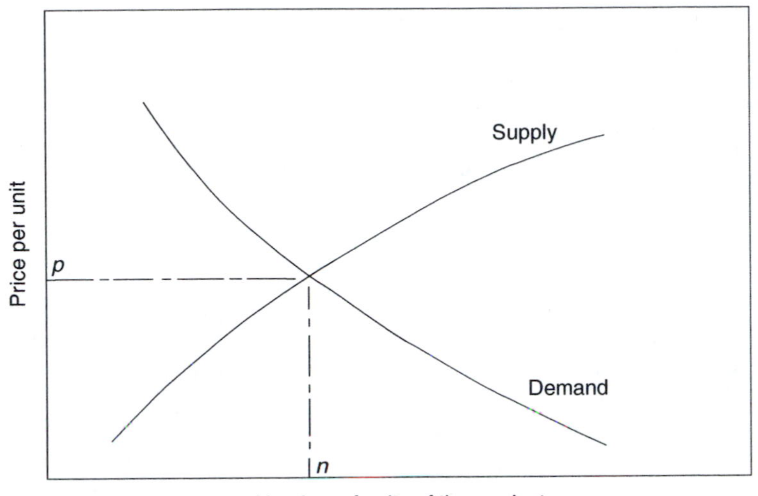

FIGURE 1.5 Schematic representation of typical supply and demand curves.

competitive economically as well as technically. There are numerous examples of products that exhibit excellent performance but possess little economic merit, and consequently have not been adopted commercially. In almost all cases, the value of engineering products lies in their utility measured in economic terms. This means that the selling price has to be low enough to be competitive but sufficiently high for the company to make a profit, as illustrated in Figure 1.6.

1.7.1 Cost of Product Engineering

Figure 1.6 shows the elements of the cost of product engineering, which covers the direct cost of labor and materials required to prepare the design and drawings, to perform development work and experiments, and to make tools or dies for the different parts of the product. The cost of product engineering is usually charged to the number of products to be sold within a given period. Each department has direct-labor and overhead rates per hour for each class of work charged to the product.

1.7.2 Actual Manufacturing Cost

As shown in Figure 1.6, the actual manufacturing cost covers direct materials that form part of the product in measurable quantities. The quantities of direct materials required as well as sizes are normally recorded on bill of materials or on process sheets. The quantities should be gross, which include necessary allowance for waste and scrap. For example, the quantity of bar stock required for a part should be indicated in units of length to cover the length of the part plus the length of the cutoff.

Indirect materials include materials used in quantities too small to be readily identified with units of product. Examples include cutting oils, solders, and adhesives.

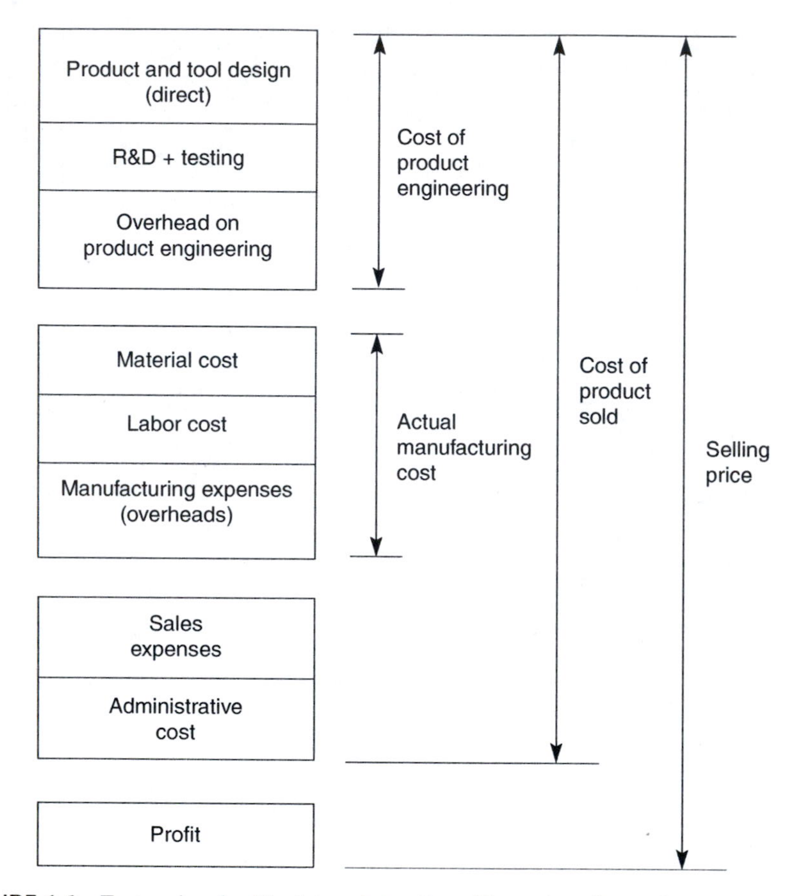

FIGURE 1.6 Factors involved in determining the selling price of a product.

Such items are considered as shop supplies. A credit should be included to account for recycled scrap or by-products that are sold elsewhere. Labor may be defined as the employment costs, wages, and other associated payments of the employees whose effort is involved in the fabrication of the product.

Manufacturing expense, or overhead cost, covers all the other costs associated with running a manufacturing company. Examples of manufacturing expense include salaries and fringe benefits of company executives, accounting personnel, purchasing department, and R&D. Other overhead expenses include depreciation, taxes, insurance, heating, light, power, and maintenance. Chapter 8 discusses the cost of materials and processes in more detail.

1.7.3 Sales Expense and Administrative Cost

The sales expense covers the cost of distribution as well as the cost of advertising and marketing the product. Administrative expenses can be treated in a similar manner to manufacturing overheads.

1.7.4 Selling Price

The cost of product sold, as shown in Figure 1.6, is the total cost of product engineering, manufacturing, selling, and administration. An amount of profit is then added to this total cost to arrive at the selling price. The net profit to the company that results from selling a certain number of units of the product per year can be calculated as

$$\text{Net profit} = (\text{number of units} \times \text{profit per unit}) - \text{income taxes}$$

Income taxes are usually calculated as a certain percent of the total profits.

1.8 PLANNING FOR RETIREMENT OF THE PRODUCT AND ENVIRONMENTAL CONSIDERATIONS

The final step in the product development cycle is disposal of the product when it has reached the end of its useful life. In the past, little attention was paid to this stage and the retired product was just abandoned or used for landfill. However, increasing environmental awareness, increasing cost of energy and raw materials, and tighter legislation have made it necessary for product development teams to consider reuse, ease of scrap recovery, recycling, and disposal costs when designing new products.

1.8.1 Recycling of Materials

Table 1.6 compares the unit energy for production of some metals from the ore (primary metal) and from scrap (secondary metal). The table shows the energy savings that result from recycling, especially in the case of nonferrous metals. Figure 1.7 shows how recycling fits in the total materials cycle.

TABLE 1.6
Energy Used for the Production of Some Engineering Materials

	Primary Metal		Secondary Metal	
Metal	**GJ/t**	**%**	**GJ/t**	**% of Primary Metal**
Mg	370	100	10	2.7
Al	350	100	15	4.3
Ni	150	100	15	10
Cu	120	100	20	17
Zn	70	100	20	29
Pb	30	100	10	33
Steel	35	100	15	43

Note: Data based on Metals Handbook, Desk Edition, ASM, Metals Park, Ohio, 1985, p. 31.5.

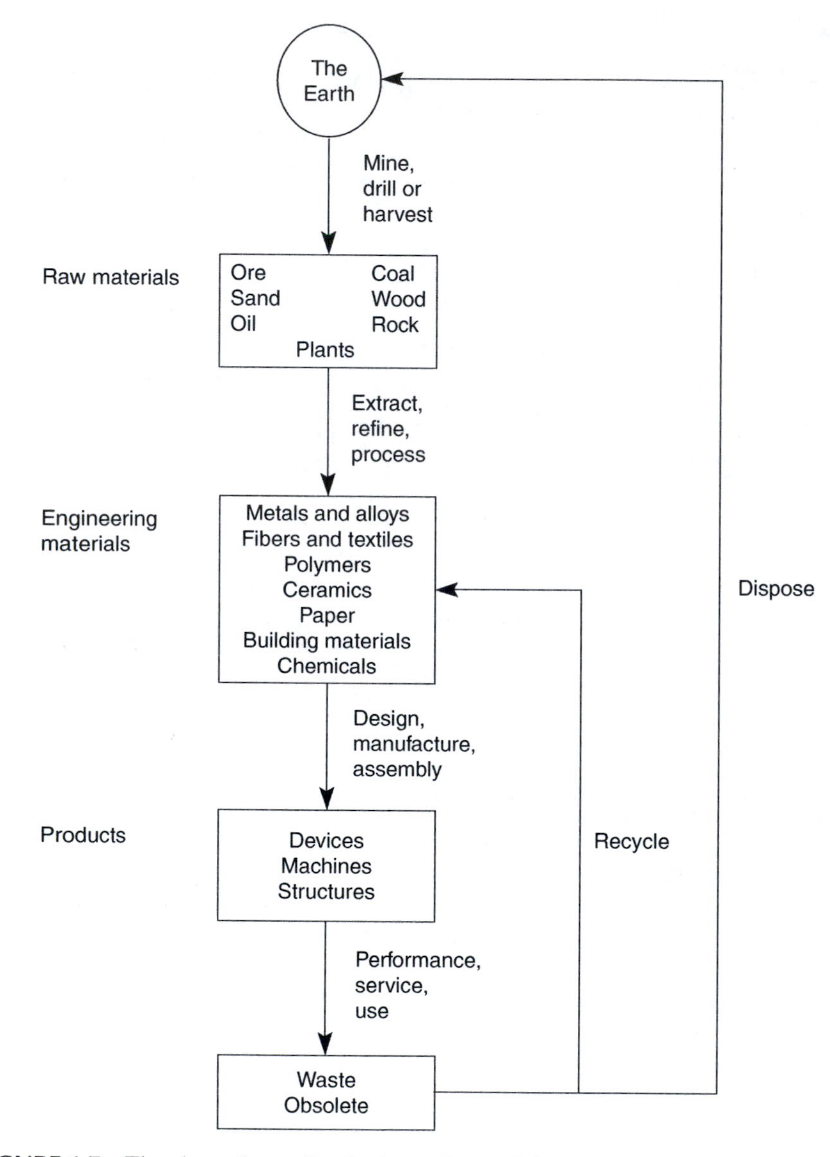

FIGURE 1.7 The place of recycling in the total material cycle.

1.8.2 Sources of Materials for Recycling

The processes involved in recycling materials depend on their type and source. For example, in the manufacturing industries the sources of metallic scrap are mainly turnings and borings from the machine shop, punchings and skeletons from the press shop, trimmings from the forging shop, and sprues and gates from the foundry. Another source for recycling is materials recovered from used products, such as worn-out machinery or discarded packages. Salvaging materials from the latter source is more complex. This is because the different materials have to be collected from various sources and sorted to separate the different materials, which are sent to the appropriate secondary manufacturer.

1.8.3 The Infrastructure for Recycling Packaging Materials

Recycling of materials involves a series of interrelated activities. This is called recycling infrastructure, starting with the individual consumer and ending with material producer mill. For recycling to work effectively, the various activities involved must be coordinated. Collection can be an expensive step, especially in the case of plastic containers that occupy a large volume but a small fraction of the total weight of recycled materials. The method of collection depends on the community and includes curbside programs, voluntary drop-off or buyback centers, and resource recovery facilities.

Curbside collection follows the traditional form of refuse pickup, while voluntary drop-off centers involve consumers bringing the recyclable items. Resource recovery facilities are repositories for all postconsumer solid waste in which recyclables are sorted and the waste is used either to create fuel (refuse-derived fuel) or energy by burning directly at the facility (waste to energy).

1.8.4 Sorting

Identification and sorting of the various materials for recycling is not always easy, and lack of proper identification is often a source of contamination and loss of quality in recycled materials. Identification of materials in the recycling industry is usually based on the following:

- Color: For example, copper is reddish, brasses are light yellow, and zinc alloys are bluish or dark gray.
- Weight: For example, the specific gravity of aluminum and its alloys is about 2.7, steels about 7.8, copper about 8.7, and lead about 11.3.
- Magnetic properties for separating steels from other materials.
- Spark testing: For example, when subjected to high-speed grinding wheel carbon steels produce heavy dense sparks that are white in color with bursts that depend on the carbon content, but nickel-base alloys emit thin short sparks that are dark red in color.

To help consumers in identifying the various packaging materials so that they may participate in sorting, the symbols shown in Figure 1.8 are printed or embossed into the containers.

1.8.5 Scrap Processing

After sorting, recyclable materials are sold to secondary material producers or intermediate processors for remelting and processing into finished products. The processes and equipment used for processing scrap vary widely depending on its nature, which can range from aluminum cans and plastic bottles to large machinery and motorcars.

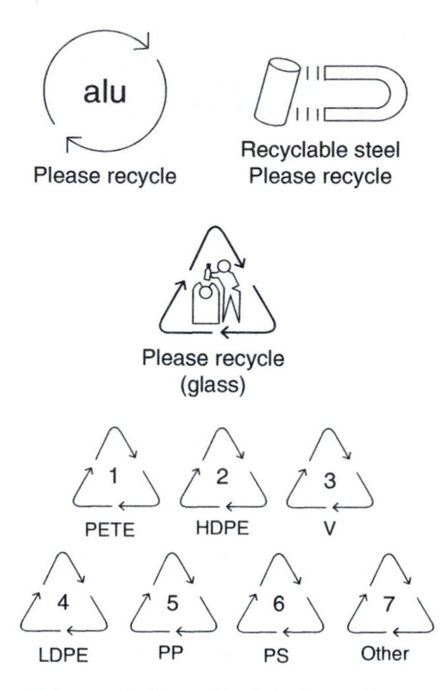

FIGURE 1.8 Examples of the symbols used to help in sorting materials for recycling.

1.8.6 Recyclability of Materials

Recyclability of a material is defined as the ease with which it can be collected, separated, reprocessed, and reused for manufacturing a new product. Generally, metals and alloys have high recyclability since they can be remelted and reused with little or no loss of quality. Glass also has high recyclability as cullet, or broken glass, and easily substitutes for primary materials. However, rubber has lower recyclability since vulcanizing is an irreversible process that precludes reprocessing in the original form. Economic issues related to recycling are discussed in Section 8.12.

1.9 THE PRODUCT LIFE CYCLE

The product life is defined as the length of time between its appearance on the market for the first time and the time when the company decides to stop its production. The life of a product may vary from a few months to several years depending on its nature, competition, economic and social climates, as well as political decisions. Examples of products that normally have a relatively short life cycle are found in the clothing and toy industries where the main determining parameters are fashion and consumer desires. Other examples are found in the electronic and computer industries where technological development is relatively fast and competition pressures are high. In contrast, machines for power generation and production as well as similar heavy equipment have a relatively long life cycle, which is determined mostly by the relatively slow advances in their well-established technology.

The life cycle of a successful product can generally be represented by a curve similar to that shown in Figure 1.9. The sales are low in the introduction stage and

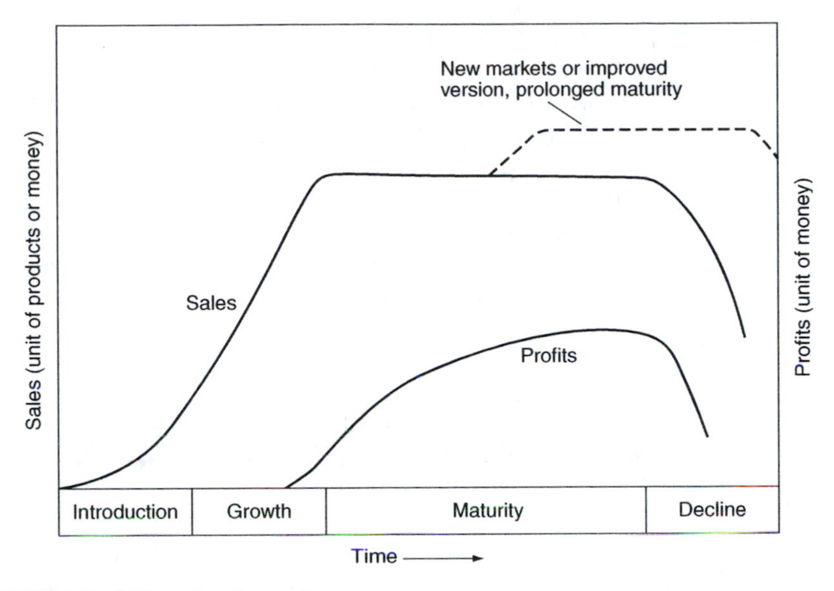

FIGURE 1.9 Life cycle of a product.

gradually increase as the product gradually gains the acceptance of an increasing number of consumers. As the product reaches maturity, the production rates and sales volume reach the design values. While the initiation and growth stages should be as short as possible, the maturity stage should be prolonged. That is because during this stage production rates are most efficient, the investment used in the product development is recovered, and most profits are made.

As shown in Figure 1.9, there is a time lag between sales and profit. Heavier investments in product development will take longer to recover, which means a longer time lag. To prolong the maturity stage, efforts should be made to develop new markets by adopting new marketing strategies. Introducing an improved version of the product can also prolong its life cycle. This can be achieved by introducing design modifications or employing new materials or manufacturing processes. This could improve efficiency or extend the use of the product to new applications and environments.

Eventually, social change, appearance of other competitive products, or technological advances make the product less competitive causing the sales to decrease and the product to reach the decline stage. When the sales volume causes the production to reach uneconomical rates, the product is normally discontinued, thus ending its life cycle.

1.10 SUMMARY

1. Ideally, product development is performed by an interdisciplinary team with representatives from different segments of an industrial enterprise including engineering design, materials and manufacturing, finance, legal, sales, and marketing. This is because in addition to satisfying the technical requirements, a successful product should also be esthetically pleasing,

safe to use, economically competitive, and compliant with legal and environmental constraints.

2. The activities involved in product development include concept development and feasibility study, system-level design, detail design and selection of materials and processes, testing and refinement, launching the product, and selling the product.
3. The selling price of a product is determined by the cost of product engineering (design, R&D, testing and refinement), manufacturing cost (material and labor costs), sales expense and administrative cost, income taxes, and net profit.
4. Because materials industry consumes a considerable amount of energy in making its products, it is no longer acceptable that engineering materials are used and then discarded in landfills. Better alternatives include waste reduction and recycling. Efficient recycling needs an infrastructure for sorting and scrap processing.
5. The life cycle of a product consists of introduction stage, growth stage, maturity stage in which the production rates and sales volume reach the design value, and decline where sales decrease, leading to the end of the life cycle of the product. Because most profits are made during the maturity stage, it should be prolonged by developing new markets and introducing design modifications.

1.11 REVIEW QUESTIONS

1.1 In the example of the Greenobile car given in Case Study 1.1, compare concept A with a new concept D, which is battery driven. What are the major technological problems that have to be solved before concept D becomes a mass-produced product?

1.2 Select a product that you would like to produce if you had your own company. Follow a procedure similar to that outlined for the Greenobile car in Case Study 1.1 in deciding on the selection criteria, setting the specifications, developing the concepts, arriving at the relative importance of the selection criteria, and finally selecting an optimum concept.

1.3 Survey the market and identify two different concepts for devices to open corked bottles. Compare the concepts on the basis of (a) the force needed to remove the cork, (b) whether the cork can be reused to reseal the bottle, (c) appearance and esthetic qualities of the device, and (d) cost of the device. Which concept would you select and why? Suggest ways to improve the concept that you did not choose to make it more competitive.

1.4 Carry out the analysis in question 1.3, but for devices that help in removing jar lids.

1.5 A dairy producer near a large city is considering different expansion schemes for their fresh milk department. Among the various alternatives are (a) doorstep delivery and (b) delivery to supermarkets in the city. Compare the feasibility of each alternative scheme.

1.6 The above mentioned dairy producer is also considering different materials for packaging their product. Among the alternative materials are reusable glass bottles, plastic containers, and multilayer carton packages. Compare the advantages and limitations of each material. Which material will be most suitable for (a) doorstep delivery and (b) supermarket delivery?
1.7 A group of new engineering graduates are thinking of starting a small business in manufacturing educational toys. What are the steps that they need to take before they can start production? Assume a reasonable development time for the project and draw a Gantt chart to show the time relations between the different activities.
1.8 What is the sequence of events involved in building, decorating, and furnishing a house with a small garden? If the total time for building, decorating, and furnishing the house is 18 months, draw a Gantt chart to show the time relations between the different events.
1.9 Explain what is meant by sustainable development and show how it is influenced by recycling of materials.
1.10 Compare the use and recyclability of teacups made of china, glass, melamine, and Styrofoam.
1.11 Compare the use of porcelain and plastics in making soup dishes.
1.12 Discuss the use of composite laminates in fruit juice packaging.
1.13 Collect 10 simple items that are in daily use:
- Draw a neat sketch of the item showing its main dimensions.
- Identify/guess the material out of which they are made.
- What is the expected life of each item?
- Are there any other uses for any of the items after its useful life?
- Determine the amount of energy that may be saved by recycling.
- Classify the items according to the ease with which they can be recycled as excellent, very good, good, fair, or poor.
- Suggest changes that may improve the recyclability of the items.

BIBLIOGRAPHY AND FURTHER READING

Ashby, M., *Materials Selection in Mechanical Design*, 3rd Ed., Butterworth-Heinemann, Amsterdam, 2005.

Ashby, M. and Johnson, K., *Materials and Design: The Art and Science of Material Selection in Product Design*, Butterworth-Heinemann, Amsterdam, 2002.

Beakley, G.C., Evans, D.L., and Keats, J.B., *Engineering: An Introduction to a Creative Profession*, 5th Ed., Macmillan, New York, 1986.

Berry, D., Recyclability & Selection of packaging materials, *JOM*, **44**, December, pp. 21–25, 1992.

Boothroyd, G., Dewhurst, P., and Knight, W., *Product Design for Manufacture and Assembly*, Marcel Dekker, New York, 1994.

Dieffenbach, J.R. and Mascarin, A.E., *JOM*, **45**, June, pp. 16–19, 1993.

Dieter, G., *ASM Handbook: Materials Selection and Design*, Vol. 20, ASM, Materials Park, OH, 1997.

Gray, C.L. and Hippel, F., The fuel economy of light vehicles, *Sci. Am.*, **244**, May, pp. 36–47, 1981.

Henstock, M.E., *Design for Recylability*, The Inst. of Metals, London, 1988.

Humphreys, K.K. and Katell, S., *Basic Cost Engineering*, Marcel Dekker, New York, 1981.

Kerzner, R., *A Systems Approach to Planning, Scheduling, and Controlling*, Van Nostrand Reinhold, New York, 1992.

Kutz, M., *Handbook of Materials Selection*, Wiley, New York, 2002.

Kutz, M., *Mechanical Engineers Handbook: Materials and Mechanical Design*, Wiley, New York, 2006.

Ludema, K.C., Caddell, R.M., and Atkins, A.G., *Manufacturing Engineering: Economics and Processes*, Prentice Hall, London, 1987.

Pugh, S., *Total Design: Integrated Methods for Successful Product Development*. Addison-Wesley, Reading, MA, 1991.

Ray, M.S., *Elements of Engineering Design*, Prentice Hall Int., Englewood Cliffs, NJ, 1985.

Sanders, R.E. Jr., Trageser, A.B., and Rollings, C.S., Recycling of lightweight aluminum containers, paper presented at the 2nd International Symposium—Recycling of Metals and Engineered Materials, J.H.C. Van Linden, Editor. The Minerals, Metals & Materials Soc., 1990, pp. 187–201.

Spinner, M., *Elements of Project Management: Plan, Schedule, and Control*, Prentice Hall, Englewood Cliffs, NJ, 1992.

Turner, R., *Project Management*, McGraw-Hill, New York, 1993.

Ulrich, K.T. and Eppinger, S.D., *Product Design and Development*, McGraw-Hill, New York, 1995.

Wiest, J.D. and Levy, F.K., *A Management Guide to PERT/CPM*, 2nd Ed., Prentice Hall, Englewood Cliffs, NJ, 1987.

Part I

Performance of Materials in Service

The level of performance of a component in service is governed not only by the inherent properties of the material used in making it, but also by the stress system acting on it and the environment in which it is operating. Unacceptably low level of performance can be taken as an indication of failure. Part I of this book discusses the different types of failure and how to prevent, or at least delay, such failures by selecting appropriate materials. Failure of engineering components occurs by several mechanisms, which can be arranged in order of importance as follows:

1. Corrosion, which can be defined as the unintended destructive chemical or electrochemical reaction of a material with its environment, represents about 30% of the causes of component failures in engineering applications.
2. Fatigue, which occurs in materials when they are subjected to fluctuating loads, represents about 25% of the causes of component failures in engineering applications.
3. Brittle fractures are accompanied by a small amount of plastic deformation and usually start at stress raisers, such as large inclusions, manufacturing defects, and sharp corners or notches. They represent about 15–20% of the causes of failure of engineering components.
4. Ductile factures are accompanied by larger amount of plastic deformation and normally occur as a result of overload. They represent about 10–15% of the causes of failure of engineering components.
5. Creep and stress rupture, thermal fatigue, high-temperature corrosion, and corrosion fatigue occur as a result of a combination of causes including high

temperature, stress, and chemical attack. They represent about 10–15% of the causes of failure of engineering components.
6. Other minor causes of failure include wear, abrasion, erosion, and radiation damage.

Failure under mechanical loading is discussed in Chapter 2 and environment degradation in Chapter 3.

The types of failure mentioned can occur as a result of a variety of causes, which can be arranged in order of importance as follows:

1. Poor selection of materials represents about 40% of the causes of failure of engineering components. Failure to clearly identify the functional requirements of a component could lead to the selection of a material that only partially satisfies these requirements. As an example, a material can have adequate strength to support the mechanical loads but its corrosion resistance is insufficient for the application.
2. Manufacturing defects, as a result of fabrication imperfections and faulty heat treatment, represent about 30% of the causes of failure of engineering components. Incorrect manufacturing could lead to the degradation of an otherwise satisfactory material. Examples include decarburization and internal stresses in a heat-treated component. Poor surface finish, burrs, identification marks, and deep scratches due to mishandling could lead to failure under fatigue loading.
3. Design deficiencies constitute about 20% of the causes of failure of engineering components. Failure to evaluate working conditions correctly due to the lack of reliable information on loads and service conditions is a major cause of inadequate design. Incorrect stress analysis, especially near notches, and complex changes in shape could also be a contributing factor.
4. Exceeding design limits, overloading, and inadequate maintenance and repair represent about 10% of the causes of failure of engineering components. If the load, temperature, speed, voltage, etc. are increased beyond the limits allowed by the factor of safety in design, the component is likely to fail. As an example, if an electrical cable carries a higher current than the design value, it overheats and this could lead to melting of the insulating polymer and then short-circuit. Subjecting the equipment to environmental conditions for which it was not designed also falls under this category. An example is using a freshwater pump for pumping seawater. In addition, when maintenance schedules are ignored and repairs are poorly carried out, service life is expected to be shorter than anticipated in the design.

In spite of the efforts to avoid failure, components do fail in service and it is the responsibility of the manufacturer to find out why and how to avoid similar failures in the future. Failure analysis is best carried out by interdisciplinary teams consisting of designers, materials and manufacturing engineers, as well as service personnel. Failure analysis techniques are described in Chapter 2, selection of materials to resist failure is discussed in Chapter 4, material-related design deficiencies are discussed in Chapter 6, and manufacturing defects in Chapter 7.

2 Failure under Mechanical Loading

2.1 INTRODUCTION

With an increasing pressure for higher performance, cheaper products, and lighter components, manufacturers are using materials closer to their limits of performance, with subsequent higher probability of failure. For example, using a stronger material allows the designer to reduce the cross-sectional area and possibly the weight of a component, but will also increase the tendency for buckling as the slenderness of the component increases. Stronger materials are also likely to exhibit lower toughness and ductility with an increasing tendency for catastrophic brittle fracture.

This chapter begins by defining the different types of failures under mechanical loading, and then gives a brief description of each type and how it takes place under service conditions. The chapter ends with a brief review of some experimental methods and analytical techniques of failure analysis. The objectives of the chapter are to

1. examine the relationships between material properties and failure under static loading,
2. discuss the different types of fatigue loading and factors affecting the fatigue strength of materials,
3. review the categories of elevated-temperature failures, and
4. describe some experimental and analytical techniques of failure analysis.

2.2 TYPES OF MECHANICAL FAILURES

Failure under mechanical loading can take place either as a result of permanent change in the dimensions of a component, which results in an unacceptably low level of performance, or as a result of actual fracture. The general types of mechanical failures encountered in practice are as follows:

1. Yielding of the component material under static loading. Yielding causes permanent deformation, which could result in misalignment or hindrance to mechanical movement.
2. Buckling takes place in slender columns when they are subjected to compressive loading, or in thin-walled tubes when subjected to torsional loading.
3. Creep failure takes place when the creep strain exceeds allowable tolerances and causes interference of parts. In extreme cases, failure can take place through rupture of the component subjected to creep. In bolted joints

and similar applications, failure can take place when the initial stressing has relaxed below allowable limits, so that the joints become loose or leakage occurs.

4. Failure due to excessive wear takes place in components where relative motion is involved. Excessive wear can result in unacceptable play in bearings and loss of accuracy of movement. Other types of wear failure are galling and seizure of parts.
5. Failure by fracture due to static overload. This type of failure can be considered as an advanced stage of failure by yielding. Fracture can be either ductile or brittle.
6. Failure by fatigue fracture due to overstressing, material defects, or stress raisers. Fatigue fractures usually take place suddenly without apparent visual signs.
7. Failure due to the combined effect of stresses and corrosion usually takes place by fracture due to cracks starting at stress concentration points, for example, caustic cracking around rivet holes in boilers.
8. Fracture due to impact loading usually takes place by cleavage in brittle materials, for example, in steels below brittle–ductile transition temperature and plastics below glass transition temperature.

Of the types of mechanical failures mentioned, the first four do not usually involve actual fracture, and the component is considered to have failed when its performance is below acceptable levels. However, the latter four types involve actual fracture of the component, and this could lead to unplanned load transfer to other components and perhaps other failures. The following sections discuss the types of failures mentioned in the foregoing list, with the exception of wear failures, which is discussed in Section 3.8, and the combined effect of stress and corrosion, which is discussed in Section 3.4.

2.3 FRACTURE TOUGHNESS AND FRACTURE MECHANICS

It is now recognized that all engineering materials contain potential sites for cracks in the form of discontinuities, heterogeneities, flaws, inclusions, or microstructural defects that can be classified as follows:

- Microstructural features such as oxide or sulfide inclusions, large carbide and intermetallic precipitates, and inhomogeneous distribution of alloying elements leading to hard or soft spots
- Processing defects such as shrinkage and gas pores in castings, slag inclusions and similar welding defects, laps and stringers in forgings, contaminants in powder metallurgy parts, and decarburization in heat treatment
- Damage during service such as surface pits due to corrosion, cracks at discontinuities due to fatigue loading, surface damage due to wear and fretting, and internal voids and cracks due to creep

Such cracks cause high local stresses at their tips, and these stresses depend on the geometry and size of the flaw and the geometry of the component. The ability of

TABLE 2.1
Nondestructive Methods of Crack Detection

Method	Applications and Standard Covering the Practice
Visual examination. The naked eye or a magnifying glass is used to locate and measure cracks.	Surface cracks
Penetrant test. Liquids that enter surface discontinuities by capillary action are first applied to the surface and then wiped off. A developer is then applied to help delineate the areas where the liquid has penetrated.	Defects open to the surface of metallic and nonmetallic materials (ASTM E 165)
Radiographic examination. X-rays and γ-rays are used to penetrate materials and are then caught on a sensitized film. Cavities or inclusions absorb the rays differently from the rest of the material and are delineated on the developed film.	Radiographs show the size and shape of discontinuities (ASTM E 94)
Magnetic particle method. A liquid containing iron powder is first brushed on the surface, and the part is then placed in a strong magnetic field. The particles pile up at discontinuities.	Detects surface cracks in magnetic materials (ASTM E 109 and E 138)
Ultrasonic tests. Ultrasonic vibrations that are transmitted through the material are reflected back at an internal discontinuity earlier than when reaching the opposite surface. The difference between the reflected waves is used to locate the position of the discontinuity.	Internal defects in ferrous and nonferrous metals and alloys (ASTM E 127)
Eddy current inspection. A coil is excited to induce eddy currents in the component to be inspected. In turn, this excitation induces a current in the coil. The presence of defects affects the induction of the component, which affects the current in the coil.	Used for the inspection of surface and subsurface defects in electrically conducting materials

a particular flaw or stress concentrator to cause fracture depends on the fracture toughness of the material, which can be qualitatively defined as the resistance of that material to the propagation of an existing flaw or crack. Therefore, to predict the fracture strength of a component, both the severity of the stress concentration and the fracture toughness of the material must be known.

2.3.1 Crack Detection

To ensure that the severity of stress concentration remains within safe limits, quality control techniques of crack detection are used to determine the crack size, orientation, and distribution in components. Table 2.1 gives a sample of the commonly used nondestructive methods of crack detection.

2.3.2 Fracture Toughness of Materials

Quantitative prediction of the fracture strength can be made using fracture mechanics techniques, and in the simple case of glass, Griffith showed that

$$\sigma_f = \left(\frac{2Ev_s}{\pi a}\right)^{1/2} \tag{2.1}$$

where

σ_f = Fracture stress in MPa (or psi)
a = Crack length for edge cracks and 1/2 crack length for center cracks (this is measured in meters or inches)
E = Young's modulus in MPa (or psi)
υ_s = Energy required to extend the crack by a unit area in J/m^2 (or in lb/in.^{2})

For glass, υ_s is simply equal to the surface energy. However, this is not the case with metals due to the plastic deformation that occurs at the tip of the propagating crack. In the latter case, the fracture toughness is proportional to the energy consumed in the plastic deformation. Because it is difficult to accurately measure this energy, the parameter called the stress intensity factor, K_I, is used to determine the fracture toughness of most materials. The stress intensity factor, as the name suggests, is a measure of the concentration of stresses at the tip of the crack under consideration.

For a given flawed material, catastrophic fracture occurs when the stress intensity factor reaches a critical value, K_C. The relationship between stress intensity factor, K_I, and the critical intensity factor, K_C, is similar to the relationship between stress and tensile strength. The value of K_I is the level of stress at the crack tip and is material independent. However, K_C is the highest value for K_I that the material can withstand without fracturing; it is material and thickness dependent.

The reason why K_C is thickness dependent is that lateral constraint imposed on the material ahead of a sharp crack in a thick plate gives rise to a triaxial state of stress, which reduces the apparent ductility of the material. Thus, the fracture strength is less for thick plates compared with thinner plates, although the inherent properties of the material have not changed. As the thickness increases, K_C decreases and reaches a minimum constant value, K_{IC}, when the constraint is sufficient to give rise to plane-strain conditions, as shown in Figure 2.1.

The thickness, t, at which plane-strain conditions occur is related to the fracture toughness, K_{IC}, and yield strength of the material, YS, according to the relationship:

$$t = 2.5(K_{IC}/\,YS)^2 \tag{2.2}$$

The critical stress intensity factor for plane-strain conditions, K_{IC}, is found to be a material property that is independent of the geometry. In an expression similar to Griffith's, the fracture stress σ_f can be related to the fracture toughness, K_{IC}, and the flaw size, $2a$:

$$\sigma_f = \frac{K_{IC}}{Y(\pi a)^{1/2}} \tag{2.3}$$

where Y is a correction factor that depends on the geometry of the part, that is, thickness, width W, and the flaw size $2a$ for center crack and a for edge crack. For the case

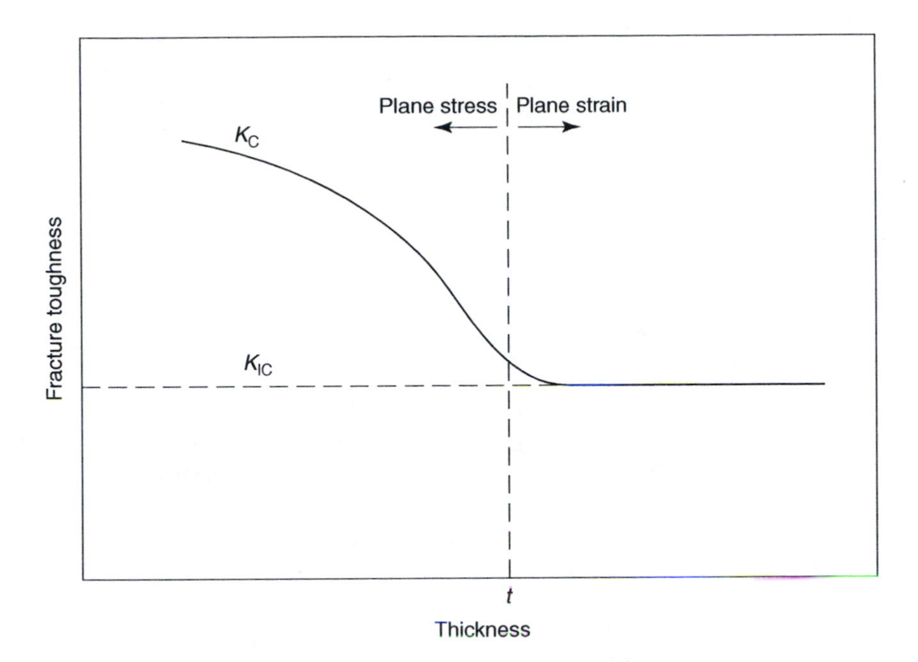

FIGURE 2.1 Effect of thickness on fracture toughness behavior.

of thick plates, as a/W decreases to 0, that is, plane strain, Y decreases to 1. From Equation 2.3, it can be shown that the units of K_{IC} are MPa $(m)^{1/2}$ or psi $(in.)^{1/2}$.

As K_{IC} is a material property, the designer can use it to determine the flaw size that can be tolerated in a component for a given applied stress level. Conversely, the designer can determine the stress level that can be safely used for a flaw size that may be present in a component. Examples 2.1 and 2.2 illustrate the use of fracture toughness in design.

Design Example 2.1: Critical Crack Length

Consider a wide plate containing a crack of length $2a$ extending through the thickness. If the fracture toughness of the material is 27.5 MPa $(m)^{1/2}$ and the yield strength is 400 MPa, calculate the fracture stress σ_f and compare it to the yield strength σ_y for different values of crack lengths.

Assume $Y = 1$.

Solution

Using Equation 2.3, σ_f can be calculated for different crack lengths. The results are given in Table 2.2.

With the smallest crack, the yield strength is reached before catastrophic failure occurs. However, longer cracks cause fracture before yielding.

TABLE 2.2
Variation of Fracture Stress with Crack Length

a (mm)	1	2	4	6	8	10
σ_f (MPa)	490.6	346.9	245.3	200.3	173.5	155.2
σ_f/σ_y	1.23	0.87	0.61	0.50	0.43	0.39

Design Example 2.2: Using Fracture Toughness in Material Selection

Problem

Ti-6 Al-4 V (K_{IC} = 60 MPa $(m)^{1/2}$) and aluminum AA7075 alloy (K_{IC} = 24 MPa $(m)^{1/2}$) are widely used in making light-weight structures. If the available nondestructive testing (NDT) equipment can only detect flaws larger than 4 mm in length, can we safely use either of these alloys for designing a component that will be subjected to a stress of 400 MPa?

Solution

From Equation 2.3 and taking $Y = 1$,

For Ti-6 Al-4 V: $\sigma_f = 400 = 60/(\pi a)^{1/2}$ $2a = 14$ mm
For AA7075 $\sigma_f = 400 = 24/(\pi a)^{1/2}$ $2a = 2.3$ mm

The preceding figures show that the critical crack can be detected in the titanium alloy but not in the aluminum alloy. Titanium alloy can be used safely but not the aluminum alloy.

Fracture toughness data are available for a wide range of materials and some examples are given in Section 4.5. Fracture toughness data can also be easily established for new materials using standardized testing methods, for example, ASTM Standard E399. Because of possible anisotropy of microstructure, it is important to orient the test specimen to correspond to the actual loading conditions of the part in service.

Fracture toughness, like other material properties, is influenced by several factors including strain rate or loading rate, temperature, and microstructure, as discussed in more detail in Section 4.5. Also, increasing the yield and tensile strengths of the material usually causes a decrease in K_{IC}. The use of fracture mechanics in design is discussed in Section 6.4.

Fracture toughness is widely accepted as a design criterion for high-strength materials where ductility is limited. In such cases, the relationship between K_{IC}, applied stress, and crack length governs the conditions for fracture in a part or a structure. This relationship is shown schematically in Figure 2.2. If a particular combination of stress and flaw size in a structure reaches the K_{IC} level, fracture can occur. Thus there are many combinations of stress and flaw size that may cause fracture in a structure made of a material having a particular value of K_{IC}. The figure shows that

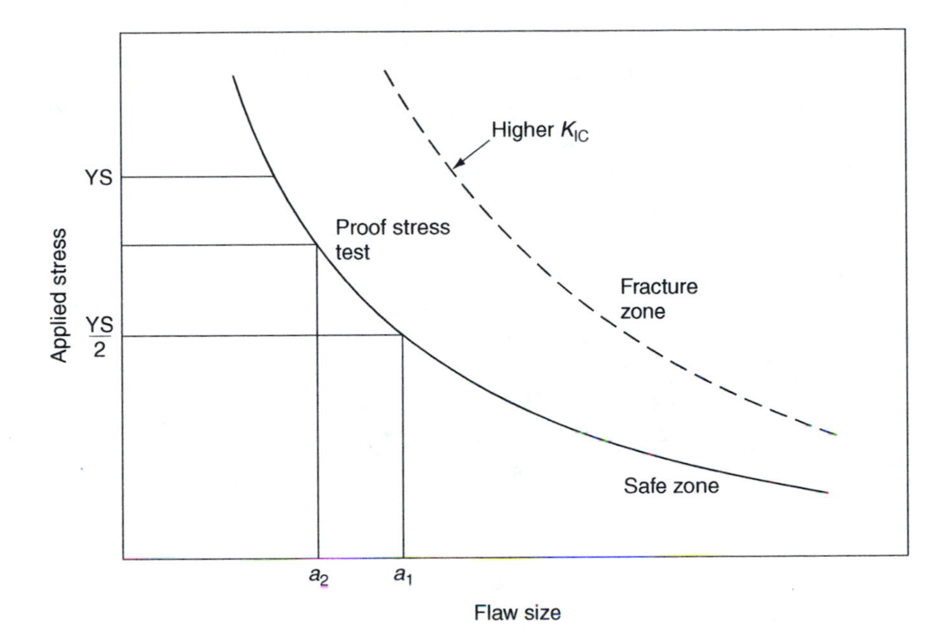

FIGURE 2.2 Schematic relationship between stress, flaw size, and fracture toughness.

materials with higher K_{IC} values tolerate larger flaws at a given stress level or higher stress levels for a given flaw size.

Figure 2.2 also shows that if a material of known K_{IC} is selected for a given application, the size of the flaw that will cause fracture can be predicted for the anticipated applied stress. If the design stress of a part is taken as 0.5YS, the critical flaw length would be (a_1). Therefore, provided that no defect of size greater than (a_1) is present, failure should not occur on loading. If in a proof test the part is loaded to a stress above the expected service stress and the test was successful, then a flaw of size greater than a_2 could not have existed. During service life, crack growth of the order of ($a_1 - a_2$) could be tolerated before failure.

From Equation 2.3 and Figure 2.2, it can be shown that the maximum allowable flaw size is proportional to $(K_{IC}/YS)^2$, where K_{IC} and YS are measured at the expected service temperature and loading rate. Thus the ratio (K_{IC}/YS) can be taken as an index for comparing the relative toughness of structural materials. Higher values of (K_{IC}/YS) are more desirable as they indicate tolerance to larger flaws without fracture, as discussed in Section 4.5. The sensitivity of the NDT techniques used to detect manufacturing defects that approach the critical size in the part or structure is determined by the value of the allowed flaw size.

2.4 DUCTILE AND BRITTLE FRACTURES

Machine and structural elements often fail in service as a result of either ductile or brittle fracture. The terms ductile and brittle are usually used to indicate the extent of macroscopic or microscopic plastic deformation that precedes fracture. For example, materials with plastic strains of less than 2% at fracture are considered brittle. The terms ductile and brittle are also related to fracture toughness, and materials with K_{IC} less than 15 MPa $(m)^{1/2}$ are considered brittle. Impact toughness, which is a measure of the energy needed for fracture, can also be used as an indication and materials that absorb less than 15 ft lb (20.3 J) are considered brittle.

2.4.1 Ductile Fractures

Service failures that occur solely by ductile fracture are relatively infrequent and may be a result of errors in design; incorrect selection of materials; improper fabrication techniques; or abuse, which arises when a part is subjected to load and environmental conditions that exceed those of the intended use.

The following case study illustrates an approach to failure analysis and the type of solution that may be available to the engineer who tries to solve the problem.

Case Study 2.3: Ductile Fracture of a Ladder

Problem

An aluminum ladder, which is 3 m long and made of four T-sections and hollow cylindrical rungs, is shown in Figure 2.3. The ladder failed when a man weighing 100 kg climbed half way up when it was leaning against a wall at an angle of 15°. Although this was the first time the man used the ladder, his wife, who weighed 60 kg, had used it many times earlier. As a result of failure, T-sections S2 and S3 suffered severe plastic deformation and buckling caused by bending, while T-sections S1 and S4 cracked just under the rung where the man was standing (Figure 2.3).

Analysis

Investigation showed that large reduction in area accompanied the fracture, and chemical analysis showed that the T-sections were made of AA 6061 alloy. The hardness of the alloy was in the range 25–30 RB in most areas, but was about 20 RB in section S2. These hardness values correspond to T4 temper condition of the AA 6061 alloy.

It is expected that the weakest section S2, which was on the tension side during loading, has yielded causing the load to be redistributed and section S3 to yield. This, in turn, caused sections S1 and S4 to be overloaded in tension.

Solution

As failure is caused by overload during normal use, it is recommended that a stronger material be used. It would be sufficient to change the temper condition from T4 to T6. The AA 6061 T6 has a hardness of 45–55 RB and yield strength about twice that of the AA 6061 T4.

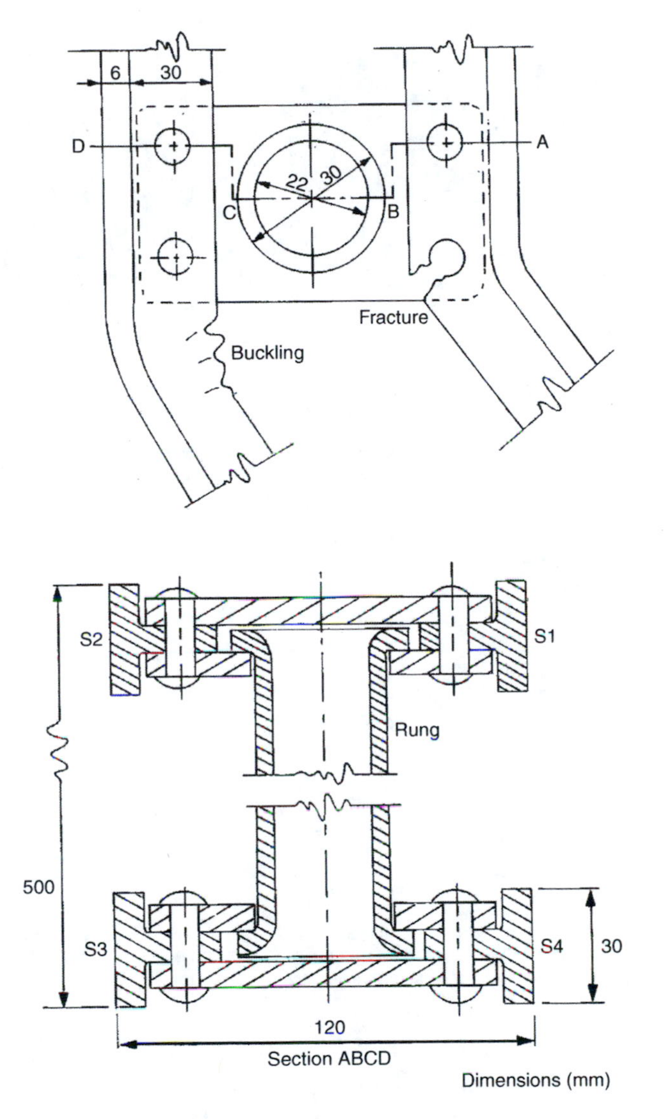

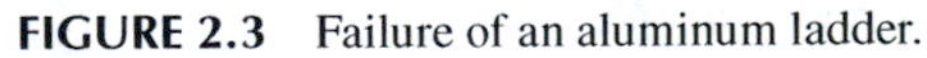
FIGURE 2.3 Failure of an aluminum ladder.

2.4.2 Brittle Fractures

Brittle fractures are usually initiated at stress raisers, such as large inclusions, cracks or surface defects, and sharp corners or notches. The single most frequent initiator of brittle fracture is the fatigue crack, which accounts for more than 50% of all brittle fractures in the manufactured products. Brittle fractures are insidious in character because they may occur under static loading at stresses below the yield strength and without warning.

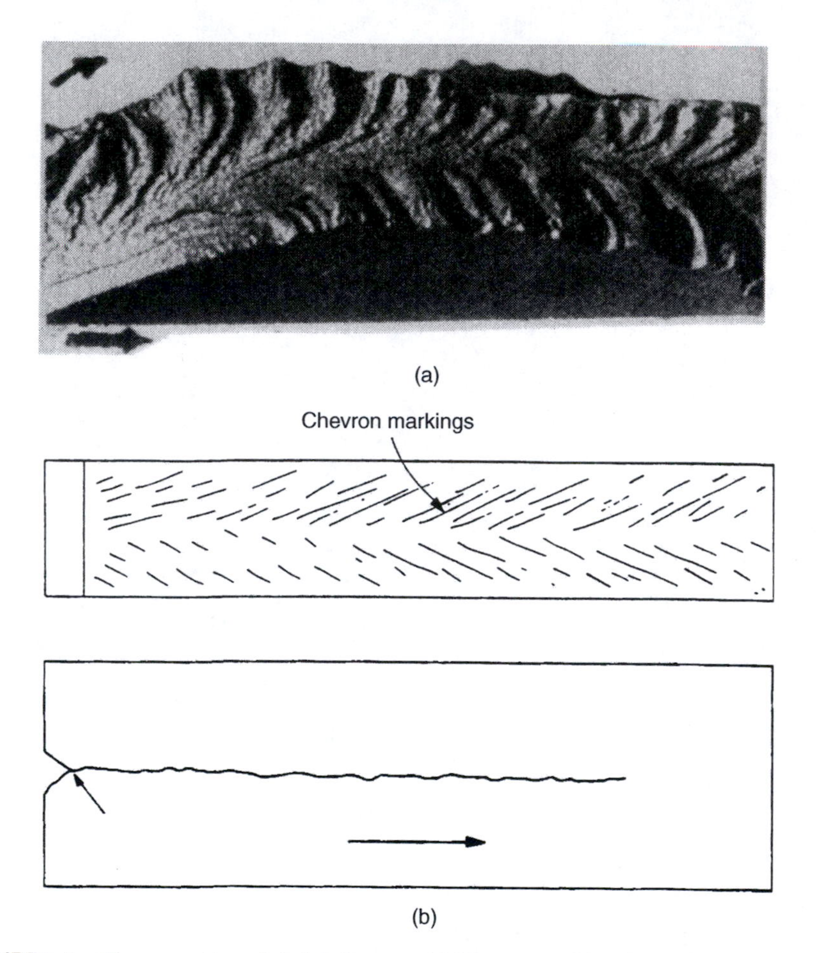

FIGURE 2.4 Chevron patterns in brittle fracture. (a) Chevron markings in steel. (From Rollason, E.C., *Metallurgy for Engineers*, Edward Arnold, London, 1977.) (b) Schematic representation.

Once started, the brittle fracture will run at high speed, reaching 1200 m/s in steel, until total failure occurs, or until it runs into conditions favorable for its arrest. The risk of occurrence of brittle fracture depends on the notch toughness of the material under a given set of service conditions. A characteristic feature of brittle fracture surfaces is the chevron pattern, which consists of a system of ridges curving outward from the center line of the plate, as shown in Figure 2.4. These ridges, or chevrons, may be regarded as arrows with their points on the center line and invariably pointing toward the origin of the fracture, thus providing an indication of its propagation pattern. This feature is useful in the analysis of service failures.

2.4.3 Ductile–Brittle Transition

The temperature at which the component is working is one of the most important factors that influence the nature of fracture. Brittle fractures are usually associated

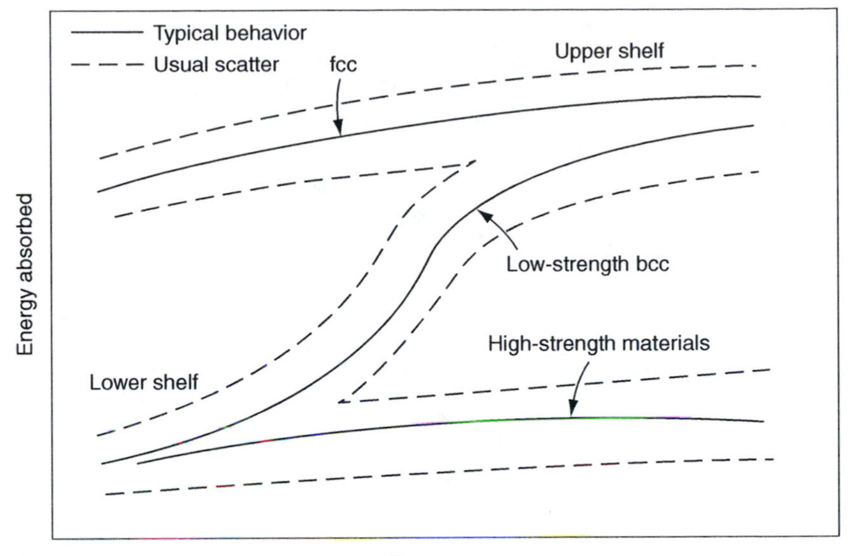

FIGURE 2.5 Schematic representation of the effect of temperature on the energy absorbed in fracture.

with low temperature, and in some steels conditions may exist where a difference of a few degrees, even within the range of atmospheric temperatures, may determine the difference between ductile and brittle behavior. This sharp ductile–brittle transition is only observed in body-centered cubic (bcc) and hexagonal close-packed (hcp) metallic materials and not in face-centered cubic (fcc) materials, as illustrated schematically in Figure 2.5.

The most widely used tests for characterizing the ductile-to-brittle transition are the Charpy, ASTM standards A23 and A370, and Izod. The temperature at which the material behavior changes from ductile to brittle is called the ductile–brittle transition temperature (T_c), and may be taken as the temperature at which the fractured surfaces exhibit 50% brittle fracture appearance. In V-notch Charpy experiments, the transition temperature can be set at a level of 20.3 J (15 ft lb) or at 1% lateral contraction at the notch. The transition temperature based on fracture appearance always occurs at a higher value than if based on a ductility or energy criterion. Therefore, the fracture appearance criterion is more conservative.

The rate of change from ductile-to-brittle behavior depends on the strength, chemical composition, structure, and method of fabrication of the material. The state of stress and the speed of loading also influence the nature of fracture. A state of triaxial tensile stresses, such as those produced by a notch, can be the cause of brittle fracture. The notches in a component can be due to shape changes, processing defects, or corrosion attack. Materials that behave normally under slowly applied loads may behave in a brittle manner when subjected to sudden applications of load, such as shock or impact. The ductile-to-brittle transition also shifts to higher temperatures as the rate of loading increases.

In the case of steels, the shift in transition temperature depends on the strength and can be as high as 68°C (155°F) in steels of yield strength of 280 MPa (about 40 ksi). The shift in transition temperature between static and impact loading decreases with increasing strength and becomes negligible at yield strengths of about 900 MPa (about 130 ksi).

In applying the Charpy V-notch (CVN) results to industrial situations, it should be borne in mind that the shock conditions encountered in the test may be too drastic. Many industrial components operate successfully in extreme cold without special consideration for notch toughness values or transition temperature. However, where stress concentration and rate of strain are high and service temperatures are low, special design and fabrication precautions should be taken and materials with low transition temperatures should be selected.

2.4.4 Design and Manufacturing Considerations

The design and fabrication precautions that should be taken to avoid brittle fracture include the following:

1. Abrupt changes in section should be avoided to avoid stress concentrations, and thickness should be kept to a minimum to reduce triaxial stresses.
2. Welds should be located clear of stress concentrations and of one another, and they should be easily accessible for inspection.
3. Whenever possible, welded components should be designed on a fail-safe basis. This concept is discussed in Section 5.2.

A useful relation between plane-strain fracture toughness (K_{IC}) and the upper-shelf CVN impact energy was suggested for steels of yield strengths (YS) higher than about 770 MPa (ca. 110 ksi) by Rolfe and Barson (1977) as

$$(K_{IC}/\mathrm{YS})^2 = (5/\mathrm{YS})\,(\mathrm{CVN} - \mathrm{YS}/20) \tag{2.4}$$

where K_{IC} is in ksi $(\mathrm{in.})^{1/2}$, YS in ksi, and CVN in ft lb.

2.5 FATIGUE FAILURES

Generally, fatigue fractures occur as a result of cracks, which usually start at some discontinuity in the material or at other stress concentration locations, and then gradually grow under repeated application of load. As the crack grows, the stress on the load-bearing cross section increases until it reaches a high level enough to cause catastrophic fracture of the part. This sequence is reflected in the fracture surfaces that usually exhibit smooth areas that correspond to the gradual crack growth stage and rough areas that correspond to the catastrophic fracture stage, as shown in Figure 2.6. The smooth parts of the fracture surface usually exhibit beach marks, which occur as a result of changes in the magnitude of the fluctuating fatigue load. Another feature of fatigue fractures is that they lack macroscopic plastic deformation and, in this respect, they resemble brittle fractures. The following case study is used to

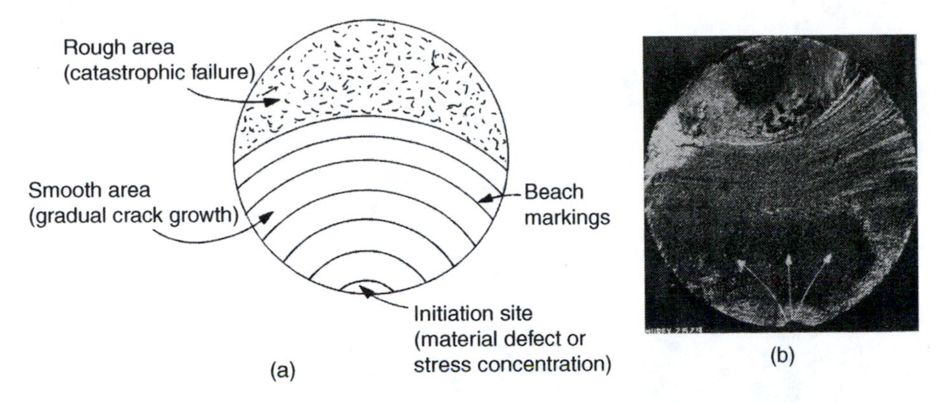

FIGURE 2.6 General appearance of a fatigue fracture surface. (a) Schematic representation; (b) fatigue fracture of automobile axle shaft. (From Rollason, E.C., *Metallurgy for Engineers*, Edward Arnold, London, 1977.)

illustrate one of the frequent causes of fatigue fracture and to show one of the solutions that may be used to solve the problem.

Case Study 2.4: Fatigue Failure of a Pressure Line

Problem

The steel pressure line of a hydraulic pump in a power-generation unit started leaking at the exit line flange assembly shown in Figure 2.7. The source of leakage was found to be a crack in the fillet weld.

Analysis

Investigation of the working conditions showed that although the pressure in the line was within the design limits, excessive vibrations existed in the 2-m-long tube, which was not sufficiently supported by the flexible hose at its end. This caused the line to act as a cantilever beam with maximum forces at the flange.

It is concluded that the crack in the fillet weld took place as a result of fatigue loading caused by the vibrations in the line.

Solution

The corrective action taken was to change the design to move the weld from the area of high stress concentration, as shown in Figure 2.7. The line was also adequately supported at the point where it joined the flexible hose to minimize vibrations.

2.5.1 Types of Fatigue Loading

The simplest type of fatigue loading is the alternating tension-compression without a static direct stress (Figure 2.8a). In this case the stress ratio, defined as $R = \sigma_{min}/\sigma_{max}$,

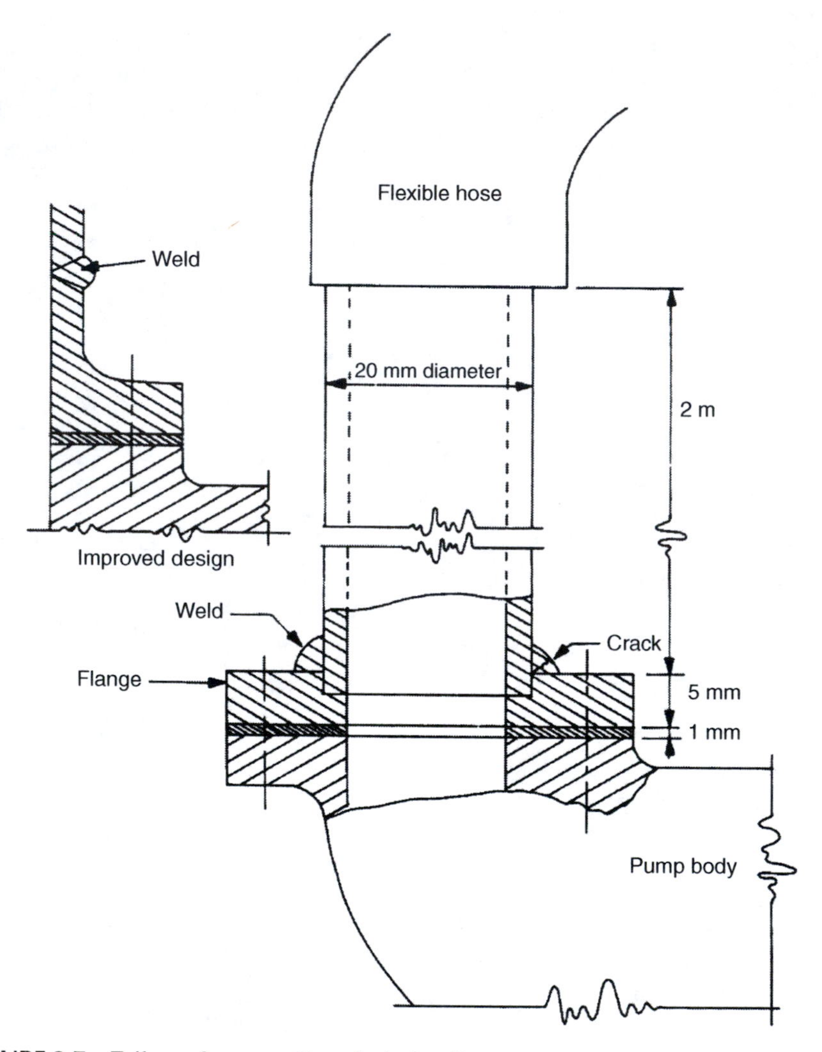

FIGURE 2.7 Failure of pressure line of a hydraulic pump.

is −1. If a static mean stress σ_m is superimposed on the alternating stress, then the stress varies between the limits of

$$\sigma_{max} = \sigma_m + \sigma_a, \quad \text{and} \quad \sigma_{min} = \sigma_m - \sigma_a,$$

as shown in Figure 2.8b. A special case is the pulsating stress with $R = 0$.

Under actual service conditions, parts may be subjected to more than one form of load, for example, alternating torsion with static tension. Many other combinations are known to be met in different applications. However, most of the available fatigue test results are for the simple alternating stresses, that is, $R = -1$. Such results are usually presented as *S–N* curves, as shown in Figure 2.9. In this case, *S* is

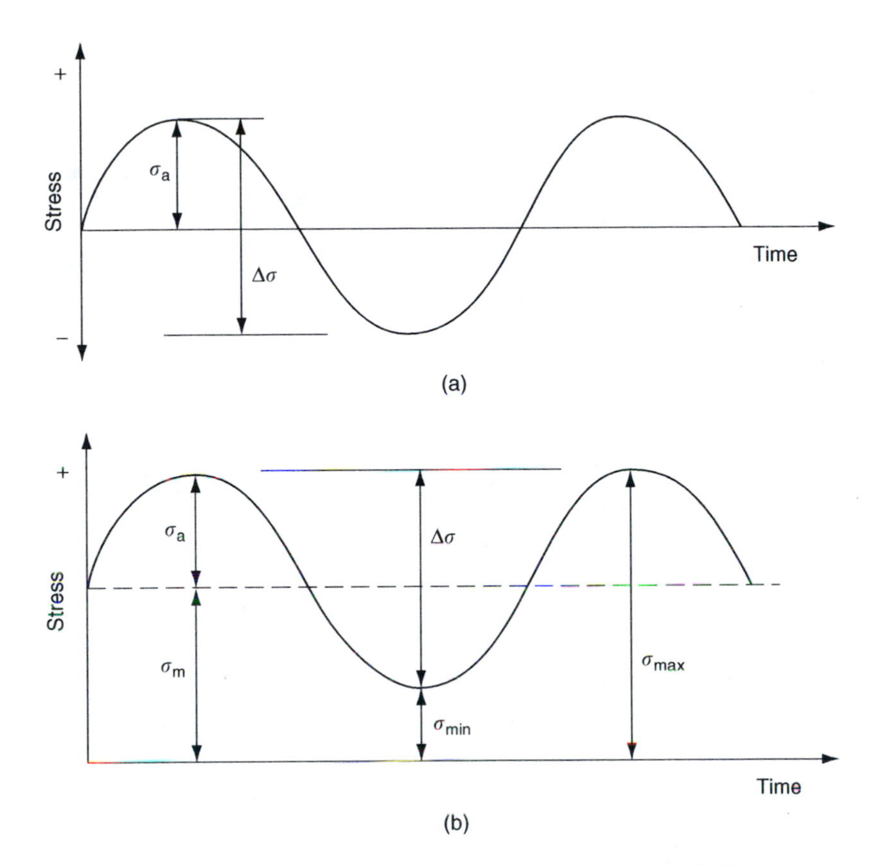

FIGURE 2.8 Types of fatigue loading. (a) Alternating stress, $R = -1$; (b) fluctuating stress.

the alternating stress and N is the number of cycles to failure. Although most available S–N curves and endurance limit values are based on laboratory experiments and controlled test conditions, fatigue results always show larger scatter than other mechanical properties. The standard deviation for endurance limit results is usually in the range of 4–10%, but in the absence of statistical values, an 8% standard deviation can be assumed. The reported endurance results can be taken to correspond to 50% survival reliability. Statistical variation of material properties is discussed in Section 5.7.

2.5.2 Fatigue Strength

Some materials, such as steels and titanium, exhibit a well-defined fatigue or endurance limit below which no fatigue fracture occurs (Figure 2.10). The figure also shows that other materials, such as aluminum alloys, do not have such a limit, and their S–N curves continue to decrease at high number of cycles. For these materials, fatigue strength is reported for a specified number of cycles. As this number of cycles is not standardized, the reported fatigue strength values are subject to large variations depending on whether the strength is taken at $N = 10^6$, 10^7, or 10^8 cycles.

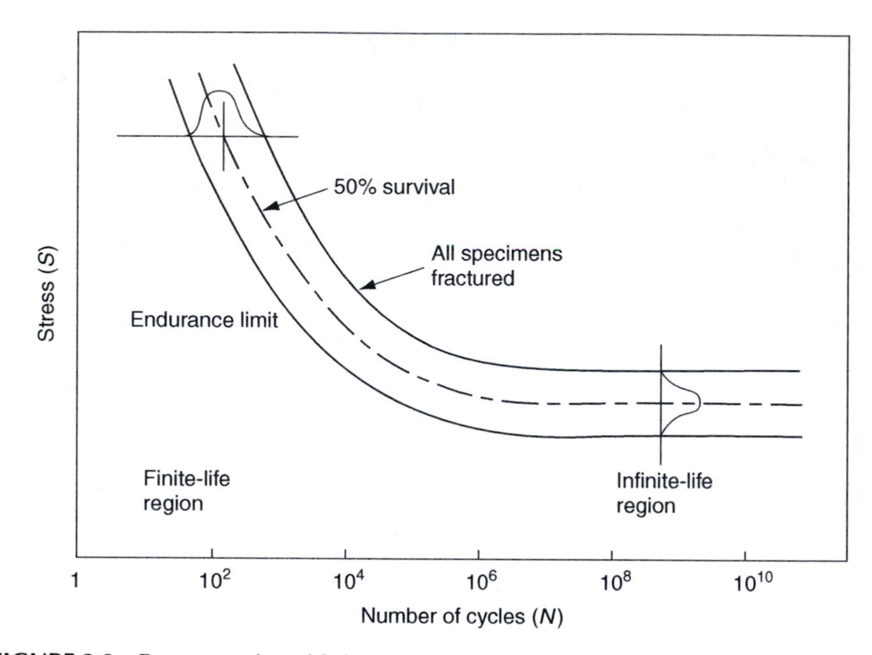

FIGURE 2.9 Representation of fatigue test results on the *S–N* curve.

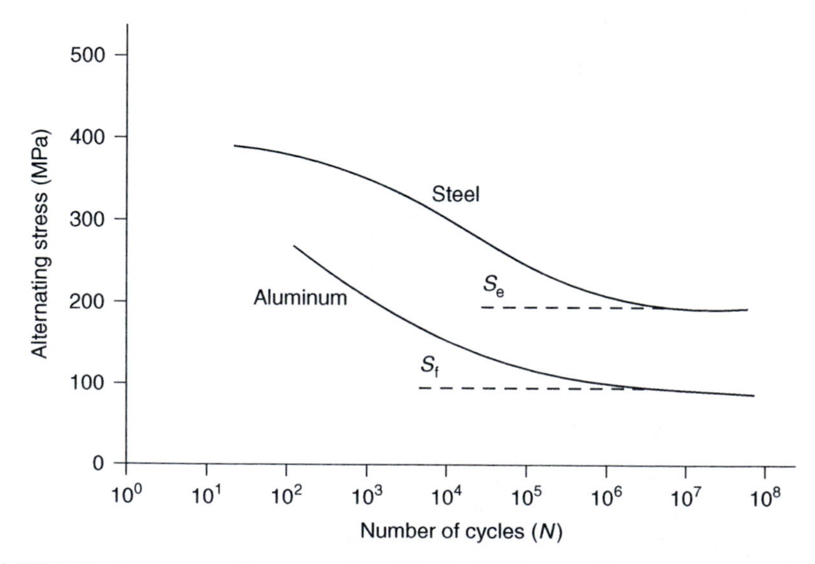

FIGURE 2.10 Fatigue results as represented by the *S–N* curves. Steel exhibits an endurance limit and its curve levels off at S_e. Aluminum does not exhibit an endurance limit, and its curve continues to decline for all values of *N*.

Therefore, it is necessary to specify the number of cycles for which the strength is reported. When the material does not exhibit a well-defined fatigue or endurance limit, it is only possible to design for a limited life.

The endurance limit, or fatigue strength, of a given material can usually be related to its tensile strength, as shown in Table 4.7. The endurance ratio, defined as endurance limit/tensile strength, can be used to predict fatigue behavior in the absence of endurance limit results. As the table shows, the endurance ratio of most ferrous alloys varies between 0.4 and 0.6.

An important limitation of *S–N* curves and the endurance limit results is that they are usually determined for relatively small specimens under controlled conditions and simple loading systems. In addition, these results do not distinguish between crack initiation life and crack propagation life. These disadvantages limit their use in designing large structural components where crack like defects can exist in the material or as a result of manufacturing. Under such conditions, it is the rate of fatigue-crack propagation that determines the fatigue life of the part.

2.5.3 Crack Initiation

Even if the nominal stresses acting on the part are below the elastic limit, local stresses may exceed the yield stress as a result of stress concentration or material discontinuity. As a result, cyclic plastic deformation takes place on favorably oriented slip planes, leading to local strain hardening and eventual crack nucleation.

It can be shown that the local stress range at the site of crack nucleation, $\Delta\sigma_{max}$, can be related to the range of the stress intensity factor, ΔK_I, by the following relationship:

$$\Delta\sigma_{max} = \Delta\sigma K_t = \left(\frac{2}{\pi^{1/2}}\right)\left(\frac{\Delta K_I}{r^{1/2}}\right) \tag{2.5}$$

where

r = Notch-tip radius
$\Delta\sigma$ = Range of applied nominal stress
K_t = Stress concentration factor

Experience shows that $\Delta K_I/(r)^{1/2}$ is the main parameter that governs fatigue-crack initiation in a benign environment. In the case of steels, there is a fatigue-crack initiation threshold, $[\Delta K_I/(r)^{1/2}]$, below which fatigue cracks do not initiate. The value of this threshold increases with increasing strength and with decreasing strain hardening exponent.

2.5.4 Crack Propagation

The performance of most parts and structures under fatigue loading is more dependent on their resistance to crack propagation than to crack nucleation. This is because microcracks are known to nucleate very early in the lives of parts, and notched high-strength materials may have propagating cracks effectively throughout their

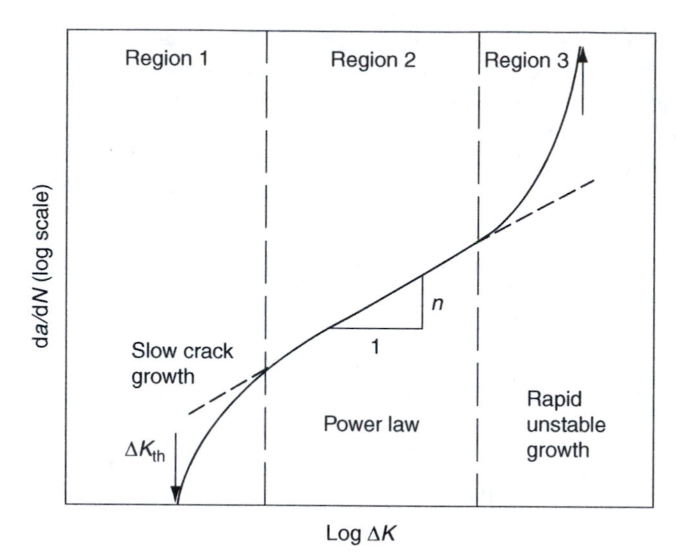

FIGURE 2.11 Schematic illustration of the effect of the range of the stress intensity factor (ΔK_I) on fatigue-crack growth rate (da/dN).

service lives. Initially, the crack propagates along the slip plane along which it nucleated, stage I, and then turns on to a plane perpendicular to the direction of the maximum tensile stress, stage II. Stage I may account for more than 90% of the life of a smooth ductile part under light loads or may be totally absent in a sharply notched, highly stressed part.

Experience based on experimental data shows that the fatigue-crack propagation behavior is controlled primarily by the stress intensity factor range, ΔK_I, and can be divided into three regions, as shown in Figure 2.11. In region 1, fatigue cracks grow extremely slowly or not at all.

The crack growth in region 2 can be represented by a power law, which is usually called the Paris relationship:

$$\frac{da}{dN} = C\,(\Delta K_I)^n \tag{2.6}$$

where

a = Crack length
N = Number of cycles
da/dN = Crack growth per cycle
C and n = Experimentally determined constants that depend on material properties and environment

In region 3, an increase in growth rate leads to rapid unstable growth as K_C or K_{IC} is approached.

The use of the mentioned principles to select materials and to design parts that will resist fatigue fracture is discussed in Sections 4.6 and 6.5, respectively.

2.6 ELEVATED-TEMPERATURE FAILURES

The effect of service environment on material performance at elevated temperature can be divided into the following three main categories:

1. Mechanical effects such as creep and stress rupture
2. Chemical effects such as oxidation
3. Microstructural effects such as grain growth and overaging

Although oxidation and creep can directly lead to the failure of a part in service, the microstructural changes can lead to weakening of the material, and therefore, can indirectly lead to failure. Many of the strengthening mechanisms that are effective at room temperature become ineffective at elevated temperatures. Generally, nonequilibrium structures change during long-term high-temperature service and this leads to lower creep strength. Thus, materials that depend on their fine grains for strengthening may lose this advantage by grain growth, and materials that have been strain-hardened by cold working may recover or anneal. Structures that have been precipitation hardened to peak values may overage, and steels that have been hardened and tempered may overtemper.

2.6.1 CREEP

A major factor that limits the life of components in service at elevated temperatures is creep. Creep is defined as the time-dependent deformation, which occurs under the combined effect of stress and elevated temperature, normally in the range of 35–70% of the melting point of the material expressed in absolute temperature. Creep occurs as a result of the motion of dislocations within the grains, grain boundary rotation, and grain boundary sliding. It is sensitive to grain size, alloying additions, microstructure of the material, and service conditions.

When creep reaches a certain value, fracture occurs. Creep fracture (also called stress rupture) usually takes place at strains much less than the fracture strains in tension tests at room temperature.

In most practical cases, the strain that is suffered by a component under creep conditions can be divided into the stages shown schematically in Figure 2.12. Following an initial instantaneous deformation, creep takes place at a decreasing strain rate, which is the slope of the curves in Figure 2.12, during the primary or transient stage. This is followed by the secondary creep or steady-state stage where the strain rate is constant under constant stress conditions. This steady-state creep rate is often a high-temperature design parameter and may be required to be lower than a specified value to ensure a minimum life of a component in service. At a given temperature, the steady-state creep rate ($\varepsilon^{\cdot}$) can be expressed as a function of the applied stress (σ) as follows:

$$\varepsilon^{\cdot} = B\sigma^{m} \tag{2.7}$$

where B and m are experimental constants and depend on the material and operating temperature.

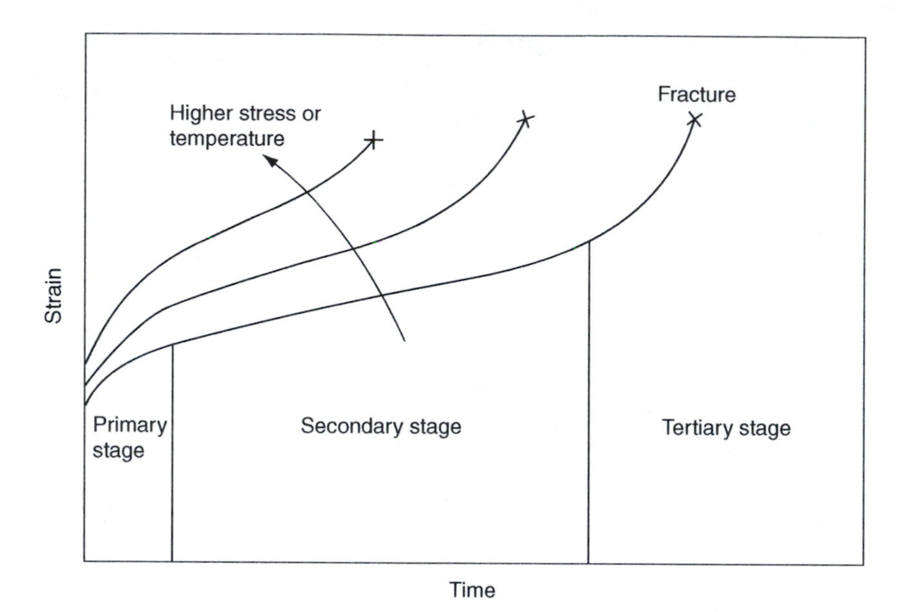

FIGURE 2.12 Schematic creep curve under tensile loading.

At the end of the steady-state stage, tertiary creep starts and the strain rate increases rapidly with increasing strain and fracture finally occurs. Tertiary creep can be caused by

1. reduction of the cross-sectional area of the component due to cracking or necking;
2. oxidation and other environmental effects, which reduce the cross-sectional area; and
3. microstructural changes that weaken the material such as coarsening of precipitates.

Under certain conditions, some materials may not exhibit all the mentioned stages of creep. For example, at high stresses or high temperatures, the primary stage may not be present, with secondary creep or even tertiary creep starting soon after load application. Another example is the case where fracture occurs before the tertiary stage is reached, as in the case of some low-ductility cast alloys. Creep ductility is an important factor in materials selection. Although the permissible creep strain in practice is usually of the order of 1%, selecting materials with higher creep ductility means a higher safety margin.

2.6.2 Combined Creep and Fatigue

In many high-temperature applications in practice, the applied loads are cyclic and could lead to a combined creep–fatigue failure. Under these conditions, the life of a component is determined by the initiation and growth of a creep or a fatigue crack. At high load frequencies and relatively lower temperatures, crack growth is independent of the frequency or temperature. This is because the material just ahead of the crack does not suffer any time-dependent processes, such as oxidation or creep relaxation. Under these conditions, the mechanism of crack growth is essentially the same as room temperature fatigue.

At low frequencies and relatively high temperatures, crack growth is affected by time-dependent processes. A mixture of the two extreme cases of behavior is expected at intermediate temperatures and load frequencies.

2.6.3 Thermal Fatigue

Another form of elevated-temperature failures is thermal fatigue. Stresses and strains induced in a component due to thermal gradients can cause failure if repeated a sufficient number of times. Faster changes in temperature, lower thermal conductivity of the material, higher elastic constant, higher thermal expansion coefficient, lower ductility, and thicker component sections often account for shorter service life. Ceramic materials are particularly prone to thermal fatigue in view of their limited thermal conductivity and brittleness.

In high-temperature applications, the environment plays an important role in determining the performance of components. Selecting the material that will resist the environment, controlling the environment, or protecting the surface is essential for prolonged service. Examples of aggressive environments are those that contain vanadium compounds, sulfur compounds, or salt. A vacuum environment may be more harmful than air if some of the alloy constituents evaporate at high temperatures.

2.7 FAILURE ANALYSIS: EXPERIMENTAL METHODS

When a component fails in service, it is important that the source of failure is located to identify the responsible party and to avoid similar failures in future designs. Owing to the complexity of most failure cases, it is useful to follow a systematic approach to the analysis such as the following:

1. Gathering background information about the function, source, fabrication, materials used, and service history of the failed component is an important step.
2. Site visits involve locating all the broken pieces, making visual examination, taking photographs, and selecting the parts to be removed for further laboratory investigation. Macroscopic, microscopic, chemical analysis, nondestructive, and destructive tests are normally used to locate possible material and manufacturing defects. Presence of oxidation and corrosion products, temper colors, surface markings, etc. can also provide valuable clues toward failure mode identification.
3. Based on the gathered information it should be possible to identify the origin of failure, direction of crack propagation, and sequence of failure. Presence of secondary damage not related to the main failure should also be identified.
4. The final step in failure analysis usually involves writing a report to document the findings and to give the conclusions. This report usually includes the background information and service history of the failed part, description of

the specimens examined and procedure of examination, information about materials and comparison with specifications, manufacturing methods, causes of failure, and how to avoid such failure in the future.

The following case study illustrates the use of the experimental method in failure analysis of a welded component.

Case Study 2.5: Failure of Welded Alloy Steel Component

Problem

A component made of alloy steel, which was manufactured by welding, failed next to the fusion zone. What factors could have contributed to this failure?

Analysis

The first step is to ensure that the failure zone does not have obvious cavities or cracks and that the load did not exceed the design limit, and that the weld was not placed in a stress concentration zone. Assuming that there are no obvious weld defects and that the design parameters are correct, the next step is to look for the less obvious materials and manufacturing defects. The following are questions that need to be answered:

1. What was the grade of the welding electrodes? Could it have introduced hydrogen in the weldment? (Look for the electrode number and specifications and whether it is a low-hydrogen grade.)
2. What is the composition of the alloy steel and what is its hardenability? Was there martensitic structure in the fracture zone? (Find out the designation number, look for chemical analysis, estimate hardenability, and perform hardness tests.)
3. What was the welding procedure? Was appropriate preheating and post-welding heating applied? (Look for the records and process sheets of the weld.)
4. Was there severe grain growth in the heat-affected zone where fracture occurred? (Perform microscopic examination.)
5. Did the parent metal have inclusions that could have caused stress concentration? (Perform microscopic examination.)

Answers to these questions would be helpful in identifying the cause of failure.

2.8 FAILURE ANALYSIS: ANALYTICAL TECHNIQUES

Several analytical techniques and computer-based methods have been developed to help the engineer in solving failure problems. Identification of failure mode is not only important for determining the cause of failure, but also a powerful tool for reviewing the design. The following discussion gives a brief review of some of the analytical techniques that have been developed for systematic identification of failure modes. Reference should be made to the original publications for more details.

2.8.1 Fault Tree Analysis

Fault tree analysis (FTA) is an analysis technique that is widely used in organizing the logic in studying reliability, critical failure modes, safety, availability, or advantages of design redundancy. When any of these issues is selected for analysis it is considered to be the "top event" that forms the main trunk from which logic branches develop. The analysis proceeds by determining how the top event can be caused by individual or combined lower-level events. As the fault tree grows, its logic is separated into successively smaller events until each element is sufficiently basic to be treated independently of other events. This separation is recorded using AND and OR gates in addition to other standard symbols, as shown in Figure 2.13.

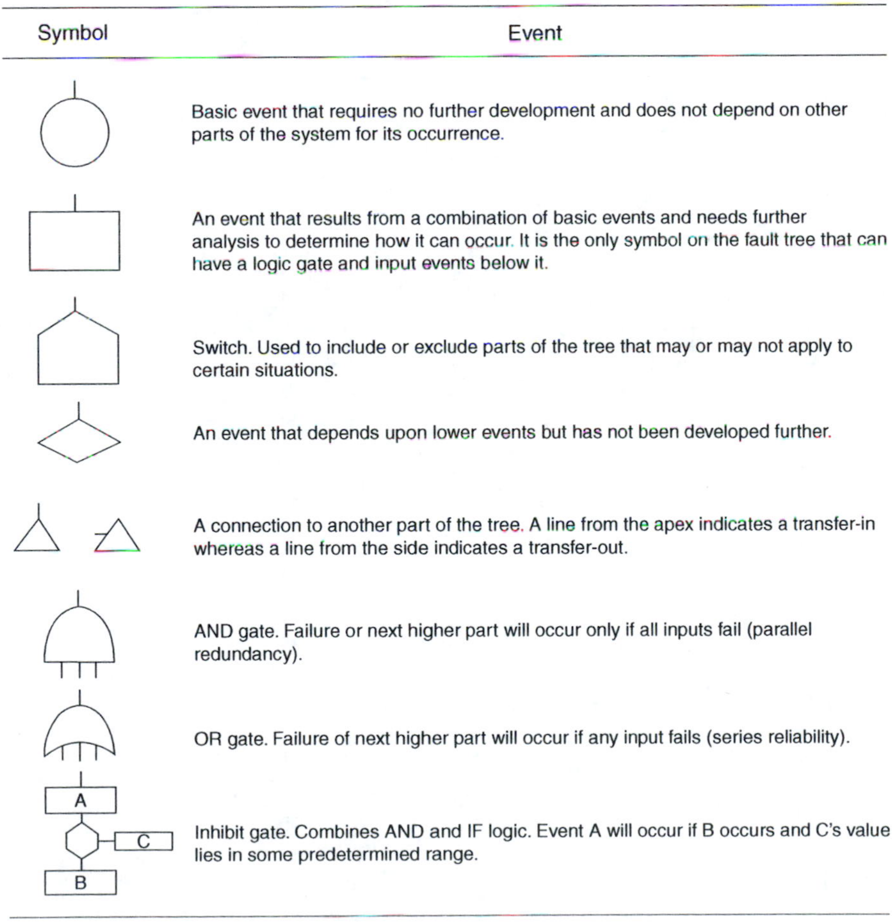

Symbol	Event
	Basic event that requires no further development and does not depend on other parts of the system for its occurrence.
	An event that results from a combination of basic events and needs further analysis to determine how it can occur. It is the only symbol on the fault tree that can have a logic gate and input events below it.
	Switch. Used to include or exclude parts of the tree that may or may not apply to certain situations.
	An event that depends upon lower events but has not been developed further.
	A connection to another part of the tree. A line from the apex indicates a transfer-in whereas a line from the side indicates a transfer-out.
	AND gate. Failure or next higher part will occur only if all inputs fail (parallel redundancy).
	OR gate. Failure of next higher part will occur if any input fails (series reliability).
A, B, C	Inhibit gate. Combines AND and IF logic. Event A will occur if B occurs and C's value lies in some predetermined range.

FIGURE 2.13 Some standard symbols used in FTA.

Normally, an engineering system's failure tree would have a large number of branches, gates, and elements that need a computer routine to keep track of the analysis. Many FTA programs are commercially available and can be used for generating and evaluating large failure trees.

Case Study 2.6: Application of FTA

Problem
Use FTA to analyze the possible causes of failure in a gearbox.

Solution
Figure 2.14 gives a simple analysis of the common causes of failure that are encountered in a mechanical gearbox.

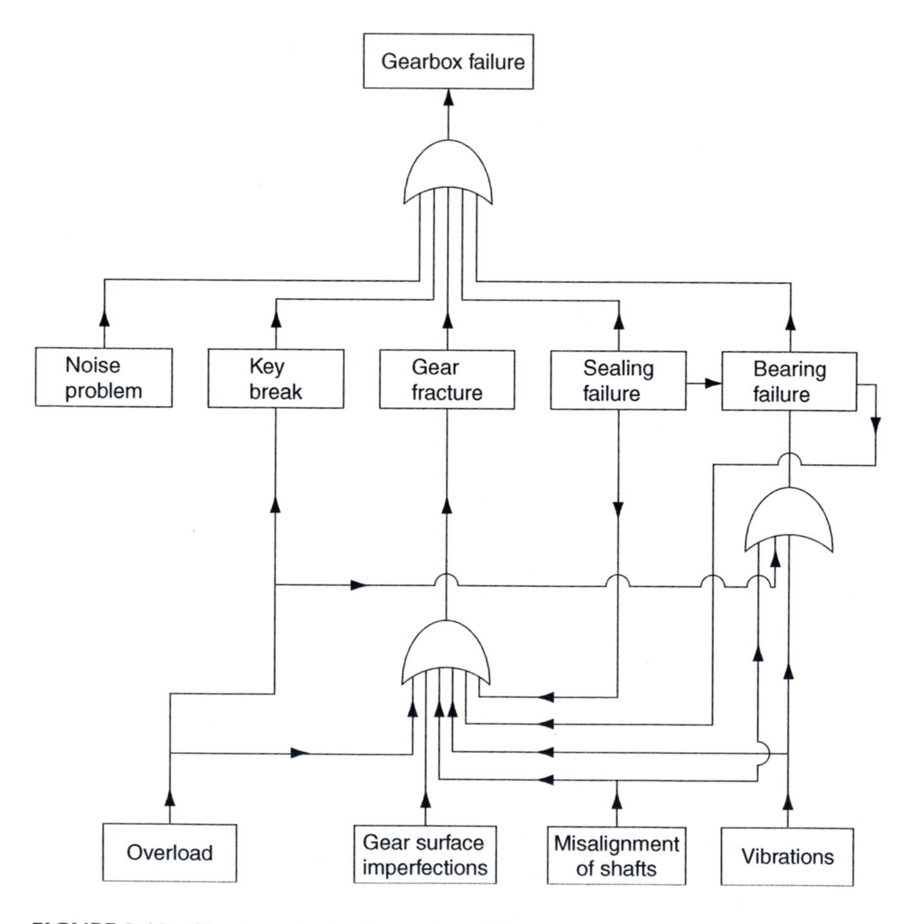

FIGURE 2.14 Simple analysis of a gearbox failure.

In addition to providing a qualitative view of the impact of each element on the system, FTA can be used to quantify the top event probabilities from reliability predictions of different events. In these calculations, an AND gate multiplies probabilities of failure and an OR gate acts additively. Two events that are connected by an OR gate will have larger contribution to the failure of the system than similar two events that are connected by an AND gate, as shown in Figure 2.15. Comparing events 1 and 2 with events 3 and 4 shows that the former events have much larger contribution to the higher event and need more accurate assessment and more care in design.

In performing an FTA, care should also be taken to identify common mode, or common cause, failures. These can lead to the failure of all paths in a redundant configuration, which practically eliminates its advantage. Examples of sources of common mode failures include

1. failure of a power or fuel supply that is common to the main and backup units;
2. failure of a changeover system to activate redundant units; and
3. failure of an item causing an overload and failure of the next item in series or the redundant unit.

An indirect source of failure that should be identified in performing FTA is the enabling event. This event may not necessarily be a failure in itself, but could cause

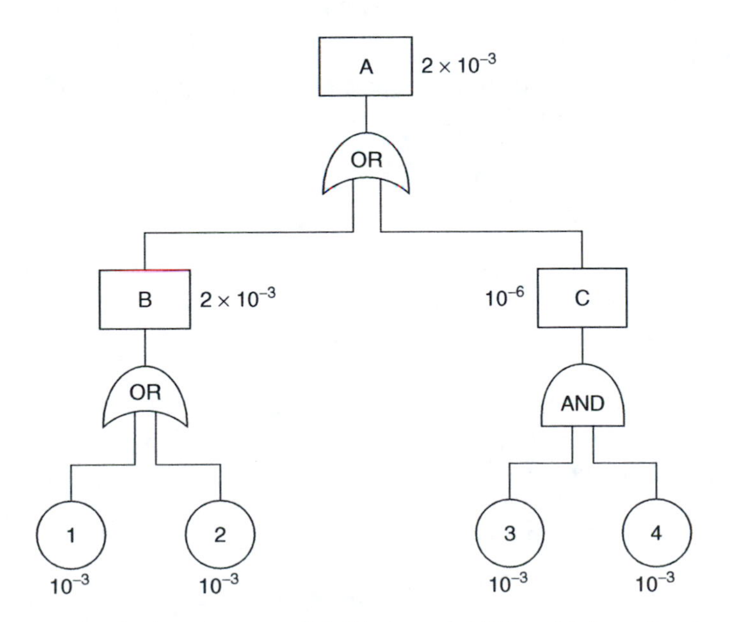

FIGURE 2.15 Sensitivity of system reliability to probabilities of failure of various components. Numbers beside each event represent probabilities.

a higher level failure event when accompanied by a failure. Examples of enabling events include

1. redundant system being out of action due to maintenance,
2. warning system disabled for maintenance, and
3. setting the controls incorrectly or not following standard procedures.

Beside their use in design and reliability assessment, FTA can also be used as a tool in troubleshooting, failure analysis, and accident investigations.

2.8.2 Failure Logic Model

The materials failure logic model (MFLM) proposed by Marriott and Miller (1982) is based on the assumptions that

1. material failure can be modeled as a logic sequence of elementary go/no-go events, and
2. each material failure mechanism can be characterized by a logic expression, which serves to identify that mechanism regardless of context.

The following case study is given as an example to illustrate the use of MFLM.

Case Study 2.7: The Use of MFLM in Failure Analysis

Problem

Consider a welded low-carbon steel pressure vessel, which failed during commissioning at less than operating load.

Solution

The failure event can be described as

$$F = A.B.(C1 + C2).D.E.G.H. \quad (2.8)$$

where

A = Low-alloy steel
B = Heat treatment defect resulting in brittle structure
C1 = Welding defect
C2 = Residual stress from welding
D = Presence of corrosive environment
E = High residual stresses as a result of inappropriate post weld heat treatment
G = Failure of nondestructive tests to detect initial defect
H = Failure to detect incorrect heat treatment of material
( ).( ) = Boolean AND operator
( ) + ( ) = Boolean OR operator

In this case, either of the following logic events could have been sufficient to cause failure:

$$F1 = A.B.C1.G.H. \quad (2.9)$$

This means that the initial defect, combined with the brittle structure, constituted a major risk.

$$F2 = A.B.C2.D.E.H. \quad (2.10)$$

This means that stress corrosion cracking is likely to lead to crack growth even in the absence of initial defect.

2.8.3 Failure Experience Matrix

Collins and Daniewicz (2006) introduced the failure experience matrix as a means of storing failure information for mechanical systems. The matrix is three dimensional, as shown in Figure 2.16, with the axes defined as follows:

1. *Failure modes*. This axis covers the different types of failure, for example, fatigue, corrosion, and wear.
2. *Elemental mechanical functions*. This axis covers all the different functions that are normally performed by mechanical elements. Examples

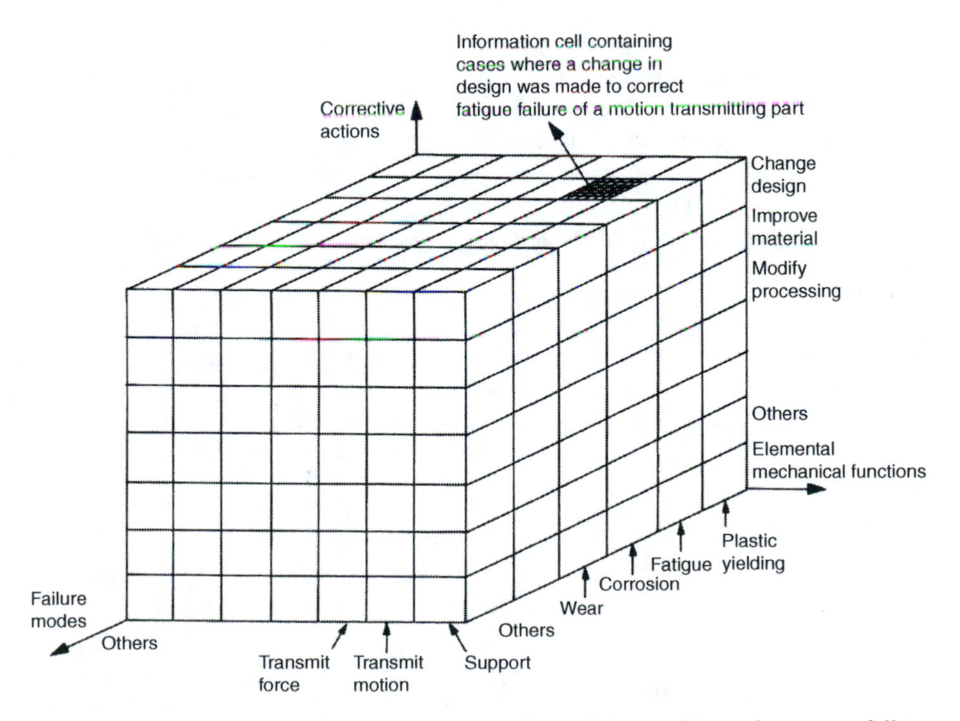

FIGURE 2.16 Part of failure experience matrix, which can be used to store failure information.

include supporting, force transmitting, shielding, sliding, fastening, liquid storing, pumping, and damping.

3. *Corrective actions.* This axis gives any measure or combination of steps taken to return a failed component or system to satisfactory performance. Examples of corrective actions include design change, change of material, improved quality control, change of lubricant, revised procurement specifications, and change of vendor.

This system can be computerized and could be of help to engineers in designing critical components. If the function of the component is entered, the system will give the most likely modes of failure and the corrective actions needed to avert them.

2.8.4 Expert Systems

Another technique that was proposed by Weiss (1986) uses expert systems for failure analysis. A logic program is written using LOGLISP language, which is a combination of logic, that is, predicate calculus and resolution; and LISP, which is the language usually used for artificial intelligence.

The principal ingredients for failure analysis are symptom–cause relationships, and facts and rules about the system under consideration. For example, the presence of a neck is a symptom of a failure due to tensile overload. Symptoms can be related to loading, for example, tension, torsion, bending, and fatigue; and failure mode, for example, neck, dimples, shearlips, beach marks, or cleavage.

Based on the observed symptoms, the expert system program gives the possible causes of failure. Introducing more than one symptom for a given failure reduces the number of possible causes of failure. For a realistic failure analysis case, the expert system needs to interface with a material database and probably a finite element stress analysis program. More information about expert systems is given in Section 9.7.

2.8.5 Failure Prevention at the Design Stage

Anticipating the different ways by which a product could fail while still at the design stage is an important factor that should be considered when selecting a material or a manufacturing process for a given application. The possibility of failure of a component can be analyzed by studying on-the-job material characteristics, the stresses and other environmental parameters that will be acting on the component, and the possible manufacturing defects that can lead to failure. The various sections of the analysis include the following:

1. *Environmental profile.* This provides a description of the expected service conditions that include operating temperature and atmosphere, radiation, presence of contaminants and corrosive media, other materials in contact with the component and the possibility of galvanic corrosion, and lubrication.
2. *Fabrication and process flow diagram.* Such flow diagrams provide an account of the effect of the various stages of production on the material properties and of the possibility of quality control. Certain processes can lead to undesirable directional properties, internal stresses, cracking, or

structural damage, which can lead to unsatisfactory component performance and premature failure in service.

3. *Failure logic models*. These models describe all possible types of failure and the conditions that can lead to those failures. In addition to its use in failure analysis, the MFLM described earlier can be interfaced with a computer-aided design system to aid the designer in the identification of potential failures.

2.9 SUMMARY

1. Causes of failure of engineering components can usually be attributed to design deficiencies, poor selection of materials, manufacturing defects, exceeding design limits and overloading, or inadequate maintenance.
2. The general types of mechanical failure include yielding, buckling, creep, wear, fracture, stress corrosion, and failure under impact loading.
3. Fracture toughness is defined as the resistance of materials to the propagation of an existing crack and is a function of the critical stress intensity factor K_{IC}. Fracture toughness is widely used as a design criterion for high-strength materials where ductility is limited. Higher values of K_{IC}/YS, where YS is the yield strength, are more desirable as they indicate tolerance to larger flaws without fracture.
4. Brittle fractures of metals are usually associated with low temperatures and usually take place at stress raisers such as sharp corners, surface defects, inclusions, or cracks. Steels and other materials that have bcc lattice are prone to brittle fracture at temperatures below the ductile–brittle transition.
5. Fatigue failures account for the largest number of mechanical failures in practice and occur in components that are subjected to fluctuating loads. The fatigue strength of most steels is usually about 0.4–0.6 times the tensile strength.
6. Creep is a major factor at high temperatures and can cause fracture at strains much less than the fracture strains in tensile tests. When the applied creep load is fluctuating, failure takes place by a combination of creep and fatigue.
7. Thermal fatigue takes place as a result of repeated changes in temperature. Faster changes in temperature, lower thermal conductivity, higher elastic constant, higher thermal expansion coefficient, lower ductility, and thicker sections encourage thermal fatigue. Ceramics are particularly prone to this type of failure.
8. Several experimental and analytical techniques are available for the analysis of failure and for predicting its occurrence at the design stage.

2.10 REVIEW QUESTIONS

2.1 K_{IC} for the aluminum alloy used in making a structure is 45 MPa $(m)^{1/2}$. If the structure contains a crack 2.5 mm long, what is the applied stress that will cause fracture? Assume $Y = 1$.

2.2 If the available NDT equipment can detect internal cracks 1 mm in length or longer, determine whether or not alloy AA 7475-T651 with YS = 462

MPa and $K_{IC} = 47$ MPa $(m)^{1/2}$ can be safely used to make a component that will be subjected to a tensile stress of 390 MPa. Assume $Y = 2$.

2.3 Explain the difference between alternating stress and fluctuating stress cycles. Which one of these loading modes is encountered in the motorcar rear axle and the connecting rod of an internal combustion engine?

2.4 Why is fatigue failure a potentially serious problem in many welded steel structures? What are the best ways of avoiding such failure?

2.5 A manufacturer of sports equipment is considering the possibility of using fiber-reinforced plastics in making racing bicycle frames. It is expected that fatigue failures of the joints could be a problem in this case. Describe a design and testing program that can solve this problem.

2.6 Use FTA to analyze the possible causes of failure of a motorcar to start.

2.7 Maraging 300 and AISI 4340 ($T = 260°C$) steels are being considered for making a structure. If the available NDT equipment can only detect flaws greater than 3 mm in length, can we safely use either of these alloys for designing a component that will be subjected to a stress of 600 MPa? Use the information in Table 4.6 and take $Y = 1$.

BIBLIOGRAPHY AND FURTHER READING

Blinn, M.P. and Williams, R.A., Design for fracture toughness, in *Materials Selection and Design, ASM Handbook*, Vol. 20, Dieter, G.E., Editor. ASM International, Materials Park, OH, 1997, pp. 533–544.

Bowman, K., *Introduction to Mechanical Behavior of Materials*, Wiley, New York, 2003.

Boyer, H.E. and Gall, T.L., *Metals Handbook, Desk Edition*, ASM, Metals Park, OH, 1985.

Brooks, C.R. and Choudhury, A., *Failure Analysis of Engineering Materials*, McGraw-Hill, New York, 2001.

Colangelo, V.J. and Heiser, F.A., *Analysis of Metallurgical Failures*, Wiley, New York, 1987.

Collins, J.A. and Daniewicz, S.R., Failure modes: performance and service requirements for metals, in *Handbook of Materials Selection*, Kutz, M., Editor. Wiley, New York, 2002, pp. 705–773.

Collins, J.A. and Daniewicz, S.R., Failure modes: performance and service requirements for metals, in *Mechanical Engineers' Handbook: Materials and Mechanical Design*, 3rd Ed., Kutz, M., Editor. Wiley, Hoboken, NJ, 2006, pp. 860–924.

Cook, N.H., *Mechanics and Materials for Design*, McGraw-Hill, New York, 1985.

Courtney, T.H., *Mechanical Behavior of Materials*, 2nd Ed., McGraw-Hill College, Blacklick, OH, 1999.

Das, A.K., *Metallurgy of Failure Analysis*, McGraw-Hill, New York, 1997.

Dieter, G., *ASM Metals Handbook Vol. 20: Materials Selection and Design*, ASM International, Materials Park, OH, 1997.

Dowling, N., *Mechanical Behavior of Materials*, 3rd Ed., Prentice-Hall, New York, 2006.

Farley, J.M. and Nickols, R.W., *Non-Destructive Testing*, Pergamon Press, London, 1988.

Flinn, R.A. and Trojan, P.K., *Engineering Materials and Their Applications*, 4th Ed., Houghton Mifflin Co., Boston, MA, 1990.

Hosford, W.F., *Mechanical Behavior of Materials*, Cambridge University Press, London, 2005.

Huffman, D.D., Metals Handbook, 8th Ed., Vol. 11, Failure Analysis and Prevention, ASM International, Materials Park, OH, 1988.

Jones, D.R.H., *Failure Analysis Case Studies II*, Pergamon Press, Oxford, 2001.

Kutz, M., *Handbook of Materials Selection*, Wiley, New York, NJ, 2002.

Kutz, M., *Mechanical Engineers' Handbook: Materials and Mechanical Design*, 3rd Ed., Wiley, Hoboken, NJ, 2006.

Marriott, D.L. and Miller, N.R., Materials failure logic models, *Trans. ASME, J. Mech. Design*, 104, 628–634, 1982.

Parker, A.P., *The Mechanics of Fracture and Fatigue*, E.&F.N. Spon Ltd., London, 1981.

Rolfe, S.T. and Barson, J.M., *Fracture and Fatigue Control in Structures*, Prentice-Hall, Englewood Cliffs, NJ, 1977.

Rollason, E.C., *Metallurgy for Engineers*, 4th Ed., Edward Arnold, London, 1977.

Schaffer, J.P., Saxena, A., Antolovich, S.D., Sanders, T.H., Jr., and Warner, S.B., *The Science and Design of Engineering Materials*, McGraw-Hill, Boston, MA, 1999.

Tawancy, H.M., Ul Hamid, A., and Abbas, N.M., *Practical Engineering Failure Analysis*, Marcel Dekker, New York, 2004.

Weiss, V., Towards failure analysis expert systems, *ASTM Standardization News*, April, pp. 30–34, 1986.

Wulpi, D.J., *Understanding How Materials Fail*, ASM, Metals Park, OH, 1985.

3 Environmental Degradation

3.1 INTRODUCTION

Engineering materials, to varying degrees, are susceptible to degradation by the environment in which they serve. Environmental degradation can be classified into three main categories:

1. Corrosion and oxidation
2. Wear
3. Radiation damage

Corrosion may be defined as the unintended destructive chemical or electrochemical reaction of a material with its environment. Metallic, polymeric, and ceramic materials are susceptible to attack from different environments and although the corrosion of metals is electrochemical in nature, the corrosion of other materials usually involves chemical reaction.

Oxidation represents a direct chemical reaction between the material and oxygen. There are various mechanisms for building up an oxide layer on the material surface. For some metals, such as pure aluminum, the oxide layer is strong and impervious and provides protection against further oxidation. For others, such as plain carbon steels, the oxide layer is weak and porous and is not protective.

The nature, composition, and uniformity of the environment and the attacked surface can greatly influence the type, rate, and extent of attack. In addition, externally imposed changes, and changes that occur as a result of corrosion and oxidation processes themselves are known to influence the type and rate of attack. Corrosion and oxidation frequently lead to failure of engineering components or render them susceptible to failure by some other mechanism. The rate and extent of corrosive attack that can be tolerated in a certain component depends on the application. For example, in many structural applications some uniform corrosion or oxidation can be allowed, while in food-processing equipment, for instance, even a minute amount of metal dissolution is not tolerated.

Wear is another form of material degradation that is usually mechanical, rather than chemical, in nature. Material is removed from the surface by the mechanical action of another solid or liquid. The rate of wear is usually accelerated in the presence of corrosion.

Radiation can cause damage to all types of materials. The nature of damage varies with the nature of radiation. Damage by ultraviolet radiation is principally encountered in polymers. In nuclear reactors, damage to the construction materials occurs by the bombardment of neutrons. Although damage takes place on the atomic scale, it generally leads to large-scale changes in strength and ductility.

The goal of this chapter is to illustrate how engineering materials degrade as a result of environmental attack and the measures that can be taken to prevent or delay the harmful effects of degradation. The main objectives are to

1. review the electrochemical principles of corrosion and to use them to describe the different types of corrosion in metallic materials,
2. examine the combined effect of stress and corrosion on the behavior of materials in service,
3. discuss the corrosion of polymers and ceramics,
4. examine how metallic materials oxidize in service,
5. describe widely used methods of corrosion control, and
6. discuss the different types of wear and radiation damages.

3.2 ELECTROCHEMICAL PRINCIPLES OF CORROSION

In the case of metallic materials, where corrosion takes place by electrochemical attack, the corroding metal is the anode in a galvanic cell and the cathode can be another metal, a conducting nonmetal or an oxide, as shown in Figure 3.1. The reaction can be written as

$$M \rightarrow M^{n+} + ne^- \tag{3.1}$$

where M stands for the metal atom, which emits n electrons and becomes a positive ion.

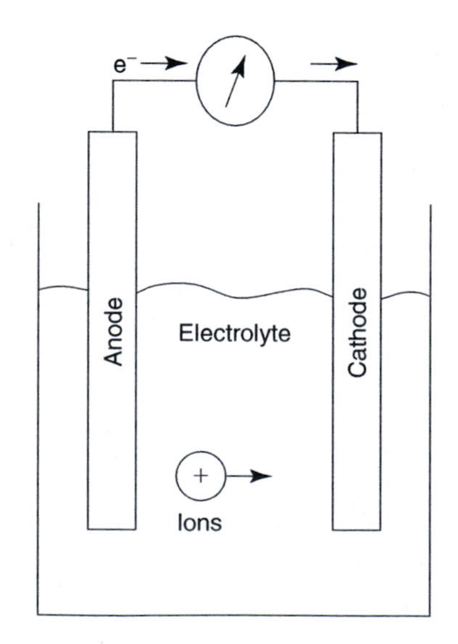

FIGURE 3.1 Electrochemical cell.

In oxygen-free liquids, such as stagnant water or HCl, the cathode-reduction reaction results in the evolution of hydrogen, usually called hydrogen electrode.

$$2H^+ + 2e^- \rightarrow H_2 \tag{3.2}$$

In aerated water, oxygen is available and an oxygen electrode is formed.

$$O_2 + 2H_2O + 4e^- \rightarrow 4OH^- \tag{3.3}$$

This reaction enriches the electrolyte in OH^- ions that react with the metal ions, M^{n+}, to form a solid product. For example, Fe^{2+} combines with two OH^- ions to form $Fe(OH)_2$, or rust.

When two metals are placed in a galvanic cell, one of them assumes the role of the anode and the other assumes the role of the cathode based on their relative tendency to ionize. For example, iron becomes the anode when placed with copper in a galvanic cell because of its stronger tendency to ionize. However, iron becomes the cathode when placed with zinc in the galvanic cell because of the stronger tendency of zinc to ionize. Table 3.1 ranks some common metals and alloys in order of their tendency to ionize in seawater. This galvanic series is a useful guide to design engineers in predicting the relative behavior of electrically connected metals and alloys in marine applications. Figure 3.2 shows examples of galvanic corrosion, in which dissimilar metals are unwisely placed in contact in the presence of an electrolyte.

Corrosion by galvanic action can also take place in a single electrode as a result of local variations in metal composition or ion concentration in the electrolyte. For example, if a piece of iron is immersed in oxygenated water, ferric hydroxide, $Fe(OH)_3$, will form at microscopic local anodes as shown in Figure 3.3. In this case, the anodic reaction is

$$Fe \rightarrow Fe^{2+} + 2e^- \tag{3.4}$$

The reaction at the local cathode is

$$O_2 + 2H_2O + 4e^- \rightarrow 4OH^- \tag{3.5}$$

The overall reaction is obtained by adding the two reactions 3.4 and 3.5 to give

$$2Fe + 2H_2O + O_2 \rightarrow 2Fe^{2+} + 4OH^- \rightarrow 2Fe(OH)_2 \tag{3.6}$$

The ferrous hydroxide is further oxidized to ferric hydroxide to give the red color of iron rust

$$2Fe(OH)_2 + H_2O + \frac{1}{2}O_2 \rightarrow 2Fe(OH)_3 \tag{3.7}$$

TABLE 3.1
Position of some Metallic Materials in the Galvanic Series Based on Seawater

Protected, noble, or cathodic end

Platinum

Gold

Graphite

Titanium

Silver

Chlorimet 3 (61 Ni, 18 Cr, 18 Mo)
Hastelloy C (62 Ni, 17 Cr, 15 Mo)
Inconel 625 (61 Ni, 21.5 Cr, 9 Mo, 3.6 Nb)

Incoloy 825 (21.5 Cr, 42 Ni, 3 Mo, 30 Fe)
Type 316 stainless steel (passive)
Type 304 stainless steel (passive)
Type 410 stainless steel (passive)

Monel alloy 400 (66.5 Ni, 31.5 Cu)

Inconel alloy 600 (passive) (76 Ni, 15.5 Cr, 8 Fe)
Nickel 200 (passive) (99.5 Ni)

Leaded tin bronze G, 923, cast (87 Cu, 8 Sn, 4 Zn)
Silicon bronze C65500 (97 Cu, 5 Al)
Admiralty brass C44300, C44400, C44500 (71 Cu, 28 Zn, 1 Sn)

Chlorimet 2 (66 Ni, 32 Mo, 1 Fe)
Hastelloy B (60 Ni, 30 Mo, 6 Fe, 1 Mn)

Inconel 600 (active)
Nickel 200 (active)

Naval brass C46400 to C46700 (60 Cu, 39.25 Zn, 0.75 Sn)
Muntz metal C28000 (60 Cu, 40 Zn)

Tin
Lead

Type 316 stainless steel (active)
Type 304 stainless steel (active)

Lead–tin solder (50 Sn, 50 Pb)

Cast irons
Low-carbon steels

Aluminium alloy 2117 (2.6 Cu, 0.35 Mg)
Aluminium alloy 2024 (4.5 Cu, 1.5 Mg, 0.6 Mn)

Aluminium alloy 5052 (2.5 Mg, 0.25 Cr)
Aluminium alloy 3004 (1.2 Mn, 1 Mg)
Aluminium 1100, commercial-purity aluminium (99 Al min, 0.12 Cu)

Galvanized steel
Zinc

Magnesium alloys
Magnesium

Corroded, anodic, least noble end

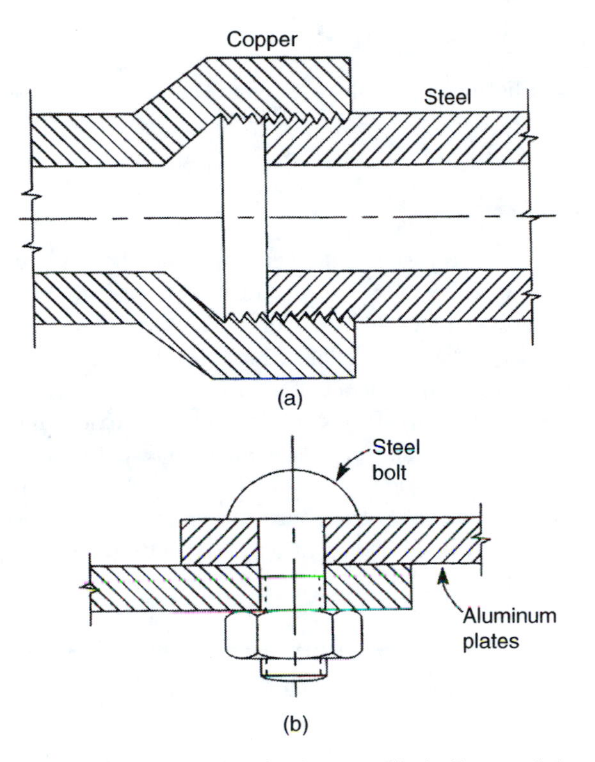

FIGURE 3.2 Examples of galvanic corrosion between dissimilar metals in contact with one another in the presence of an electrolyte.

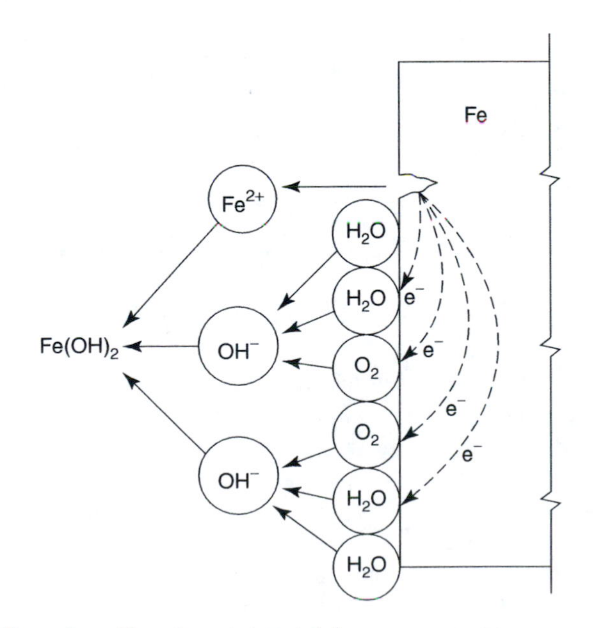

FIGURE 3.3 Corrosion of iron in water containing oxygen.

3.3 TYPES OF ELECTROCHEMICAL CORROSION

Corrosion of metallic materials may occur in a number of forms, which differ in appearance.

3.3.1 General Corrosion

General, or atmospheric corrosion of metals is probably the most commonly encountered, and the most significant in terms of economic losses, form of corrosion. When a metal is exposed to the atmosphere, its surface is covered with a thin layer of condensed or adsorbed water, even at relative humidity less than 100%, and this layer can act as the electrolyte. The presence of industrial contaminants in the atmosphere increases the corrosion rate. Examples are dust, sulfur dioxide, and ammonium sulfate. Sodium chloride is also an impurity, which is present in marine atmospheres, and it increases the corrosion rate.

General corrosion does not usually lead to sudden or unexpected failure, but gradual reduction in thickness needs to be taken into account during the design stage, as illustrated in Example 3.1.

Design Example 3.1: Effect of General Corrosion on Service Life

Problem

The rate of corrosion of a steel tank is measured regularly and is approximately constant, 50 mg/dm^2/day. What is the useful life of the tank if the initial thickness is 10 mm and the minimum safe thickness is 6 mm?

Solution

Taking the density of steel as 7.8 g/cc,

The weight loss $= 0.050 \text{ g} = 10 \times 10 \times 7.8 \times t$

$t =$ the thickness loss per day $= 6.41 \times 10^{-5}$ cm $= 6.41 \times 10^{-4}$ mm

$$\text{Useful life} = \frac{(10 - 6)}{t} = 6.24 \times 10^3 \text{ days} = 17 \text{ years}$$

3.3.2 Galvanic Corrosion

When dissimilar metals are in electrical contact in an electrolyte, the less noble metal becomes the anode in the galvanic cell and is attacked to a greater extent than if it were exposed alone. The more noble metal becomes the cathode and is attacked to a lesser extent than if it were exposed alone. The severity of galvanic corrosion depends on the separation of the two metals in the galvanic series (Table 3.1). In most cases, metals from one group can be coupled with one another without causing a substantial increase in the corrosion rate.

Another factor that affects the severity of galvanic corrosion is the relative areas of the anodic metal to the cathodic metal. Because the density of current is higher with small anode, a steel rivet in a copper plate will be more severely corroded than a steel plate containing a copper rivet.

Galvanic corrosion can also take place between two different areas of a structure, which is made of the same metal and immersed in the same electrolyte, if the contact areas are at different temperatures. For a steel structure in contact with dilute aerated chloride solution, the warmer area is anodic to the colder area, whereas for copper in aqueous salt solution, the warmer area is cathodic to the colder area.

If a structure, which is made of the same material, is in contact with two different concentrations of an electrolyte, concentration–cell corrosion will take place. This type of attack is known to take place in buried metals as a result of their being in contact with soils that have different chemical compositions, especially with respect to the concentration of sodium chloride, sodium sulfate, and organic acids. Differences in water contents or degrees of aeration can also be detrimental. Corrective action in such cases usually involves coating of the buried metal in asphalt, enclosing in a concrete trough, or adopting cathodic protection.

Design Example 3.2: Avoiding Galvanic Corrosion

Problem

Plain carbon steel bolts, which were used to fasten an aluminum roof truss exhibited severe corrosive attack.

1. How would you explain this problem?
2. What action would you recommend to avoid this problem?

Analysis

According to the galvanic series in Table 3.1, steel should be protected since it is higher than aluminum. However, with the tendency of aluminum to form strong nonporous oxide film, the galvanic couple is actually between Al_2O_3 and steel, with steel being the anode. The corrosion rate of the steel bolts is increased as a result of their small area compared with the cathode.

Solution

It is recommended to use high-strength aluminum alloy fasteners. Although aluminum fasteners are more expensive than steel bolts, they are expected to give a longer trouble-free service.

3.3.3 Crevice Corrosion

Crevice corrosion occurs within confined spaces or crevices formed when components are in close contact. A crevice at a joint between two metallic surfaces or between a metallic and nonmetallic surface provides conditions for concentration–cell corrosion, as shown in Figure 3.4. The area of the metallic surface just inside the crevice becomes anodic and suffers faster attack than other surfaces as a result of the oxygen

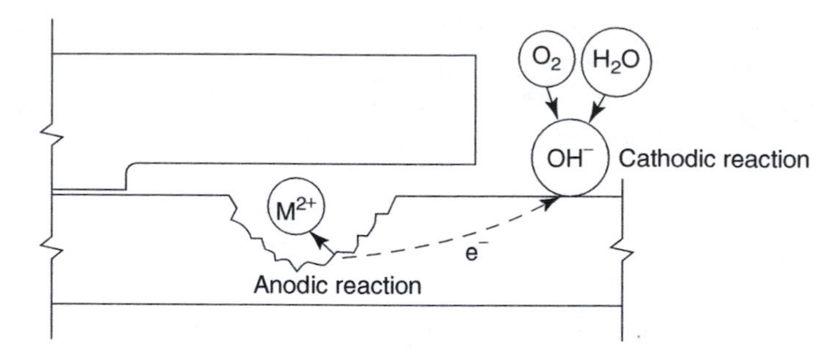

FIGURE 3.4 Crevice corrosion as a result of differences in concentration.

concentration difference in the two locations. For crevice corrosion to occur, the crevice must be wide enough to allow liquid to enter but sufficiently narrow to keep the liquid stagnant.

Crevice corrosion occurs under gaskets, rivets and bolts, and under porous deposits. This type of corrosion is called poultice corrosion when it occurs beneath the shielded areas caused by mud splashes and road debris thrown by motorcar tires on the underside of the fenders and other parts of the car body. Crevice corrosion occurs in many metallic materials including carbon and stainless steels, and titanium, aluminum, and copper alloys.

Crevice corrosion can be minimized by taking appropriate precautions in design and manufacture of components and assemblies. These include using weldments instead of rivets, using nonabsorbent gaskets, and provision of complete drainage in vessels where stagnant solutions may accumulate.

3.3.4 Pitting Corrosion

Pitting is a form of localized attack that produces pits or holes in a metal. Pitting corrosion occurs when one area of the surface becomes anodic with respect to the rest of the surface due to segregation of alloying elements or inclusions in the microstructure. Surface deposits that set up local concentration cells, dissolved halides that produce local anodes by rupture of the protective oxide film, or mechanical ruptures in protective organic coatings, are also common sources of pitting corrosion. Differences in ion and oxygen concentrations create concentration cells, which can also initiate pits.

Failure due to pitting can occur unexpectedly because of its localized nature. Pits may also contribute to the initiation of fatigue cracks in components subjected to fatigue by acting as notches.

Selection of materials with adequate pitting resistance is among the measures that can be taken to avoid failures. For example, titanium and type 316 stainless steel have better pitting corrosion resistance than type 304 stainless steel.

3.3.5 Intergranular Corrosion

Intergranular attack is another type of localized corrosion, which takes place at grain boundaries when they become more susceptible to corrosion than the bulk of the grains.

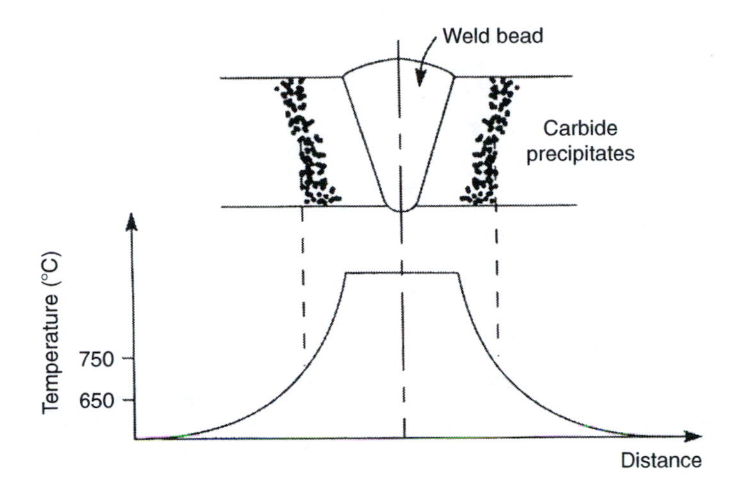

FIGURE 3.5 Corrosion of a sensitized stainless steel near the weld area.

Intergranular attack is often strongly dependent on the mechanical and thermal treatment given to the alloy. For example, unstabilized stainless steels are susceptible to intergranular corrosion when heated in the temperature range of 550–850°C (1000–1550°F). In this sensitizing temperature range, chromium combines with carbon to form chromium carbides, which precipitate at the grain boundaries, and this depletes the neighboring areas of chromium. In many corrosive environments, the chromium-depleted areas are attacked. Intergranular attack may take place in welded joints in the areas that were heated to the sensitizing temperature range, as shown in Figure 3.5. Dissolving chromium carbides by solution heat treatment at 1060–1120°C (about 1950–2050°F) followed by water-quenching eliminates sensitization. Susceptibility of stainless steels to sensitization can also be reduced by reducing the carbon content to less than 0.03% as in the case of extra-low-carbon grades, for example, 304L, or by adding sufficient titanium and niobium to combine with all the carbon in the steel, for example, 347 or 321 stainless steels. The following case study illustrates how intergranular corrosion occurs in practice and how to avoid it.

Design Example 3.3: Failure of an Exhaust Pipe Assembly

Failure Analysis

An exhaust pipe assembly of a racing motorcar was found to be cracked at the toe of a weld between a pipe and flange after one season of racing. Examination of the cracked surface with a magnifying lens showed signs of brittle fracture, and no measurable deformation, and microscopic examination of polished and etched samples of the fractured pipe revealed relatively large precipitates, some of which were connected by cracks at the grain boundaries in the heat-affected zone. Chemical analysis gave the following results: C = 0.15%, Cr = 18%, and Ni = 9%, which are close to AISI 302 stainless steel.

Conclusion

Failure took place as a result of intergranular corrosion as a result of Cr depletion near chromium carbide precipitates.

Recommended Action

After welding of the exhaust pipe assembly, heat to a temperature of 1100°C to dissolve chromium carbides and cool rapidly by quenching in water.

3.3.6 Selective Leaching

Some alloys are susceptible to selective leaching, or dissolution, where the less corrosion-resistant element is removed by corrosion. Common examples include dezincification, where zinc is removed from brasses, and graphitic corrosion or graphitization, where iron is removed from gray cast irons. The severest attack occurs when the dissolved material is present as a continuous phase. There may be a little change in the geometry of the component, but the mechanical properties are severely reduced and unexpected fracture could occur. Dezincification can be minimized by changing to a brass with lower zinc content (85% Cu, 15% Zn) or to cupronickel (70–90% Cu, 10–30% Ni).

3.4 COMBINED ACTION OF STRESS AND CORROSION

During their service life, components may be subjected to corrosion in addition to the normal stresses they are designed to bear. Experience has shown that the corrosion rate of some materials is accelerated as a result of stress. In some cases, chemical attack does not take place in the absence of stresses.

3.4.1 Stress Corrosion Cracking

Stress corrosion cracking (SCC) occurs in some alloys as a result of the combined effect of tensile stresses and chemical attack. The stresses involved in SCC can either be due to normal service loads, or due to residual stresses resulting from manufacturing and assembly processes. Examples of manufacturing processes, which could lead to residual stresses include casting, welding, cold forming, and heat treatment.

Normally, a threshold stress is required for SCC to occur, and shorter lives are expected with higher stresses. This threshold stress may be as low as 10% of the yield stress and is not usually a practical design stress. Susceptibility of an alloy to SCC is often a function of the content of major alloying elements, such as nickel and chromium in stainless steels and zinc in brasses. Increasing the strength is also known to increase the susceptibility to SCC.

Another important factor that affects the occurrence of SCC is the environment. The presence of certain ions, even in small concentrations, can be detrimental to some alloys but not to others. For example, stainless steels crack in chloride environments but not in ammonia-containing environments, whereas brasses crack in ammonia-containing environments but not in chlorides.

In the presence of certain chemicals and under the influence of stress, some plastics can fail by gradual cracking. This is known as environmental stress cracking. For example, polyethylenes suffer environmental stress cracking in detergents and oils.

SSC may be reduced or prevented by using one or more of the following methods:

1. Lowering the stress below the threshold value by eliminating residual stresses and reducing externally applied stresses.
2. Eliminating the critical environmental species.
3. Selecting the appropriate alloy. For example, carbon steels, rather than stainless steels, are often used in the construction of heat exchangers used in contact with seawater. This is because carbon steels are more resistant to SCC although they are less resistant to general corrosion than stainless steels.
4. Applying cathodic protection, as discussed in Section 3.7.
5. Adding inhibitors to the system, as discussed in Section 3.7.
6. Applying protective coatings, as discussed in Section 4.9.
7. Introducing residual compressive stresses in the surface by processes such as shot peening to avoid residual tensile stresses, which increase SCC.

3.4.2 Corrosion Fatigue

Corrosion fatigue is caused by the combined effects of fluctuating stresses and corrosive environment. Unfavorable environments cause fatigue cracks to be initiated within fewer cycles and increase the crack growth rate, thus reducing the fatigue life. For example, the fatigue strength of smooth samples of high-strength steel in saltwater can be as little as 12% of that in dry air as shown in Table 6.4. Under saltwater conditions, the smooth surface is attacked creating local stress raisers that make the initiation of fatigue cracks much easier. Saltwater also increases crack growth rate in steels. However, austenitic stainless steels and aluminum bronzes retain about 75% of their normal fatigue strength when tested in seawater.

Under corrosion fatigue conditions, the frequency of the stress cycle, the shape of the stress wave, the stress ratio, as well as the magnitude of the cyclic stress and the number of cycles affect the fatigue life. Generally, corrosion fatigue strength decreases as the stress frequency decreases, because this allows more time for interaction between the material and environment. This effect is most important at frequencies of less than 10 Hz. The temperature, pH, and aeration of the environment affect the corrosion fatigue life.

Many of the methods used to reduce or eliminate SCC can also be used to combat corrosion fatigue. Among the possible methods are the following:

1. Reducing the applied stress by changing the design and eliminating tensile residual stresses
2. Introducing residual compressive stresses in the surface, to avoid tensile stresses, which increase corrosion fatigue
3. Selecting the appropriate materials

4. Using corrosion inhibitors
5. Applying protective coatings

3.4.3 Erosion Corrosion

Erosion corrosion can be defined as the acceleration of the rate of corrosion in a metallic material under wear and abrasion conditions. Metallic surfaces that have been subjected to erosion corrosion are characterized by the appearance of grooves, valleys, pits, and other means of surface damage, which usually occur in the direction of motion.

The increased corrosion rate in carbon steel pipes conveying sand slurries and similar sludge is believed to be due to the removal of the surface rust by the abrasive action of the hard suspended particles, thus allowing easy access of dissolved oxygen to the corroding surface.

3.4.4 Cavitation Damage

Cavitation damage occurs in a metalic surface where high-velocity liquid flow and pressure changes exist, as in the case of pump impellers and ship propellers. The damage is caused as a result of the formation and collapse of air bubbles or vapor-filled bubbles in the liquid near the surface. It has been shown that localized pressures as high as 400 MPa (60,000 psi) can be generated as a result of collapsing vapor bubbles. Such pressures are capable of removing surface films or even tearing metal particles away from the surface. Cavitation damage increases the corrosion and wear rates.

3.4.5 Fretting Corrosion

Fretting corrosion takes place between mating surfaces that are subjected to sliding and vibrations, as in the case of shafts and bearings. The damage appears as grooves or pits surrounded by corrosion products, which have been torn loose by the wearing action. Damage is accelerated as more debris accumulate and act as an abrasive between the two surfaces.

3.5 CORROSION OF PLASTICS AND CERAMICS

Plastics and ceramics do not corrode in the same way as metals since they are electrical insulators. They do not suffer any of the damage caused by corrosion in metals. In many cases plastic parts can be used to insulate metal parts from corrosive interaction. However, plastics and ceramics can suffer chemical reaction, dissolution, or absorption, depending on the type of material and the nature of solution.

3.5.1 Corrosion of Plastics

Plastics are only slightly affected by the atmosphere, but can be affected by sunlight. Plastic coatings degrade and crack as they lose their plasticizers and cross-link by oxygen. Ultraviolet radiation from sunlight accelerates this degradation.

TABLE 3.2
Relative Chemical Stability of Selected Polymeric Materials

Material	Water Absorption[a]	Weak Acids	Strong Acids	Weak Alkalis	Strong Alkalis	Organic Solvents
Thermoplastics						
Fluoroplastics	0	5	5	5	5	5
Polyethylene	0.01–0.02	5	2	5	5	5
Polyvinylidene chloride	0.04–0.10	5	3	5	5	3
Vinylchloride	0.45	5	3	5	3	2
Polycarbonate	—	5	3	5	3	2
Acrylics	0.03	3	3	5	3	1
Polyamides (nylon)	1.5	3	1	5	3	3
Acetals	—	2	1	2	1	5
Polystyrene	0.04	3	1	3	1	1
Cellulose acetate	3.80	3	1	3	1	2
Thermosets						
Epoxy	0.10	5	3	5	5	3
Melamine	0.30	5	1	5	1	5
Silicones	0.15	5	3	3	2	3
Polyesters	0.01	3	1	2	1	2
Ureas	0.60	3	1	2	1	2
Phenolics	0.07–1.00	2	1	2	1	3

Note: 1. Poor: rapid attack; 2. Fair: temporary use; 3. Good: reasonable service; 4. Very good: reliable service; and 5. Excellent: unlimited service.

[a] After 24 h of immersion (wt.%).

Plastics are generally resistant to water but there is a small percentage of water absorption except for teflon. Polyethylene, acrylics, and polyester are less absorbent than others (Table 3.2).

Plastics show wide variation in their resistance to chemicals, but most of them are resistant to weak acids and alkalis. Strong acids, strong alkalis, and organic solvents attack certain plastics, as shown in Table 3.2. An important rule in predicting the performance of plastics in organic solvents is that "like dissolves like." For example, straight-chain polymers tend to dissolve in straight-chain solvents such as ethyl alcohol, whereas those with benzene rings tend to dissolve in benzene and other aromatic solvents. Table 3.3 gives the effect of some chemicals on the strength of selected plastics. Increasing the molecular weight and crystallinity decreases the attack by organic solvents. The most resistant plastics are teflon, polyethylene, and vinyl.

3.5.2 Corrosion of Ceramics

Ceramics are only slightly affected by the atmosphere. The main danger is the effect of water as it enters the cracks or joints and expands on freezing. Salt in water aggravates this problem. Acids associated with air pollution could also cause damage.

TABLE 3.3
Relative Tensile Strength in Polymers after 24 h Exposure to Chemicals at 93°C

	Polyphenylene Sulfide	Nylon 6/6	Polycarbonate	Polysulfone
Hydrochloric acid (37%)	100	0	0	100
Sulfuric acid (30%)	100	0	100	100
Sodium hydroxide (30%)	100	89	7	100
Gasoline	100	80	99	100
Chloroform	87	57	0	0
Ethylene chloride	72	65	0	0
Phenol	100	0	0	0
Ethyl acetate	100	89	0	0

Source: Data based on *Guide to Engineering Materials, Advanced Materials and Processes*, Vol. 138, ASM International, Materials Park, OH, 1988.

Nonporous ceramics are widely used for water containment, as in the case of glass-lined and enameled steel tanks, water pipes, etc.

There are wide differences in the resistance of ceramics to chemicals. Fused silica and borosilicate glasses are very resistant, but soda-lime-silica glasses are slowly attacked by alkalis. Glasses are attacked by hydrofluoric acid (HF). Organic solvents have no effect on ceramics.

3.6 OXIDATION OF MATERIALS

Many materials, metallic and nonmetallic, combine with oxygen during service, especially at elevated temperatures. In the case of metals, the high-temperature oxidation is particularly important in the design of gas turbines, rocket engines, and high-temperature petrochemical equipment. Unlike electrochemical corrosion, oxidation does not require an electrolyte as part of the process.

3.6.1 Oxidation of Metals

In metals, oxidation often starts rapidly and continues until an oxide film or scale is formed on the surface. After this stage, the rate of further oxidation depends on the soundness of the oxide film. The degree of protection provided by the oxide film to the metal surface depends on several factors including the following:

1. The ratio of volume of the oxide layer to the volume of the metal used in forming it should be close to unity to avoid excessive internal stresses in the oxide layer.
2. The film should strongly adhere to the metal surface to avoid peeling off.

3. The coefficient of expansion of the oxide should be close to that of the metal to avoid excessive internal stresses in the oxide layer on cooling from the oxidation temperature.
4. The film should have low conductivity and low diffusion coefficients for metal ions and oxygen to prevent or reduce further oxidation of the metal.
5. The film should have good, high-temperature plasticity, high melting point, and low vapor pressure to avoid cracking and ensure high-temperature stability.

If the oxide film is porous and allows continuous access of oxygen to the metal surface, oxidation will continue until all the material is oxidized. Examples of such metals include sodium and potassium. The rate at which oxidation occurs, as represented by the oxide film thickness, T, in such cases can be given by a linear relationship:

$$T = Kt \tag{3.8}$$

where K is a constant and t the exposure time.

A parabolic relationship is observed when diffusion of ions or electrons through a nonporous oxide film is the controlling factor, as in the case of iron, copper, and nickel:

$$T = (Kt)^{1/2} \tag{3.9}$$

A logarithmic relationship is observed when the oxide film is dense, impervious, and exceptionally protective against further oxidation, as in the case of aluminum and chromium. The degree of protection provided by the oxide film increases as its thickness increases, and oxidation practically stops after a critical thickness is reached.

$$T = K \log (ct + 1) \tag{3.10}$$

where c is a constant.

The ratio of the volume of the oxide film to the volume of the metal used in making it is called the Pilling–Bedworth (P–B) ratio. It may be calculated from the relationship:

$$\text{P–B ratio} = \frac{\text{volume of oxide produced by oxidation}}{\text{volume of metal consumed by oxidation}} = \frac{Wd}{Dw} \tag{3.11}$$

where W and w are the molecular weights of the oxide molecules and metal atoms, respectively and D and d are the densities of the oxide and metal, respectively.

When the P–B ratio is much less than unity, the oxide is nonprotective because it will be porous and cracked. However, if it is much greater than unity, the oxide may crack off because of the volume difference. The following example illustrates how the P–B ratio is calculated.

Design Example 3.4: Calculation of the P–B Ratio

Problem

Compare the P–B ratios of the oxidation of Al, Mg, and W and use the results to explain the behavior of their oxide layers.

Solution

The oxidation reactions can be represented by

$$4Al + 3O_2 \rightarrow 2Al_2O_3$$

$$2Mg + O_2 \rightarrow 2MgO$$

$$2W + 3O_2 \rightarrow 2WO_3$$

The molecular weights of Al, Mg, and W are 26.98, 24.32, and 183.85 and those of their oxides are 101.96, 40.32, and 231.85, respectively. The densities of Al, Mg, and W are 2.7, 1.74, and 19.25 and those of their oxides are 3.7, 3.58, and 7.3, respectively.

$$\text{P–B ratio for Al} = \frac{(101.96 \times 2.7)}{(2 \times 26.98 \times 3.7)} = 1.379$$

$$\text{P–B ratio for Mg} = \frac{(40.32 \times 1.74)}{(24.32 \times 3.58)} = 0.806$$

$$\text{P–B ratio for W} = \frac{(231.85 \times 19.25)}{(183.85 \times 7.3)} = 3.325$$

The P–B ratio for Al is close to unity and is, therefore, expected to provide better protection than either that for Mg, which is less than unity, or W, which is higher than unity. Experience substantiates these observations.

In many cases, the compositions and characteristics of oxide films can be changed by adding alloying elements to the base metal. For example, chromium, aluminum, and silicon are added to iron to modify its normally porous oxide layer and make it more protective. Materials for elevated-temperature service can also be protected against oxidation by applying protective coatings.

3.6.2 Oxidation of Plastics

Most plastics and rubbers oxidize in the presence of oxygen. Rubbers are especially susceptible to oxidation and the process is called ageing. The reaction of oxygen with rubber initially reduces elasticity and increases hardness. This is because oxygen diffuses into the structure and provides additional cross-linking. As ageing proceeds, the rubber degrades, and eventually loses most of its strength. The rate of ageing depends on the temperature, type of atmosphere, material composition, and method of manufacture.

Oxygen may also cause depolymerization, or chain scission, permitting small molecules to escape as a gas, or cause charring or even burning of the polymer at high temperatures. Polymers based on silicon rather than carbon are more resistant to oxidation and can be used at higher temperatures.

3.6.3 Oxidation of Ceramics

Most oxide ceramics are not significantly affected by oxygen, even at high temperatures. Carbides and nitrides can oxidize, which limits their use at high temperatures.

3.7 CORROSION CONTROL

Corrosion can be prevented or at least controlled by several methods. The selection of the method of control is usually influenced by the cost. For example, it may be more economical to make a certain component out of a less-expensive but less-resistant material and periodically replace it, than to make it from a more-resistant material, which is also more expensive. The deleterious effects of the different types of corrosion can be eliminated or at least reduced by adopting one or more of the following preventive measures:

1. Using galvanic protection, as discussed in this section
2. Using corrosion inhibitors, as discussed in this section
3. Selecting the appropriate material, as discussed in Section 4.8
4. Using protective coatings, as discussed in Section 4.9
5. Observing certain design rules, as discussed in Section 6.7

3.7.1 Galvanic Protection

Galvanic protection methods are used to protect metallic structures and can be divided into cathodic protection and anodic protection. Cathodic protection is achieved by supplying electrons to the component to be protected. For example, the corrosion of a steel structure in an acidic environment involves the following reactions:

$$Fe \rightarrow Fe^{2+} + 2e^- \tag{3.12}$$

$$2H^+ + 2e^- \rightarrow H_2 \tag{3.13}$$

Supplying electrons to the steel from another source will suppress the corrosion reaction in Equation 3.12 and increase the hydrogen evolution in Equation 3.13.

Electrons for cathodic protection can be supplied by

1. connecting the steel structure to a more anodic metal, sacrificial anode, such as Zn, Mg, or Al, as shown in Figure 3.6a. The sacrificial anode will

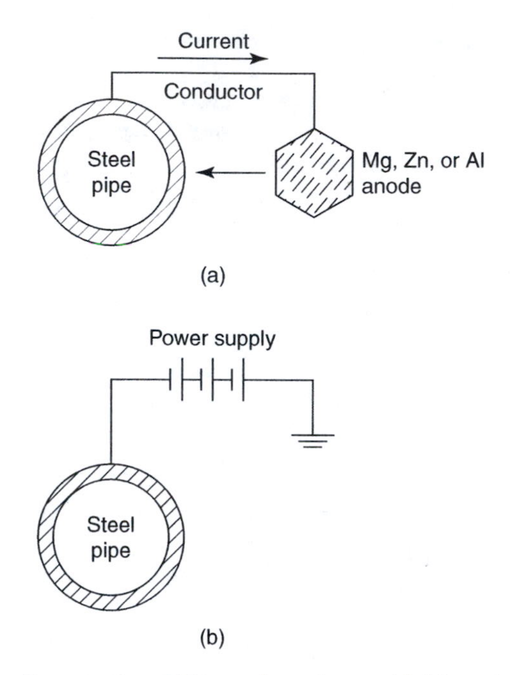

FIGURE 3.6 Cathodic protection. (a) Protection using sacrificial anode, (b) protection using impressed voltage to oppose the electrochemical potential difference.

corrode instead of the steel and is replaced periodically. This method is commonly used in protecting pipelines, ship hulls, and marine structures; and

2. using an external power supply to provide a voltage, which will oppose the one caused by electrochemical reaction, as shown in Figure 3.6b. This voltage stops the flow of electrons needed for the corrosion reaction to proceed.

Anodic protection can only be used for metallic materials that passivate and is based on the formation of a protective passive film on the surface to be protected by externally impressed anodic current. A potentiostat is used to control the anodic current, which is normally very small. The high installation cost is a disadvantage of this method. The following example illustrates the use of sacrificial anode in corrosion protection.

Design Example 3.5: Calculating the Life of a Sacrificial Anode

Problem

A steel ship hull is protected by a magnesium sacrificial anode weighing 3 kg. The current produced by the anode was found to be 2 A. Estimate the expected life of the Mg anode.

Solution

The weight of metal w lost in uniform corrosion can be estimated from Faraday's equation:

$$w = \frac{(ItM)}{(nF)} \tag{3.14}$$

where

w = Weight of metal corroded in grams
I = Corrosion current in amperes
t = Time in seconds
n = Number of electrons/atoms involved in the process
M = Atomic mass of the corroding metal in gram per mol
F = Faraday's constant = 96,500 As/mol

The corrosion reaction of Mg is

$$Mg \rightarrow Mg^{2+} + 2e^{-2}$$

From Equation 3.14:

$$\text{The expected life of the anode} = t = \frac{(wnF)}{(IM)} = \frac{(3000 \times 2 \times 96500)}{(2 \times 24.31)}$$

$$= 11.9 \times 10^6\,\text{s} = 4.6\ \text{months}$$

3.7.2 Inhibitors

Adding inhibitors to a system can decrease corrosion either by inhibiting the corrosion reaction at the anode or the cathode. In many cases, the role of the inhibitor is to form an impervious, insulating film of a compound either on the anode or the cathode. Most inhibitors have been developed by empirical methods and are of proprietary nature. Adsorption inhibitors are adsorbed on the surface and form a protective film. The scavenger inhibitors react to remove corrosion agents such as oxygen from solution.

A common example of inhibitors is chromate salts in motorcar radiators. The iron ions liberated at the anode surface combine with the chromate to form an insoluble film. Many oils, greases, and waxes are adsorbed on metallic surfaces and provide temporary protection.

3.8 WEAR FAILURES

Wear can be defined as the removal of surface material and reduction of dimensions and loss of weight as a result of plastic deformation and detachment of material. This phenomenon normally occurs in components whose function involves sliding or rolling contact with other surfaces. Examples of such components include sliding and rolling bearings, gears and splines, piston rings, and breaks and clutches. Wear can also take place as a result of contact with moving liquids or gases, especially when they contain hard particles. Wear in any material can occur by a variety of mechanisms, depending on the relative motion (rolling, sliding, or a combination), service environment (dry or wet and whether abrasive particles are present), and type of motion (impact, unidirectional, or oscillating). Therefore, life under wear conditions is a function of both the wear resistance of the material and the wear system variables.

The following sections describe the different types of wear, which include adhesive wear, abrasive and erosive wear, surface fatigue, corrosive wear, erosion corrosion, and fretting. Erosion corrosion and fretting are discussed in Section 3.4. Table 3.4 shows the predominant type of wear as a function of the type of motion and service environment.

TABLE 3.4
The Predominant Type of Wear as a Function of the Type of Motion and Service Environment

		Predominant Mechanism		
Type of Motion	**Environment**	**Surface Fatigue**	**Adhesive**	**Abrasive**
Pure rolling	All environments	x		
Rolling with slip	Without abrasive particles	x	x	
	With abrasive particles	x	x	x
Sliding in one direction	Dry	x	x	
	Fluid	x	x	
	With abrasive particles	x	x	x
Sliding cyclic	Dry	x	x	
	Fluid	x	x	
	With abrasive particles	x	x	x
Impact with another moving body	Dry or wet without particles	x	x	
	Dry or wet with particles	x	x	x
Impingement with fluid	Without abrasive particles	x		
	With abrasive particles	x	x	
	Streamline without particles			
	Streamline with particles		x	
	Turbulent without particles		x	
	Turbulent with particles	x	x	

Source: Based on Bayer, R.G. in *Materials Selection and Design, ASM Handbook*, ASM International, Materials Park, OH, 1997, 603–614.

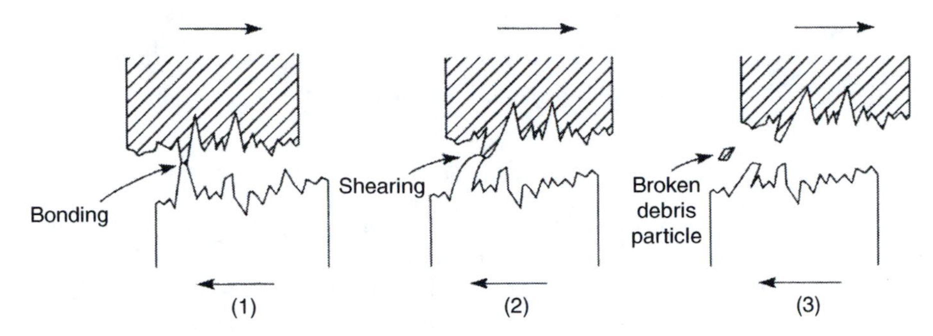

FIGURE 3.7 Schematic representation of the processes involved in adhesive wear.

3.8.1 Adhesive Wear

Adhesive wear, also known as scoring, galling, or seizing, occurs when two surfaces slide against each other under pressure. The process involves plastic deformation followed by adhesion and then fracture of surface asperities, as shown in Figure 3.7. The fractured asperities frequently form tiny abrasive particles, which contribute to further wear of the surfaces.

The volume of material (V) removed from a surface of hardness (H) under a load (P) and a sliding distance (D) can be estimated from the relationship

$$V = \frac{(kPD)}{(3H)} \tag{3.15}$$

where k is a dimensionless quantity referred to as the wear coefficient. Representative values of k are

0.045, for low-carbon steel on low carbon steel
0.0015, for copper on low-carbon steel
0.021, for stainless steel on stainless steel
0.00002, for phenol formaldehyde on phenol formaldehyde

Relationship 3.15 shows that the adhesive wear is reduced when the load is reduced or the hardness increased. It is also generally better to avoid contact between similar metals, since adhesion takes place more readily. Smoother surfaces will also wear at a slower rate.

Design Example 3.6: Designing for Adhesive Wear

Problem
A rotating steel shaft is supported by a sleeve bearing. Compare the rate of wear when the sleeve is (a) made of the same steel and (b) made of copper.

Analysis

Assuming that the hardness of steel is 54 R_c and the hardness of copper is 36 R_c, from Equation 3.15, the relative volume loss due to wear in (a) and (b) is given by

$$\frac{V_a}{V_b} = \frac{(0.045/54)}{(0.0015/36)} \approx 20$$

If the volume loss is taken as an indication of the expected life of the bearing, the mentioned figures indicate that using a copper sleeve will allow the bearing to last about 20 times longer than if a steel sleeve is used. Introducing a lubricant with the copper sleeve will allow the bearing life to be even longer.

3.8.2 Abrasive, Erosive, and Cavitation Wear

Abrasive wear results from contact with hard projections on a mating surface, or with loose hard particles that are trapped between two surfaces. Unlike adhesive wear, no bonding occurs. This type of wear is common in earth-moving equipment, scraper blades, and crushers. Materials with high hardness exhibit better resistance to abrasive wear.

Erosion usually takes place as a result of contact with a moving fluid that contains hard particles. Erosion can also be caused by liquid impingement caused by droplets carried in a fast-moving gas.

Cavitation is a related type of wear, which occurs when a liquid containing dissolved gas enters a low-pressure region. Gas bubbles precipitate from the liquid and then subsequently collapse when the pressure increases again. The collapse of the gas bubbles sends high-pressure waves that exert high pressures against the surrounding surface. Cavitation is frequently encountered in hydraulic pumps, propellers, and dams.

3.8.3 Surface Fatigue

Surface fatigue is a special type of surface damage where parts of the surface are detached under externally applied cyclic stresses. Surface or subsurface crack formation causes particles to separate from the surface causing pitting or spalling. This is an important source of failure in rolling-contact systems such as ball bearings and railway lines.

3.8.4 Lubrication

In practice, determining the cause of wear may be difficult because failure may have resulted from the combined effect of different types of wear modes. In addition, as wear progresses there may be a change in the predominant wear mode.

An important method of combating wear is lubrication. Many kinds of surface films can act as lubricants, preventing cold welding of asperities and thus reducing friction and wear. Lubricants may be solid, liquid, or gas. Liquid lubricants have the advantage of combining cooling with lubricating action. In systems, which depend on lubricants to combat wear, failure of the lubricant can lead to scuffing, galling, or even seizure. Most lubricant failures occur due to chemical decomposition, contamination, or change in properties due to excessive heat. In many cases in practice, more than one of the discussed causes can be involved in lubricant failures.

3.9 RADIATION DAMAGE

In many applications, materials are subjected to radiation fields as part of their service function. Examples of such applications include nuclear power generation, radiation therapy, and communication satellites. Radiation can either be electromagnetic waves or particles, as described in Table 3.5.

3.9.1 Radiation Damage by Electromagnetic Radiation

In the case of electromagnetic radiation, the radiation energy E increases with decreasing wavelength, λ, according to the relation:

$$E = \frac{(hc)}{\lambda} \tag{3.16}$$

where h is the Plank's constant $= 0.6626 \times 10^{-33}$ J/s and c the speed of light $= 0.2998 \times 10^{9}$ m/s.

The amount of radiation damage depends on the energy of radiation, the radiation density, time of exposure, and the bond strength of the material receiving the radiation.

Electromagnetic radiation may be absorbed by the material, thus becoming a potential source of damage, or transmitted through it, thus causing no damage.

Polymers are especially susceptible to ultraviolet (UV) radiation damage. A single UV photon has sufficient energy to break the C–C bond in many linear-chain polymers. Breaking in C–C bond reduces the molecular weight and strength, encourages cross-linking, and also provides sites for oxidation. Adding carbon black or TiO_2 to polymers, or applying an appropriate coating material reduces the damage caused by UV radiation.

TABLE 3.5
Types and Characteristics of Radiations

Category of Radiation	Characteristics
Electromagnetic	
Ultraviolet	Wavelength 1–400 nm
X-ray	Wavelength 0.001–10 nm
γ-ray	Wavelength < 0.1 nm
Particles	
α particles	Helium nucleus (two protons + two neutrons)
β particles	Positive or negative particles with a mass of an electron

3.9.2 Radiation Damage by Particles

In the case of metals, radiation damage caused by particles such as neutrons involves knocking atoms out of their normal lattice sites and creating interstitials and vacancies, which can collect to form dislocations. These point and line defects reduce electrical conductivity and ductility and increase hardness. Annealing reduces the number of vacancies and dislocation lines, and thus reduces or eliminates radiation damage.

Ceramics are also affected by particle radiation. Thermal conductivity and optical properties may be impaired. In the case of ionic bonds, the damage is similar to that in metals and can be eliminated by annealing. In the case of covalent bonds, however, the damage causes irreversible breaking of the bonds and results in modification of the structure.

3.10 SUMMARY

1. Corrosion may be defined as the unintended destructive chemical or electrochemical reaction of a material with its environment.
2. Corrosion of metallic materials may occur in a number of forms that differ in appearance. These include general corrosion, galvanic corrosion, crevice corrosion, pitting, intergranular corrosion, selective leaching, stress corrosion, corrosion fatigue, erosion corrosion, cavitation damage, and fretting corrosion.
3. Corrosion of plastics and ceramics takes place by chemical reaction. Most plastics resist a variety of chemicals but are usually attacked by organic solvents. Glasses resist most chemicals but are attacked by HF acid.
4. Corrosion can be controlled using galvanic protection, corrosion inhibitors, material selection, protective coatings, or observing certain design rules.
5. Oxidation is the reaction between the material and oxygen. Oxidation usually starts rapidly and then slows down depending on the degree of protection provided by the oxide film. The P–B ratio, which is the ratio of volume of oxide to the volume of metal consumed by oxidation, gives a good indication of the degree of protection. Metals with P–B ratios close to unity are expected to have protective oxides.
6. Wear can be defined as the removal of surface material as a result of mechanical action. Adhesive wear, also known as scoring, galling, or seizing, occurs when two surfaces slide against each other under pressure. Abrasive and erosive wear results from contact with hard particles. Most types of wear can be reduced by selecting a hard surface and using a lubricant.
7. Radiation damage can be the result of electromagnetic waves such as ultraviolet radiation, x-rays, γ-rays, or particles such as neutrons and α or β particles. The amount of damage depends on the energy of radiation, radiation density, time of exposure, and the bond strength of the material receiving radiation.

3.11 REVIEW QUESTIONS

3.1 What are the differences in the mechanism of corrosion protection by galvanizing and chrome plating when used for plain carbon steels?

3.2 Discuss the mechanisms by which plastics are attacked by the environment and explain why they do not corrode in the same way as metallic materials.

3.3 The following methods are used to protect steel sheets against corrosion: galvanizing, tinning, and coating by polymeric paints. Explain the mechanism by which steel is protected.

3.4 Compare the merits of using galvanizing and tinning in protecting steel for (a) food cans and (b) outdoor fencing.

3.5 Discuss two of the methods that can be used in practice to eliminate or reduce corrosion attack. Give examples of where each method is used.

3.6 The wall thickness of a steel tank is measured monthly and the loss in thickness is approximately the same each month, 50 mg/dm^2/day. What is the useful life of the tank, if the initial thickness is 10 mm and the minimum safe thickness is 6 mm? (Answer: 17 years).

3.7 A steel tank of 1 m diameter is made of welded AISI 1030 steel sheets of 350 MPa yield strength. The loss in thickness due to corrosion is constant and equals 50 mg/dm^2/month. If the pressure in the tank is 3.5 MPa and expected life is 10 years, what is the starting thickness of the steel sheet? Take a weld efficiency of 0.7 and a factor of safety of 2. You may treat the tank as a thin-walled container and calculate the stress in walls (σ) according to the formula $\sigma = (pr)/t$, where p is the pressure, r the radius of the tank, and t the wall thickness. What are the possible measures that can be taken to reduce the corrosion rate?

BIBLIOGRAPHY AND FURTHER READING

Bayer, R.G., Design for wear resistance, in *Materials Selection and Design, ASM Handbook*, vol. 20, Dieter, G.E., Editor. ASM International, Materials Park, OH, 1997, pp. 603–614.

Billmeyer, F.W., *Textbook of Polymer Science*, Wiley, Sussex, 1984.

Boyer, H.E. and Gall, T.L., *Metals Handbook*, Desk Edition, ASM, Metals Park, OH, 1985.

Brooks, C.R. and Choudhury, A., *Failure Analysis of Engineering Materials*, McGraw-Hill, New York, 2001.

Colangelo, V.J. and Heiser, F.A., *Analysis of Metallurgical Failures*, Wiley, New York, 1987.

Collins, J.A. and Daniewicz, S.R., Failure modes: performance and service requirements for metals, in *Handbook of Materials Selection*, Kutz, M., Editor. Wiley, New York, 2002, pp. 705–773.

Collins, J.A. and Daniewicz, S.R., Failure modes: performance and service requirements for metals, in *Mechanical Engineers' Handbook: Materials and Mechanical Design*, 3rd Ed., Kutz, M., Editor. Wiley, Hoboken, NJ, 2006, pp. 860–924.

Cook, N.H., *Mechanics and Materials for Design*, McGraw-Hill, New York, 1985.

Courtney, T.H., *Mechanical Behavior of Materials*, 2nd Ed., McGraw-Hill College, Blacklick, OH, 1999.

Crawford, R.J., *Plastics Engineering*, Pergamon Press, London, 1987.
Das, A.K., *Metallurgy of Failure Analysis*, McGraw-Hill, New York, 1997.
Dieter, G., *ASM Metals Handbook, Vol. 20, Materials Selection and Design*, ASM International, Materials Park, OH, 1997.
Dowling, N., *Mechanical Behavior of Materials*, 3rd Ed., Prentice-Hall, London, 2006.
Farag, M.M., *Materials Selection for Engineering Design*, Prentice-Hall, London, 1997.
Flinn, R.A. and Trojan, P.K., *Engineering Materials and their Applications*, 4th Ed., Houghton Mifflin Co., Boston, MA, 1990.
Fontana, M.G., *Corrosion Engineering*, 3rd Ed., McGraw-Hill, New York, 1986.
Hosford, W.F., *Mechanical Behavior of Materials*, Cambridge University Press, Cambridge, London, 2005.
Jones, D.R.H., *Failure Analysis Case Studies II*, Pergamon Press, Oxford, 2001.
Khobaib, M. and Krutenant, R.C., *High Temperature Coatings*, The Metals Society, New York, 1986.
Kutz, M., *Handbook of Materials Selection*, Wiley, New York, NJ, 2002.
Kutz, M., *Mechanical Engineers' Handbook: Materials and Mechanical Design*, 3rd Ed., Wiley, Hoboken, NJ, 2006.
Metals Handbook, *Failure Analysis and Prevention*, 8th Ed., vol. 11, ASM, Materials Park, OH, 1988.
Schaffer, J.P., Saxena, A., Antolovich, S.D., Sanders, Jr. T.H., and Warner, S.B., *The Science and Design of Engineering Materials*, McGraw-Hill, Boston, MA, 1999.
Tawancy, H.M., Ul Hamid, A., and Abbas, N.M., *Practical Engineering Failure Analysis*, Marcell Dekker, New York, 2004.
Thornton, P.A. and Colangelo, V.J., *Fundamentals of Engineering Materials*, Prentice-Hall, Englewood Cliffs, NJ, 1985.
Weiss, V., *Towards Failure Analysis Expert Systems*, ASTM Standardization News, April 1986, pp. 30–34.
Wulpi, D.J., *Understanding How Materials Fail*, ASM, Metals Park, OH, 1985.

4 Selection of Materials to Resist Failure

4.1 INTRODUCTION

In addition to design deficiencies, manufacturing defects, overloading, and inadequate maintenance, poor selection of materials is known to be a major source of failure of engineering components in service. Generally, most widely used engineering materials have established ranges of applications and service conditions that match their capabilities. Exceeding such capabilities could lead to a failure in service. However, erring on the conservative side by underutilizing the material capabilities will usually result in uneconomic products.

The discussion in this chapter starts by identifying the material properties that are required for a given type of loading or service environment. The different types of materials that are most suited for a given application are then examined. The objectives of the chapter are to

1. provide an overview of the different engineering materials that are available to the design engineer;
2. identify the material properties that are required to resist failure under mechanical loading—including static, dynamic, fatigue, and creep;
3. review the different types of materials that are most suited for resisting failure under mechanical loading;
4. identify the material properties that are required to resist failure in hostile service environments including carrion and wear; and
5. review the different types of materials that are most suited for resisting failure under corrosion and wear conditions.

4.2 CLASSIFICATION AND DESIGNATION OF ENGINEERING MATERIALS

Modern engineers have a great and diverse range of materials at their disposal. These materials can be conveniently classified into ferrous and nonferrous metals and alloys, nonmetallic organic and inorganic materials, composite materials, and semiconductors. Table 4.1 illustrates this classification and gives examples of materials within each of the classes. Appendix A gives some properties of selective widely used engineering materials.

Although there are large variations in the properties of materials within a given class, there are also similarities. For example, all metallic materials are conductors of heat and electricity, whereas all nonmetallic materials are generally insulating.

TABLE 4.1
Classification of Engineering Materials and Examples of the Different Classes

Engineering Materials					
Metallic Materials		**Composites and Semiconductors**		**Nonmetallic Materials**	
Ferrous metals and alloys	Nonferrous metals and alloys	Composite materials	Semiconductors	Organic nonmetallic	Inorganic nonmetallic

Ferrous metals and alloys. Carbon steels, high-strength low-alloy steels, high-alloy steels, stainless steels, GCI, nodular cast irons, etc.

Nonferrous metals and alloys. Light metals and alloys (Al, Mg, Ti), copper and zinc and their alloys (brasses, bronzes, zamak), low-melting point metals and alloys (Pb, Sn, Bi, Sb, Cd, In), precious metals (Au, Pt, Ag), refractory metals (W, Mo, Ta, Nb), nickel and alloys, superalloys (Fe–Ni-base, Ni-base, Co-base), etc.

Organic nonmetallic materials. Thermoplastics (polyethylene, polystyrene, vinyls, polypropylene, ABS, acrylic, nylon, acetals, polycarbonate, fluoroplastics, polyesters, polyurethane, cellulosics, PEEK, PETE, PMMA), thermosetting plastics (phenolics, epoxy, polyester, silicone, urea, melamine), elastomers (natural rubber, neoprene, butyl rubber, styrene butadiene rubber, silicone elastomers), natural materials (wood, cork, bamboo), etc.

Inorganic nonmetallic materials. Refractory ceramics (oxides, carbides, nitrides), whitewares, clay products, glasses (fused silica, soda-lime, lead-alkali glasses, borosilicates, glass ceramics), bricks, stone, concrete, etc.

Composite materials. Polymer-matrix composites (CFRP, GFRP, KFRP, CNTRP, NFRP, laminated composites, sandwich materials), metal matrix composites (SAP, aluminum–graphite composites, Al–SiC composites, TD nickel), etc.

Semiconductors. Single-crystal silicon, germanium, and gallium arsenide.

Metallic materials have relatively high elastic moduli and are usually ductile. Ceramics also have high elastic moduli, and they are hard and corrosion resistant but are generally brittle. Plastics are generally lighter than metals and ceramics, and they are corrosion resistant but they are softer and have much lower elastic moduli. Composites combine two or more material groups, which allow them to combine the attractive properties of the constituencies.

Although classification is the systematic arrangement or division of materials into groups on the basis of some common characteristics, designation is the identification of each class by a number, letter, symbol, name, or a combination thereof. Designations are developed by professional societies and organizations and are normally based on either the chemical composition or the mechanical properties. Appendix A gives some examples. The unified numbering system (UNS) has been developed by the American Society for Testing and Materials (ASTM), the Society of Automotive Engineers (SAE) and several other technical societies, trade associations, and the U.S. government agencies. The UNS number is a designation of chemical composition and consists of a letter and five numerals. The letter indicates the broad class of alloys, and the numerals define specific alloys within that class. A sample of the UNS numbering system is included in Appendix A.

It should be noted that the designation systems are not specifications but are often incorporated into specifications describing products that are made of the designated materials. A standard specification is a published document that describes the characteristics a product must have to be suitable for a certain application, as discussed in Section 5.4.

In most engineering applications, selection of materials is usually based on one or more of the following considerations:

1. Product shape: (a) sheet, strip, or plate; (b) bar, rod, or wire; (c) tubes; (d) forgings; and (e) castings or moldings
2. Mechanical properties, as ordinarily revealed by the tensile, fatigue, hardness, creep, or impact tests
3. Physical and chemical properties such as specific gravity, thermal and electrical conductivities, thermal expansion coefficient, and corrosion resistance
4. Microstructural considerations such as anisotropy of properties, hardenability of steels, grain size, and consistency of properties, that is, absence of segregations and inclusions
5. Processing considerations such as castability, formability, machinability, weldability, and moldability
6. Esthetic qualities and environmental impact
7. Cost and availability

In the following discussion, the main types of materials are evaluated and compared on the basis of their mechanical, physical, and chemical properties. Processing considerations will be discussed in Chapter 7; economic, esthetic, and environmental considerations will be discussed in Chapter 8.

4.3 SELECTION OF MATERIALS FOR STATIC STRENGTH

4.3.1 Aspects of Static Strength

Static strength can be defined as the ability to resist a short-term steady load at moderate temperatures. This resistance is usually measured in terms of yield strength, ultimate tensile strength, compressive strength, and hardness. When the material does not exhibit a well-defined yield point, the stress required to cause 0.1% or 0.2% plastic strain, that is, the proof stress, is used instead. Most ductile wrought metallic materials are equally strong in tension and compression; brittle materials, however, are generally much stronger in compression than in tension.

Although many engineering materials are almost isotropic, there are important cases where significant anisotropy exists. In the latter cases, the strength depends on the direction in which it is measured. The degree of anisotropy depends on the nature of the material and its manufacturing history. Anisotropy in wrought metallic materials is more pronounced when they contain elongated inclusions and when processing consists of repeated deformation in the same direction. Composites reinforced with unidirectional fibers also exhibit pronounced anisotropy, which can be useful if the principal external stress acts along the direction of highest strength.

4.3.2 Level of Strength

The level of strength in engineering materials may be viewed either in absolute terms or relative to similar materials. For example, it is generally understood that high-strength steels have tensile strength values in excess of 1400 MPa (ca. 200 ksi), which is also high strength in absolute terms. Relative to light alloys, however, an aluminum alloy with a strength of 500 MPa (ca. 72 ksi) would also be designated as high-strength alloy even though this level of strength is low for steels.

From the design point of view, it is more convenient to consider the strength of materials in absolute terms. From the manufacturing point of view, however, it is important to consider the strength as an indication of the degree of development of the material concerned, that is, relative to similar materials. This is because highly developed materials are often complex, more difficult to process, and relatively more expensive compared to other materials in their class. Figure 4.1 gives the strength of some materials, both in absolute terms and relative to similar materials. In a given class of materials, the medium-strength members are usually more widely used because they generally combine optimum strength, ease of manufacture, and economy. The most developed members in a given class of materials are usually

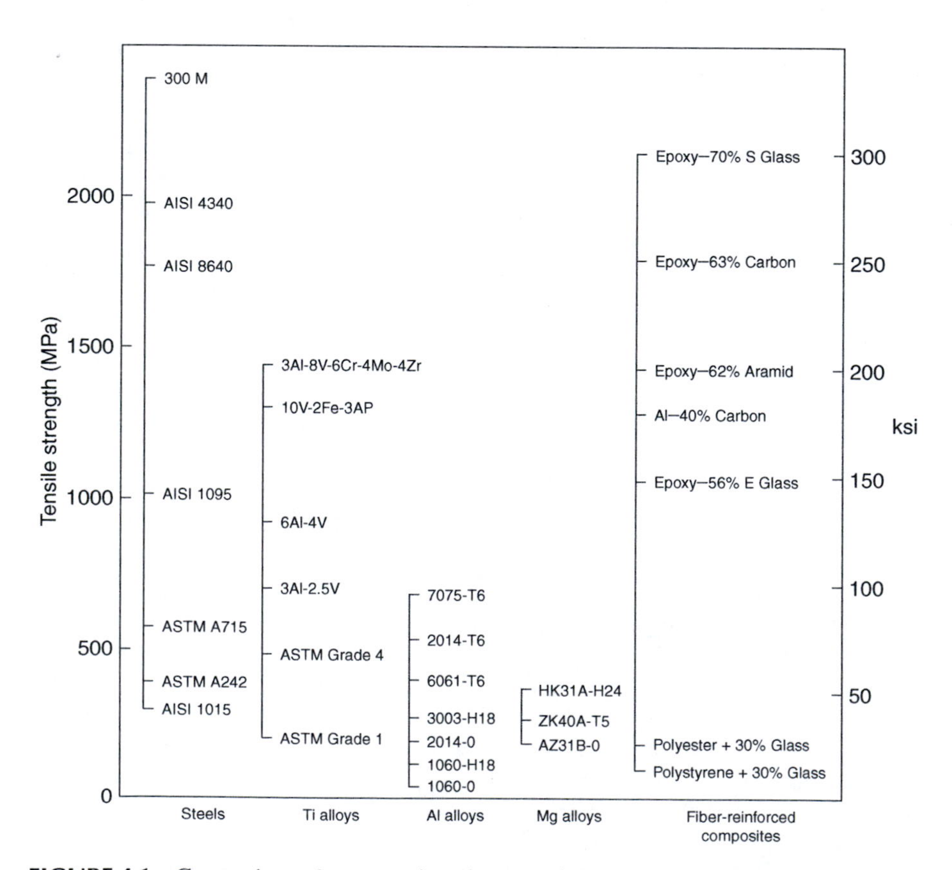

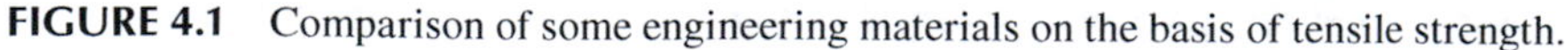
FIGURE 4.1 Comparison of some engineering materials on the basis of tensile strength.

highly specialized and, as a result, they are produced in much lower quantities. The low-strength members of a given class are usually used to meet requirements other than strength. Requirements such as electrical and thermal conductivities, formability, corrosion resistance, or cost may be more important than high strength in some applications.

Frequently, higher-strength members of a given class of materials are also more expensive. Using a stronger but more expensive material could result in a reduction of the total cost of the finished component. This is because the amount of material used would be less as a result of smaller cross section, as discussed in Section 8.5.

4.3.3 Load-Carrying Capacity

The load-carrying capacity of a given component is a function of both the strength of the material used in making it and its dimensions. This means that a lower-strength material can be used in making a component to bear a certain load, provided that its cross-sectional area is increased proportionally. However, the designer is not usually completely free in choosing the strength level of the material selected for a given part. Other factors like space and weight limitations could limit the choice. Space limitations can usually be solved by using stronger material, which will allow smaller cross-sectional area and smaller total volume of the component.

Weight limitations are encountered with many applications including aerospace, transport, construction, and portable appliances. The weight of a component is a function of both its volume and density. For example, the weight w of a tie-rod of cross-sectional area A and length l is given by

$$w = Al\rho = (L/S)l\rho \tag{4.1}$$

where

ρ = Density of the material
L = Applied tensile load
S = Working strength of the material, which is equal to the yield strength of the material divided by a factor of safety, typically in the range of 1.5–3

Equation 4.1 shows that the weight of the tie-rod can be minimized by maximizing the ratio S/ρ, which is the specific strength of the material. In selecting a material for a tie-rod from a list of candidates, S/ρ can be used as the main performance index for comparison. The optimum material in this case is the one that has the highest S/ρ. Figure 4.2 compares the materials of Figure 4.1 on the basis of specific strength. The figure shows that the light alloys—Ti, Al, and Mg—have similar specific strengths to steel, whereas fiber-reinforced composites have a clear advantage over other materials.

Following a procedure similar to Equation 4.1, performance indices of components subjected to different types of loading can be calculated, and the results are shown in Table 4.2. The table shows that although strength and density have equal influence on the performance index in cylinders under tension, compression or internal pressure, density has more influence in the case of flat plates, rectangular sections, and cylinders under torsion or bending since S has a power less than unity.

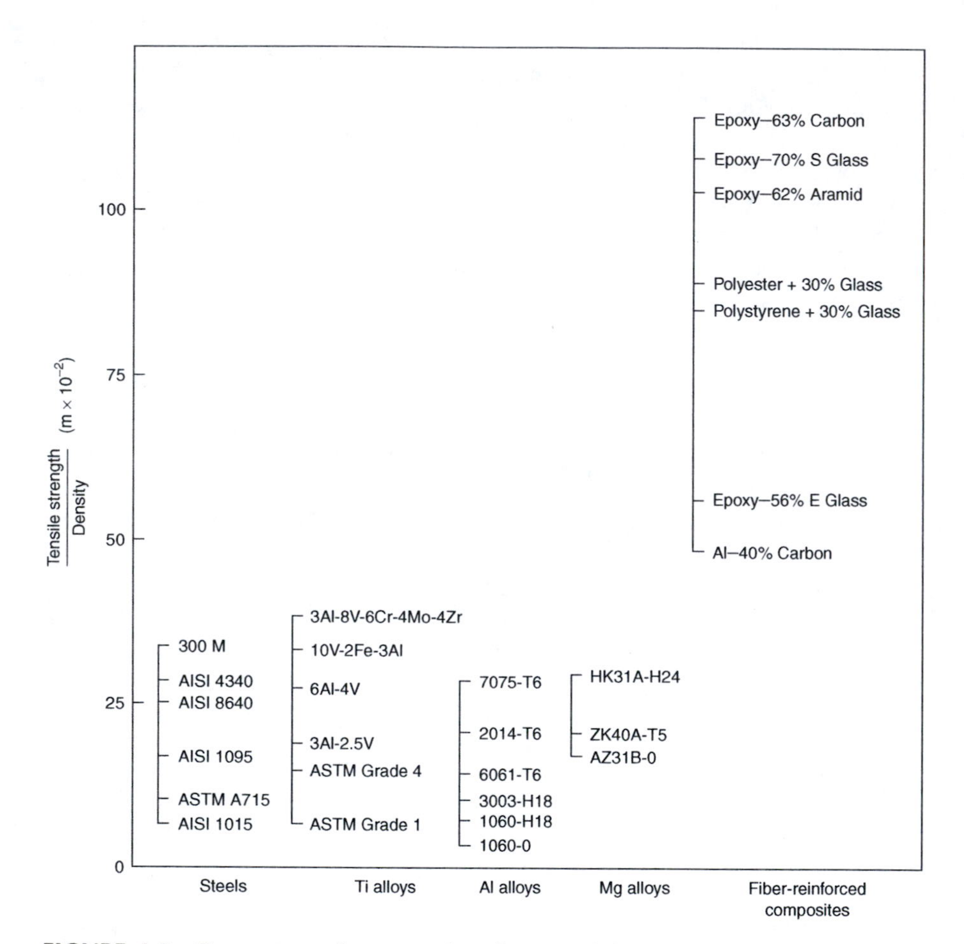

FIGURE 4.2 Comparison of some engineering materials on the basis of specific tensile strength.

TABLE 4.2
Performance Indices in Selection for Static Strength

Cross Section and Loading Condition	Performance Index
Solid cylinder in tension or compression	S/ρ
Solid cylinder in torsion	$S^{2/3}/\rho$
Solid cylinder in bending	$S^{2/3}/\rho$
Solid rectangle in bending	$S^{1/2}/\rho$
Flat plate in bending	$S^{1/2}/\rho$
Flat plate under in-plane compression	$S^{1/2}/\rho$
Thin-walled cylindrical pressure vessel	S/ρ

In cases where a component carries a compressive load, reducing the cross-sectional area by choosing a strong material could cause failure by buckling due to increased slenderness of the part. Example 4.1 illustrates this point.

Design Example 4.1: Material Selection for a Compression Element

Problem

A load of 50 kN is to be supported on a cylindrical compression element of 200 mm length. As the compression element has to fit with other parts of the structure, its diameter should not exceed 20 mm. Weight limitations are such that the mass of the element should not exceed 0.25 kg. Which of the materials given in Table 4.3 is most suited for making the compression element?

TABLE 4.3
Comparison of Compression Element Materials

Material	Strength (MPa)	Elastic Modulus (GPa)	Specific Gravity	Diameter Based on Strength (mm)	Diameter Based on Buckling[a] (mm)	Mass Based on Larger Diameter (kg)	Remarks
Steels							
ASTM A675 Grade 45	155	211	7.8	20.3	15.75	—	Reject (1)
ASTM A675 Grade 80	275	211	7.8	15.2	15.75	0.3	Reject (2)
ASTM 717 Grade 80	550	211	7.8	10.8	15.75	0.3	Reject (2)
Aluminum							
AA 2014- T6	420	70.8	2.7	12.3	20.7	—	Reject (1)
Plastics and Composites							
Nylon 6/6	84	3.3	1.14	27.5	44.6	—	Reject (1)
Epoxy–70% glass	2100	62.3	2.11	5.5	21.4	—	Reject (1)
Epoxy–62% Kevlar	1311	82.8	1.38	7.0	19.9	0.086	Accepted

Note: Reject (1) = material is rejected because it violates the limits on diameter.
Reject (2) = material is rejected because it violates the limits on weight.

[a] Assuming that the ends of the compression element are not constrained, the Euler formula can be used to calculate the minimum diameter that will allow safe use of the compression member without buckling.

Solution

Table 4.2 shows the calculated diameter of the compression element when made of different materials. The diameter is calculated on the basis of strength and buckling. The larger value for a given material is used to calculate the mass of the element. The results in Table 4.3 show that only epoxy–62% Kevlar satisfies both the diameter and weight limits.

4.4 SELECTION OF MATERIALS FOR STIFFNESS

4.4.1 EFFECT OF MATERIAL STIFFNESS ON DEFLECTION UNDER LOAD

Stiffness of a component can be defined as its resistance to deflection under load. In many cases, such resistance is a function of both material stiffness and geometry of the component. For example, when a load is placed on a beam, the beam is bent and every portion of it is moved in a direction parallel to the direction of the load. The distance that a point on the beam moves, deflection, depends on its position in the beam, the type of beam, and the type of supports. A beam that is simply supported at both ends suffers maximum deflection (y) in its middle when subjected to a concentrated central load (L). In this case the maximum deflection, y, is given by

$$y = \frac{(Ll^3)}{(48EI)} \tag{4.2}$$

where

l = Length of the beam
E = Young's modulus of the beam material
I = Second moment of area of the beam cross-section, with respect to the neutral axis

Equation 4.2 shows that the stiffness of a beam may be increased by increasing its second moment of area, which is computed from the cross-sectional dimensions, and by selecting a high-modulus material for its manufacture.

An important characteristic of metallic materials is that their elastic moduli are very difficult to change by changing the composition or heat treatment. However, the elastic moduli of composite materials can be changed over a wide range by changing the volume fraction and orientation of the constituents. Table 4.4 gives representative values of the modulus of elasticity of some engineering materials. When a metallic component is loaded in tension, compression, or bending, the Young's modulus, E, is used in computing its stiffness. When the loading is in shear or torsion, the modulus of rigidity, G, is used in computing stiffness. The relationship between these two elastic constants is given by

$$G = \frac{E}{2(1+\nu)} \tag{4.3}$$

where ν is Poisson's ratio.

TABLE 4.4
Comparison of Stiffness of Selected Engineering Materials

Material	Modulus of Elasticity E (GPa)	Density ρ (mg/m^3)	$(E/\rho)\times 10^{-5}$	$(E^{1/2}/\rho)\times 10^{-2}$	$(E^{1/3}/\rho)$
Steel (carbon and low alloy)	207	7.825	26.5	5.8	35.1
Aluminum alloys (average)	71	2.7	26.3	9.9	71.2
Magnesium alloys (average)	40	1.8	22.2	11.1	88.2
Titanium alloys (average)	120	4.5	26.7	7.7	50.9
Epoxy–73% E glass fibers	55.9	2.17	25.8	10.9	81.8
Epoxy–70% S glass fibers	62.3	2.11	29.5	11.8	87.2
Epoxy–63% carbon fibers	158.7	1.61	98.6	24.7	156.1
Epoxy–62% aramid fibers	82.8	1.38	60	20.6	146.6

The importance of stiffness arises in complex assemblies where differences in stiffness could lead to incompatibilities and misalignment between various components, thus hindering their efficiency or even causing failure. Using high-strength materials in attempts to reduce weight usually comes at the expense of reduced cross-sectional area and reduced second moment of area. This could adversely affect stiffness of the component if the elastic constant of the new strong material does not compensate for the reduced second moment of area. Another solution to the problem of reduced stiffness is to change the shape of the component cross section to achieve higher second moment of area, I. This can be achieved by placing as much as possible of the material as far as possible from the axis of bending. Figure 4.3 gives the formulas for calculating I for some commonly used shapes, and the values of I for constant cross-sectional area.

4.4.2 Specific Stiffness

In applications where both the stiffness and weight of a structure are important, it becomes necessary to consider the stiffness/weight, specific stiffness, of the structure. In the simple case of a structural member under tensile or compressive load, the specific stiffness is given by E/ρ, where ρ is density of the material. In such cases, the weight of a beam of a given stiffness can be easily shown to be proportional to ρ/E. The performance index in this case is E/ρ. This shows that the weight of the component can be reduced equally by selecting a material with lower density or higher elastic modulus. When the component is subjected to bending, however, the dependence of the weight on ρ and E is not as simple. From Equation 4.2 and Figure 4.3, it can be shown that the deflection of a simply supported beam of square cross-sectional area is given by

$$y = \frac{(Ll^3)}{(4Eb^4)} \tag{4.4}$$

where b is the breadth or width of the beam.

The weight of the beam, w, can be shown to be

$$w = lb^2\rho = \frac{l^{5/2}}{2}\left(\frac{L}{y}\right)^{1/2}\frac{\rho}{E^{1/2}} \tag{4.5}$$

Section shape	Formula for I	Value of I for different geometries		
B, H	$\frac{BH^3}{12}$	$H/B = 1$		$I = 833$
		$H/B = 2$		$I = 1650$
		$H/B = 3$		$I = 2511$
		$H/B = 4$		$I = 3333$
B, H, h, b; B, h, $\frac{b}{2}$, $\frac{b}{2}$	$\frac{BH^3 - bh^3}{12}$	$H = 19$, $h = 15$	$B = 10$, $b = 6$	$I = 4028$
		$H = 21$, $h = 17$	$B = 8$, $b = 4$	$I = 4536$
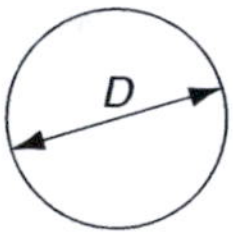 D	$\frac{\pi D^4}{64}$	$D = 11.29$		$I = 796$
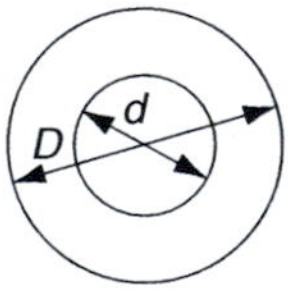 D, d	$\frac{\pi(D^4 - d^4)}{64}$	$D = 20$	$d = 16.5$	$I = 4300$

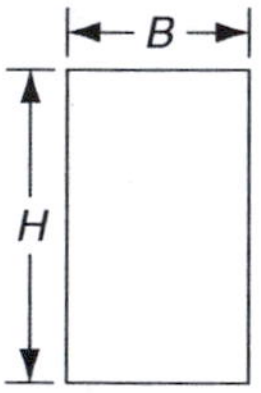

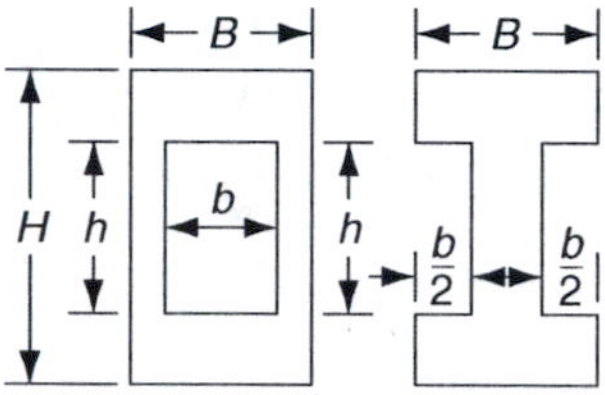

FIGURE 4.3 Effect of shape on the value of second moment of area (I) of a beam in bending. Cross-sectional area is the same in all cases and equals 100 units of area.

This shows that for a given deflection y under load L, the weight of the beam is proportional to $\rho/E^{1/2}$. The performance index in this case is $E^{1/2}/\rho$. As E in this case is present as the square root, it is not as effective as ρ in controlling the weight of the beam. It can be similarly shown that the weight of the beam in the case of a rectangular cross section is proportional to $\rho/E^{1/3}$, and the performance index is $E^{1/3}/\rho$, which is even less sensitive to variations in E, as shown in Table 4.4.

4.4.3 Effect of Material Stiffness on Buckling Strength

Another selection criterion, which is also related to the elastic modulus of the material and cross-sectional dimensions, is the elastic instability, or buckling, of slender components, struts, subjected to compressive loading. The compressive load, L_b, that can cause buckling of a strut is given by Euler formula as

$$L_b = \frac{(\pi^2 EI)}{l^2} \tag{4.6}$$

where l is the length of the strut.

TABLE 4.5
Performance Indices in Selection for Stiffness

Cross Section and Loading Condition	Performance Index
Solid cylinder in tension or compression away from the buckling limit	E/ρ
Column in compression, with failure by buckling	$E^{1/2}/\rho$
Solid cylinder in torsion	$G^{1/2}/\rho$
Simply supported beam of square cross section in bending	$E^{1/2}/\rho$
Simply supported beam of rectangular cross section in bending	$E^{1/3}/\rho$
Flat plate in bending	$E^{1/3}/\rho$
Flat plate under in-plane compression	$E^{1/3}/\rho$
Thin-walled cylindrical pressure vessel	E/ρ

Equation 4.6 shows that increasing E and I will increase the load-carrying capacity of the strut. As buckling can take place in any lateral direction, an axially symmetric cross section can be considered. For a solid round bar of diameter, D, the second moment of area, I, is given as

$$I = \frac{(\pi D^4)}{64} \tag{4.7}$$

The use of the resistance to buckling as a selection criterion is illustrated in Example 4.1.

The weight of a strut, w, is given by

$$w = l\left(\frac{\pi D^2}{4}\right)\rho = \left(\frac{2l^2 L_b^{1/2}}{\pi^{1/2}}\right)\left(\frac{\rho}{E^{1/2}}\right) \tag{4.8}$$

Equation 4.8 shows that the weight of an axisymmetric strut can be reduced by reducing ρ or by increasing E of the material, or both, with the performance index in this case being $E^{1/2}/\rho$. However, reducing ρ is more effective, as E is present as the square root. In the case of a flat panel subjected to buckling, it can be shown that the weight is proportional to $\rho/E^{1/3}$ and the performance index is $E^{1/3}/\rho$.

Following a procedure similar to the one just mentioned, the performance indices of components subjected to different types of loading can be calculated, and the results are shown in Table 4.5. The use of performance indices in making material substitution decisions is illustrated in Example 4.2

Design Example 4.2: Material Substitution for an Interior Door Panel

Problem

A motorcar manufacturer is considering material substitution of the interior door panels of one of their models as part of a weight reduction effort. The panel is 100 cm long and 50 cm wide and is currently made of polyvinyl chloride (PVC) of 3.7 mm thickness. The candidate materials under consideration include

PP + 40% glass fibers and PP + 40% flax fibers. If PVC is substituted, the new panels should have at least the same stiffness and buckling resistance.

Solution

The performance index for equal stiffness and buckling resistance of the panel is $E^{1/3}/\rho$.

The thickness of the new (t_n) and the original (t_o) panels for equal performance indices are related as $t_n = t_o (E_o/E_n)^{1/3}$, where E_o and E_n are modulus of elasticity of the original and new materials, respectively.

For the materials under consideration:

Material	E (GPa)	ρ (g/cc)	t (mm)	Weight (kg)
PVC	2	1.3	3.7	1.85
PP + 40% glass fibers	7.75	1.67	2.37	3.3
PP + 40% flax fibers	4.65	1.19	2.8	1.67

Conclusion

Panels made of PP + 40% flax fibers are lighter than the current PVC panels and can be used as a substitute.

4.5 SELECTION OF MATERIALS FOR HIGHER TOUGHNESS

It is now recognized that small cracks or discontinuities can exist in materials during their manufacture, processing, or service life, and can lead to catastrophic failure if they exceed a critical size, as discussed in Section 2.3. It was also shown that the maximum allowable flaw size is proportional to $(K_{IC}/YS)^2$, where YS is the yield strength of the material and K_{IC} fracture toughness, which is the property of a material that allows it to withstand fracture in the presence of cracks. The ratio K_{IC}/YS can be taken as performance index when comparing the relative toughness of materials. Higher values of K_{IC}/YS are more desirable as they indicate tolerance to larger flaws without fracture.

4.5.1 Metallic Materials

There is a close relationship between toughness and other mechanical properties. Within a given class of materials, there is an inverse relationship between strength and toughness, as shown in Figure 4.4 and Table 4.6.

Generally, the toughness of a material is influenced by its chemical composition and microstructure. For example, steels become less tough with increasing carbon content, larger grain size, and more brittle inclusions. The grain size of steels is affected by the elements present, especially those used for deoxidizing. Small

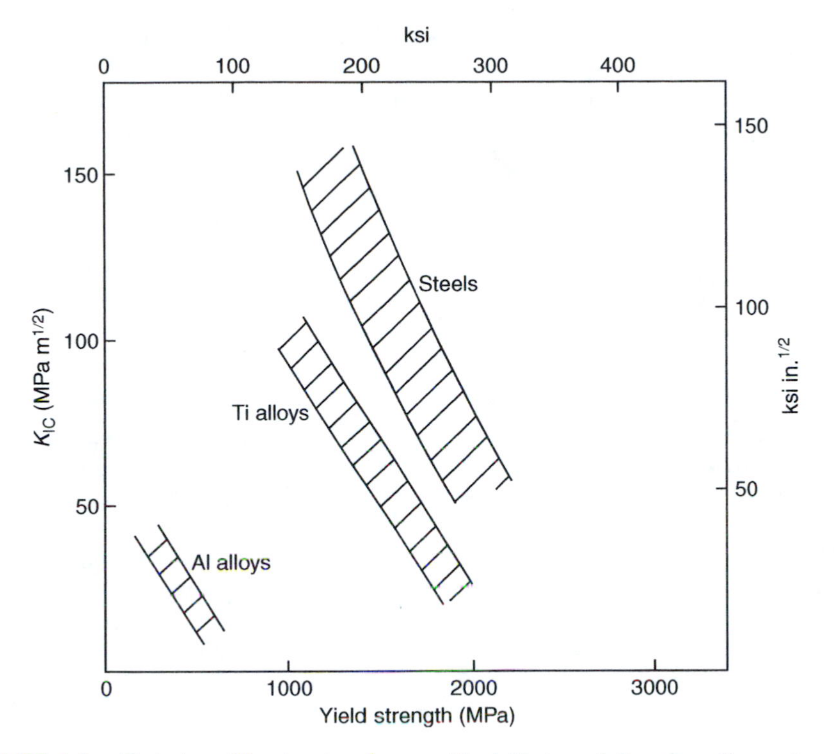

FIGURE 4.4 Variation of fracture toughness with yield strength for some alloy systems.

additions of aluminum to steel are known to promote fine grain size, which improves the toughness. Fully killed fine-grained steels also have lower ductile–brittle transition temperatures and are normally selected for applications where brittle fracture may occur. Fine grains can also be obtained in steels by using alloying elements, by controlling the rolling practice, or by normalizing treatment. A thoroughly deoxidized steel grade has fewer nonmetallic inclusions and gives better toughness. When brittle inclusions are elongated, their influence on ductility is more pronounced in the transverse and through–thickness directions.

The method of fabrication can also have a pronounced effect on toughness, and experience has shown that a large proportion of brittle fractures originate from welds or their vicinity. This can be caused by the residual stresses generated by the welding process, reduction of toughness of the heat-affected zone (HAZ), or by defects in the weld area.

The rate of load application also influences the toughness. Materials that are tough under slowly applied load may behave in a brittle manner when subjected to shock or impact loading.

Decreasing the operating temperature generally causes a decrease in toughness of most engineering materials. This is particularly important in the case of bcc materials, as they tend to go through a ductile–brittle transition as the temperature decreases. All carbon and most alloy steels are of the bcc group

TABLE 4.6
Comparison of Toughness and Strength of Some Engineering Materials

	Yield Strength		K_{IC}		K_{IC}/YS	
Materials	**MPa**	**ksi**	**MPa m$^{1/2}$**	**ksi in.$^{1/2}$**	**m$^{1/2}$**	**in.$^{1/2}$**
Steels						
Medium-carbon steel	260	37.7	54	49	0.208	1.30
ASTM A533B Q&T	500	72.5	200	182	0.400	2.51
AISI 4340 (T260°C)	1640	238	50	45.8	0.030	0.19
AISI 4340 (T425°C)	1420	206	87.4	80	0.062	0.388
Maraging 300	1730	250	90	82	0.052	0.328
Aluminum Alloys						
AA 2024-T651	455	66	24	22	0.053	0.333
AA 2024-T3	345	50	44	40	0.128	0.80
AA 7075-T651	495	72	24	22	0.048	0.306
AA 7475-T651	462	67	47	43	0.102	0.642
Titanium Alloys						
Ti-6Al-4V	830	120	55	50	0.066	0.417
Ti-6Al-4V-2Sn	1085	155	44	40	0.04	0.258
Ti-6Al-4Mo-2Sn-0.05Si	960	139	45	40	0.047	0.288
Plastics						
PMMA	30	4	1	0.9	0.033	0.225
Polycarbonate	63	8.4	3.3	3	0.052	0.357
Ceramics						
Reaction-bonded Si_3N_4	450	63.3	5	4.6	0.011	0.07
Al_2O_3	262	36.9	4.5	4.1	0.017	0.11
SiC (self-bonded)	140	19.7	3.7	3.4	0.026	0.173

and their behavior is illustrated in Figure 4.5. Figure 4.6 shows the catastrophic failure of a welded steel ship as a result of brittle fracture under reduced service temperature. As most bcc materials become too brittle to use under cryogenic temperatures, austenitic steels and nonferrous materials of fcc structures become the only possible materials, as they do not suffer ductile–brittle transition. Several aluminum-, titanium-, copper-, and nickel-base alloys are available for cryogenic applications.

An important aspect of selecting materials for toughness is the likelihood of detection of a crack before it reaches a critical size. As larger cracks can be more easily discovered, it follows that materials tolerating larger critical cracks are more advantageous. The materials listed in Table 4.6 are compared on the basis of the performance index K_{IC}/YS. The values in the table show that a material may exhibit a good crack tolerance even though its fracture toughness is modest. In general, therefore, the material selected for a given application must have such combination of K_{IC} and YS that the critical crack length is appropriate for that application and the available NDT techniques, as illustrated in Example 4.3.

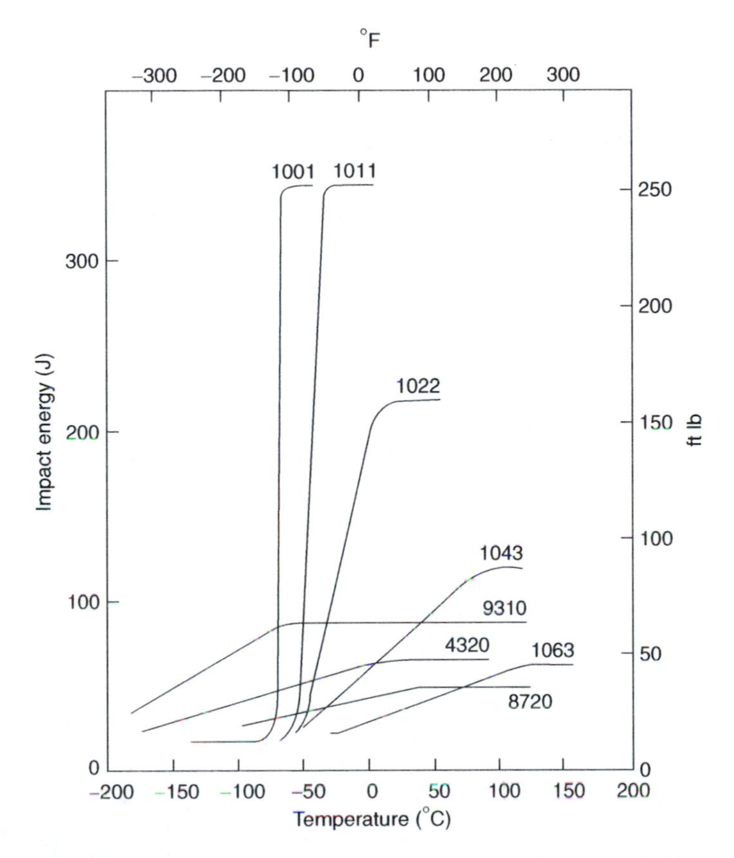

FIGURE 4.5 Effect of temperature on the notch toughness of some AISI-SAE steels.

FIGURE 4.6 Catastrophic brittle fracture of a steel liberty ship at relatively low temperatures. (From Flinn, R.A. and Trojan, P.K., *Engineering Materials and Their Applications*, Houghton Mifflin, Boston, MA, 1990.)

Design Example 4.3: Selecting a Material for a Light Tie-Rod

Problem

Aluminum AA7075-T651 and titanium Ti-6Al-4V are being considered for making a 1 m long tie-rod that will carry a tensile load of 50 kN. If the available NDT equipment can only detect flaws larger than 3 mm in length, which of these two materials can be used to make a lighter member? Ignore the factor of safety in this example and assume the following:

AA7075: yield strength σ_y = 495 MPa, K_{IC} = 24 MPa $m^{1/2}$,
specific gravity = 2.7

Ti-6Al-4V: yield strength σ_y = 830 MPa, K_{IC} = 60 MPa $m^{1/2}$,
specific gravity = 4.5

Answer

From Equation 2.3

$$\sigma_f = \frac{K_{IC}}{[Y(\pi a)^{1/2}]}$$

where

a = Half the crack length
σ_f = Fracture stress
Y = 1

σ_f for AA7075 = 338 MPa; this is lower than σ_y. For this case σ_f should be used for calculating the cross section

σ_f for Ti-6Al-4V = 845 MPa; this is higher than σ_y. For this case σ_y is used for calculating the cross section.

Cross section of the member when made from AA7075 = 148 mm^2, weight = 400 g

Cross section of the member when made from Ti-6Al-4V = 60 mm^2, weight = 270 g

Conclusion

Ti-6Al-4V can be used to make a lighter member.

4.5.2 Plastics and Composites

Although unreinforced plastics generally have lower impact strength than most metallic materials, as shown in Table 4.6, numerous techniques have been developed to improve their toughness. Examples of such techniques include the following:

1. Alloying the plastic with a rubber phase or with another higher-impact plastic. Examples of this method include toughened nylons, which are alloyed with polyolefin or other polymeric modifier. An alloy of nylon and ABS combines the characteristics of both crystalline and amorphous polymers, which result in a combination of high-flow rate, high-temperature warp resistance, good-surface appearance, chemical resistance, and toughness.
2. Copolymerization to create a tougher chemical structure. This approach is used to produce less notch-sensitive polycarbonates that retain their ductility at lower temperatures.
3. Incorporating high-impact-resistance fibers. For example, polyethylene terephthalate (PET), nylon, and polyethylene fibers have been used to replace a portion of the glass fibers in injection-molded polyesters for automotive components.

In general, thermoplastic-matrix composites are tougher than those with thermoset matrix, which is one of the reasons why the former are being developed to replace current epoxies.

4.5.3 Ceramics

The fracture of ceramics is dependent on critical flaw size, which is a function of fracture toughness (K_{IC}). With careful processing, the average flaw size can be reduced to about 30 μm, but this may still be larger than the critical flaw size. In addition, a single flaw in the material that is larger than the critical flaw size is sufficient to cause fracture. This is why toughness data for ceramic materials are often inconsistent and strength and toughness do not always respond in the same manner to changes in microstructure or interfacial properties. Table 4.6 lists typical toughness values of some ceramic materials.

An important technique to improve the toughness of ceramics like ZrO_2, Al_2O_3, and Si_3N_4, is to induce a phase transformation in the region of applied stress within the material. This absorbs energy at the tip of the advancing crack, arresting its propagation and significantly increasing both strength and toughness. Another technique is to introduce fibers to increase the toughness as a result of fiber debonding, crack deflection, or fiber pullout. Internal stresses due to differences in thermal expansion between matrix and fibers in a composite can also provide a toughening effect.

Because ceramics are sensitive to surface damage, surface modification techniques are also being developed as a means of improving their toughness.

4.6 SELECTION OF MATERIALS FOR FATIGUE RESISTANCE

In many engineering applications, the behavior of a component in service is influenced by several other factors besides the properties of the material used in its manufacture. This is particularly true for the cases where the component or structure is subjected to fatigue loading. Under such conditions, the fatigue resistance can be greatly influenced by the service environment, surface condition of the part, method of fabrication, and design details. In some cases, the role of the material in achieving satisfactory fatigue life is secondary to the mentioned parameters, as long as the material is sound and free from major flaws. For example, if the component has welded, bolted, or riveted joints, the contribution of crack initiation stage (see Section 2.5) is expected to be small, and most of the fatigue life is determined by the crack propagation stage. Experience shows that crack propagation rate is more sensitive to continuum mechanics considerations than to material properties.

Fatigue strength of metallic materials generally increases with increasing tensile strength, as shown in Table 4.7. However, the higher the strength, the higher the notch sensitivity of the material and the greater the need to eliminate coarse second phase particles and to produce a more refined, homogeneous structure. Meeting these needs could require expensive metallurgical processes or the addition of expensive alloying elements. A measure of the degree of notch sensitivity of the material is usually given by the parameter q:

$$q = \frac{(K_f - 1)}{(K_t - 1)} \tag{4.9}$$

where

K_t = Stress concentration factor, which represents the severity of the notch and is given by the ratio of maximum local stress at the notch to average stress. K_t is mainly related to the geometry of the component, as discussed in Section 5.5, and is material-independent.

K_f = Ratio of the fatigue strength in the absence of stress concentrations to the fatigue strength with stress concentration. Unlike K_t, K_f is material-dependent.

The value of q is a measure of the degree of agreement between K_f and K_t and can be taken as a fatigue notch sensitivity index. Thus, as q increases from 0 to 1, the material becomes more sensitive to the presence of stress concentrations. Generally, increasing the tensile strength of the material makes it more notch-sensitive and increases q. The value of q is also dependent on component size, and it increases as size increases. Therefore, stress raisers are more dangerous in larger components made from stronger materials.

4.6.1 Steels and Cast Irons

Steels are the most widely used structural materials for fatigue applications as they offer high fatigue strength and good processability at a relatively low cost. Steels have the unique characteristic of exhibiting an endurance limit, which enables them to perform indefinitely, without failure, if the applied stresses do not exceed this limit. Table 4.7 shows that the endurance ratio of most steels ranges between 0.4 and 0.6.

The optimum steel structure for fatigue resistance is tempered martensite, since it provides maximum homogeneity. Steels with high hardenability provide high strength with relatively mild quenching and, hence, low residual stresses, which is desirable in fatigue applications. Normalized structures, with their finer structure, give better fatigue resistance than the coarser pearlitic structures obtained by annealing.

TABLE 4.7
Comparison of Static and Fatigue Strengths of Some Engineering Materials

Material	Tensile Strength		Endurance Limit		Endurance Ratio
	MPa	ksi	MPa	ksi	
Ferrous Alloys					
AISI 1010, normalized	364	52.8	186	27	0.46
1025, normalized	441	64	182	26.4	0.41
1035, normalized	539	78.2	238	34.5	0.44
1045, normalized	630	91.4	273	39.6	0.43
1060, normalized	735	106.6	315	45.7	0.43
1060, oil Q, tempered	1295	187.8	574	83.3	0.44
3325, oil Q, tempered	854	123.9	469	68	0.55
4340, oil Q, tempered	952	138.1	532	77.2	0.56
8640, oil Q, tempered	875	126.9	476	69	0.54
9314, oil Q, tempered	812	177.8	476	69	0.59
302, annealed	560	81.2	238	34.5	0.43
316, annealed	560	81.2	245	35.5	0.44
431, quenched, tempered	798	115.7	336	48.7	0.42
ASTM 20 GCI	140	20.3	70	10.2	0.50
30 GCI	210	30.5	102	14.8	0.49
60 GCI	420	61	168	24.4	0.40
Nonferrous Alloys					
AA 2011-T8	413	59.9	245	35.5	0.59
2024, annealed	189	27.4	91	13.2	0.48
6061-T6	315	45.7	98	14.2	0.31
6063-T6	245	35.5	70	10.2	0.29
7075-T6	581	84.3	161	23.4	0.28
214 As-cast	175	25.4	49	7.1	0.28
380 Die-cast	336	48.7	140	20.3	0.42
Phosphor bronze, annealed	315	45.7	189	27.4	0.60
Phosphor bronze, hard drawn	602	87.3	217	31.5	0.36
Aluminum bronze, quarter hard	581	84.3	206	29.9	0.35
Incoloy 901, at 650°C (1202°F)	980	142.1	364	52.8	0.37
Udimet 700, at 800°C (1472°F)	910	132	343	49.7	0.38
Reinforced Plastics					
Polyester–30% glass	123	17.8	84	12.2	0.68
Nylon 66–40% glass	200	29	62.7	9.1	0.31
Polycarbonate–20% glass	107	15.5	34.5	5	0.32
Polycarbonate–40% glass	131	19	41.4	6	0.32

Inclusions in steel are harmful as they represent discontinuities in the structure that could act as initiation sites for fatigue cracking. Therefore, free-machining steels should not be used for fatigue applications. However, if machinability considerations make it essential to select a free-machining grade, leaded steels are preferable to those containing sulfur or phosphorus. This is because the rounded lead particles give rise to less structural stress concentrations than the other angular and elongated inclusions. By the same token, cast steels and cast irons are not recommended for critical fatigue applications. In rolled steels, the fatigue strength is subject to the same directionality as the static properties.

4.6.2 Nonferrous Alloys

Unlike ferrous alloys, the nonferrous alloys, with the exception of titanium, do not normally have an endurance limit. Aluminum alloys usually combine corrosion resistance, lightweight, and reasonable fatigue resistance. The endurance ratio of aluminum alloys is more variable than that of steels (Table 4.7), but an average value can be taken as 0.35.

Generally, the endurance ratio is lower for as-cast structures and precipitation-hardened alloys. Fine-grained inclusion-free alloys are most suited for fatigue applications.

4.6.3 Plastics

The viscoelasticity of plastics makes their fatigue behavior more complex than that of metals. In addition to the set of parameters that affect the fatigue behavior of metals, the fatigue behavior of plastics is also affected by the type of loading, small changes in temperature and environment, and method of sample fabrication. Because of their low thermal conductivity, hysteretic heating can build up in plastics causing them to fail in thermal fatigue or to function at reduced stress and stiffness levels. The amount of heat generated increases with increasing stress and test frequency. This means that failure of plastics in fatigue may not necessarily mean fracture. In flexural fatigue testing by constant amplitude of force, ASTM D671 sets an arbitrary level of stiffness—70% of the original modulus—as failure.

Some unreinforced plastics such as polytetrafluoroethylene (PTFE), polymethylmethacrylate (PMMA), and polyesterether ether ketone (PEEK) have fatigue endurance limits. At stresses below this level, failure does not occur. Other plastics, usually amorphous materials, show no endurance limit. In many unreinforced plastics, the endurance ratio can be taken as 0.2.

4.6.4 Composite Materials

The failure modes of reinforced materials in fatigue are complex and can be affected by the fabrication process when differences in shrinkage between fibers and matrix induce internal stresses. There is a growing body of practical experience, however, and some FRP are known to perform better in fatigue than some metals, as shown in Table 4.7. The advantage of FRP is even more apparent when compared on per weight basis. For example, because of its superior fatigue properties, glass–fiber-reinforced epoxy has replaced steel leaf springs in several motorcar models.

Generally, fiber-reinforced crystalline thermoplastics exhibit well-defined endurance limits, whereas amorphous-based composites do not. The higher strengths, higher thermal conductivity, and lower damping account for the superior fatigue behavior of crystalline polymers.

As with static strength, fiber orientation affects the fatigue strength of fiber-reinforced composites. In unidirectional composites, the fatigue strength is significantly lower in directions other than the fiber orientation. Reinforcing with continuous unidirectional fibers is more effective than reinforcing with short random fibers. Example 4.4 illustrates the use of fatigue strength in design.

Design Example 4.4: Selecting a Material for a Connecting Rod

Problem

Aluminum alloy AA 6061-T6, steel AISI 4340 oil Q and tempered, and polyester–30% glass fibers are being considered as a replacement for steel AISI 1025 normalized, in manufacturing a connecting rod to save weight. The connecting rod has a circular cross section, a length of 300 mm, and is subjected to an alternating tensile load of 60 kN. Given the following information and assuming a derating factor of 0.4 on the fatigue strength for all alternatives, select the most suitable material.

AISI 1025: Tensile strength = 440 MPa, endurance ratio = 0.41, specific gravity = 7.8

AA 6061-T6: Tensile strength = 314 MPa, endurance ratio = 0.31, specific gravity = 2.7

AISI 4340: Tensile strength = 952 MPa, endurance ratio = 0.56, specific gravity = 7.8

Polyester–30% glass: Tensile strength = 123 MPa, endurance ratio = 0.68, specific gravity = 1.45

Answer

AISI 1025: Cross section = 832 mm^2, weight = 1.947 kg

AA 6061-T6: Cross section = 1541 mm^2, weight = 1.248 kg

AISI 4340: Cross section = 281 mm^2, weight = 0.658 kg

Polyester–30% glass: Cross section = 1793 mm^2, weight = 0.780 kg

Using steel AISI 4340 gives the lightest connecting rod, with polyester–30% glass as a close second.

4.7 SELECTION OF MATERIALS FOR HIGH-TEMPERATURE RESISTANCE

4.7.1 Creep Resistance of Metals

It is shown in Section 2.6 that creep is a major factor that limits the service life of parts and structures at elevated temperatures. Experience shows that many of the methods used to improve low-temperature strength of metallic materials become ineffective as the operating temperature approaches 0.5 T_m (T_m is the absolute melting temperature expressed in degrees Kelvin or Rankine). This is because atomic mobility is sufficient to cause softening of cold-worked structures and coarsening of unstable precipitates. At these high temperatures, the differences in creep resistance from one material to another depend on the stability of the structure and on the hardening mechanism.

The most important method of improving creep strength is to incorporate a fine dispersion of stable second-phase particles within the grains. These particles can be introduced by dispersion, as in the case of thoria particles in nickel (TD nickel), or by precipitation, as in the case of precipitation-hardened nickel alloys. To minimize particle coarsening, it is the practice to make the chemical composition of the precipitates as complex as possible and to reduce the thermodynamic driving force for coarsening by reducing the interfacial energy between the precipitates and the matrix. Precipitates at the grain boundaries are important in controlling creep rupture ductility as they control grain boundary sliding, which causes premature failure.

4.7.2 Performance of Plastics at High Temperatures

The mechanical strength of plastics at high temperatures is usually compared on the basis of deflection temperature under load (DTUL), also known as heat deflection temperature. According to ASTM D-648 specification, DTUL is defined as the temperature at which a specimen deflects 0.25 mm (0.010 in.) under a load of 455 or 1820 kPa (66 or 264 psi), when heated at the rate of 2°C/min. Generally, thermosets have higher temperature resistance than thermoplastics. However, adding glass and carbon fibers, as well as mineral and ceramic reinforcements, can significantly improve DTUL of crystalline thermoplastics such as nylon, thermoplastic polyesters, polyphenylenesulfone (PPS), and fluoroplastics. For example, at 30% glass–fiber, the DTUL of nylon 6/6 at 264 psi increases from about 71°C to 249°C (160–480°F).

Although several plastics can withstand short excursions to high temperatures, up to 500°C (930°F), continuous exposure can result in a dramatic drop in mechanical properties and extreme thermal degradation.

4.7.3 Widely Used Materials for High-Temperature Applications

Because operating temperature is the single-most important factor that affects the selection of materials for elevated-temperature service, it is normal practice to classify temperature-resistant materials according to the temperature range in which they are expected to be used. Table 4.8 provides a summary of the widely used materials at the different temperature ranges, and the following description provides additional information.

4.7.3.1 Room Temperature to 150°C (300°F)

Most engineering metals and alloys, with the exception of lead, can be used in this temperature range. Several unreinforced thermoplastics are suitable for continuous service at temperatures above 100°C (212°F). In addition, fluoroplastics, polycarbonates,

TABLE 4.8
Widely Used Materials for Different Temperature Ranges

Temperature Range	Widely Used Materials
Room temperature–150°C	≤100°C thermoplastics
	≤150°C most engineering metals and alloys, FRP
150–400°C	≤200°C high-temperature plastics (polysulfones, polyphenylenesulfides, polyethersulfone, and fluoroplastics)
	≤250°C aluminum alloys, thermosetting plastics
	≤400°C plain carbon steels (short exposures), low-alloy steels (long exposures)
400–600°C	≤450°C alpha–beta titanium alloys, low-alloy steels
	≤600°C 5–12% (Cr + Mo) steels
600–1000°C	≤650°C ferritic stainless steels
	≤750°C austenitic stainless steels
	≤800°C Fe–Ni-base superalloys
	≤850°C Ni-base superalloys
	≤980°C Co-base superalloys
1000°C and above	Refractory metals (Mo, Nb, Ta, W) Ceramics

polyamides, polysulfones, polyphenylenesulfides, and the newly developed materials like PEEK and PPS can be used at temperatures up to 200°C. Several FRP, for example, nylon 6/6–glass fiber, can also serve in this temperature range.

4.7.3.2 150–400°C (300–750°F)

Plain carbon or manganese–carbon steels provide adequate properties in this temperature range, although it may be necessary to use low-alloy steels if very long service, more than 20 years, is required. High-grade cast irons can be used at temperatures up to 250°C for engine casings. Aluminum alloys can be used at temperatures up to about 250°C (480°F), although some P/M alloys have been used for short intervals at about 480°C (900°F).

High-temperature plastics can be used at temperatures up to 200°C (400°F) and will withstand temperatures up to about 300°C (500°F) for short periods. These include polysulfones, polyphenylenesulfides, polyethersulfones, and fluoroplastics. Thermoset polyamides–graphite composites can serve in the range of 260–290°C (500–550°F). New experimental plastics, like polyparaphenylene benzobisthiazole, are expected to withstand temperatures up to about 370°C (700°F) for long periods.

4.7.3.3 400–600°C (750–1100°F)

Low-alloy steels and titanium alloys are the main materials used in this temperature range. Low-alloy steels are relatively inexpensive and are used if there are no restrictions on weight. The main alloying elements that are usually added to these steels include molybdenum, chromium, and vanadium. An example of such steels is the 0.2C–1Cr–1Mo–0.25V steel, which is used for intermediate and high-pressure steam turbine rotors.

In applications at temperatures approaching 600°C (1100°F), oxidation resistance becomes an important factor in determining the performance of materials. In such cases, at least 8% chromium needs to be added to steels. Several steels are available with chromium contents in the range of 5–12%. These steels usually also contain molybdenum to improve their creep resistance.

Titanium alloys of alpha-phase structure exhibit better creep resistance than those of beta-phase structure. The alpha–beta 6Al-4V alloy is most widely used for general purposes and is limited to a maximum operating temperature of about 450°C (840°F). The near-alpha alloy 5.5Al–3.5Sn–3Zr–1Nb–0.25Mo–0.25Si can be used at temperatures up to about 600°C (1110°F).

4.7.3.4 600–1000°C (1100–1830°F)

The most widely used materials for this temperature range can be divided into the following groups:

- Stainless steels
- Fe–Ni-base superalloys
- Ni-base superalloys
- Co-base superalloys

Oxidation and hot corrosion resistance of the mentioned alloys become increasingly important with increasing operating temperature. The level of oxidation resistance in this temperature range is a function of chromium content. Aluminum can also contribute to oxidation resistance, especially at higher temperatures. Chromium is also important for hot corrosion resistance. Chromium content in excess of 20% appears to be required for maximum resistance.

When the oxidation and hot corrosion resistance of a given alloy is not adequate, protective coatings may be applied. Diffusion coatings, CoAl or NiAl, are commonly used for protection. FeCrAl, FeCrAlY, CoNiAl, or CoNiAlY overlay coatings can also be used, and they do not require diffusion for their formation.

4.7.3.4.1 Stainless Steels

The ferritic stainless steels of the 400 series are less expensive than the austenitic grades of the 200 and 300 series, see Table A.7. The ferritic grades are usually used at temperatures up to 650°C (1200°F) in applications involving low stresses. Austenitic stainless steels of the 300 series can be used at temperatures up to 750°C (1380°F). Type 316 stainless steel with 2.5% Mo is widely used and has the highest stress-rupture strength of all the 300 series alloys. The more highly alloyed compositions 19-9 DL and 19-9 DX contain 1.25–1.5 Mo, 0.3–0.55 Ti, and 1.2–1.25 W. They have superior stress-rupture strengths than the 300 series alloys. These alloys can be used at temperatures up to 815°C (1500°F). Also, the 202 and 216 Cr–Ni–Mn alloys have higher stress-rupture capabilities than the 300 series alloys.

4.7.3.4.2 Fe–Ni-Base Superalloys

Fe–Ni-base superalloys consist mainly of fcc solid–solution matrix, γ phase, strengthened by intermetallic compounds and carbide precipitates. A common precipitate is gamma prime, γ', which is $Ni_3(Al,Ti)$. Other precipitates include carbides, nitrides, and carbonitrides. Table A.16 gives the composition of a representative sample of these alloys.

4.7.3.4.3 Ni-Base Superalloys

Ni-base superalloys (see Table A.16) also consist of fcc matrix strengthened by intermetallic compound precipitates. Gamma prime, γ', is used to strengthen alloys like Waspaloy and Udimet 700. Oxide dispersions are also used for strengthening, as in the case of IN MA-754 and IN MA-6000E. Other Ni-base superalloys are essentially solid–solution strengthened in addition to some carbide precipitation, as in the case of Hastelloy X. Table A.17 gives the rupture strength of selected Ni-base superalloys.

4.7.3.4.4 Co-Base Superalloys

Co-base superalloys (Table A.16) are strengthened mainly by a combination of carbides and solid–solution hardeners. In terms of strength, Co-base alloys can only compete with Ni-base superalloys at temperatures above 980°C (1800°F). Co-base superalloys are used in the wrought form, for example, Haynes 25, and in the cast form, for example, X-40. Table A.17 gives the rupture strength of selected Co-base superalloys.

4.7.3.5 1000°C (1830°F) and above

The refractory metals Mo, Nb, Ta, and W as well as their alloys can be used for stressed applications at temperatures above 1000°C (1830°F). Table A.18 lists the composition of some commercial refractory metals and alloys. Mo-30W alloy has a melting point of 2830°C (5125°F) and excellent resistance to liquid-metal attack.

4.7.4 Niobium, Tantalum, and Tungsten

Niobium (Table A.18) can be used in contact with liquid lithium and sodium–potassium alloys at high temperatures, even above 800°C (1470°F). Addition of 1% Zr to niobium increases its resistance to embrittlement due to oxygen absorption.

Tantalum (Table A.18) can be used for structural applications at temperatures in the range of 1370–1980°C (2500–3600°F), but it requires protection against oxidation. Tantalum is also used for heat shields and heating elements in vacuum furnaces.

Tungsten (Table A.18) has the highest melting point of all materials, which makes it the obvious candidate for structural applications at very high temperatures. Molybdenum is added to tungsten to improve its machinability and rhenium is added to improve resistance to cold fracture in lamp filaments. Surface protection is an important obstacle to the widespread use of refractory metals in high-temperature oxidizing environments. Various aluminide and silicide coatings are available commercially, but they all have a maximum temperature limit of about 1650°C (3000°F).

4.7.5 Ceramics

Ceramics (Table A.24) can withstand extremely high temperatures and are being increasingly used for structural applications. Creep resistance, thermal conductivity, thermal expansion, and thermal-shock resistance are the major factors that determine the suitability of a ceramic material for high-temperature applications. Creep resistance of many ceramics is affected by intergranular phases. Because crystalline phases are generally more creep resistant than amorphous ones, it is the usual practice to reduce the amorphous intergranular phases as a means of improving creep resistance. Doping can also be used to improve the strength of grain–boundary phases, as in the case of doping Si_3N_4 with Y_2O_3 and ZrO_2. Silicon-based ceramics have lower thermal expansion coefficient, which helps in improving their thermal-shock resistance. However, this may not be an advantage if the ceramic is used as a coating on metals where a large difference in expansion may present difficulties.

Thermal-shock resistance is a function of thermal conductivity, coefficient of thermal expansion, tensile strength, and modulus of elasticity. For structural ceramics, thermal-shock resistance is dependent on both material type and processing method. For example, silicon nitride has a better thermal-shock resistance when hot-pressed than when reaction-sintered. Generally, silicon carbide and tungsten carbide have better thermal-shock resistance than zirconium oxide and aluminum oxide. Silicon nitride has good thermal-shock resistance and good oxidation resistance, which make it a feasible candidate for service temperatures of about 1200°C (2192°F) in gas turbines.

4.8 SELECTION OF MATERIALS FOR CORROSION RESISTANCE

Although corrosion resistance is usually the main factor in selecting corrosion-resistant materials, it is often difficult to assess this property for a specific application. This is because the behavior of a material in a corrosive environment can be dramatically changed by seemingly minor changes in the medium or the material itself. The main factors that can affect the behavior of the material can be classified as follows:

- Corrosive medium parameters
- Design parameters
- Material parameters

4.8.1 CORROSIVE MEDIUM PARAMETERS

Corrosive medium parameters include

1. chemical composition and presence of impurities;
2. physical state whether solid, liquid, gas, or combinations;
3. aeration, oxygen content, and ionization; and
4. bacteria content.

In the case of metallic materials, the most significant factor controlling the probability of atmospheric corrosion is whether or not an aqueous electrolyte is provided by condensation of moisture under prevailing climatic conditions. Hot and dry as well as cold and icy conditions give less attack than wet conditions. Clean atmosphere is less aggressive than industrial or marine atmospheres containing sulfur dioxide and salt, respectively. Direction of exposure to the sun, wind, and sources of pollution can also affect the rate of atmospheric corrosion.

In buried structures, increasing the porosity of the soil and the presence of water raises the rate of corrosion. In addition to allowing continuing access of oxygen to the corroding surface, porosity also encourages the activity of aerobic bacteria, which can lead to local variation in aeration, consumption of organic protection systems, and production of corrosive H_2S. In general, dry, sandy, or chalky soils of high electrical resistance are the least corrosive whereas heavy clays and saline soils are the most corrosive.

The rate of corrosion of underwater structures is affected by the amount of dissolved oxygen as well as the amount of dissolved salts and suspended matter. Since oxygen enters the water by dissolution from the air, its concentration can vary with depth and flow rate. Soft freshwater is generally more corrosive than hard water, which precipitates a protective carbonate on the corroding surface. In seawater, the presence of chloride ions increases the electrical conductivity and, therefore, the rate of corrosion. The presence of organic matter, such as bacteria or algae, in water can decrease the rate of corrosion in the covered areas but produce regions of local deaeration where accelerated attack occurs. Increasing the water temperature generally increases the rate of attack. In chemical plants, the rate of attack depends on several factors including the temperature, concentration of chemicals, fluid velocity, degree of aeration, purity of the metal, and applied stresses. In general, the attack is severest where protective or oxide films are disrupted or become locally unstable.

4.8.2 Design Parameters

The design parameters that affect the rate of corrosive attack include the following:

1. Stresses acting on the material in service
2. Operating temperature
3. Relative motion of medium with respect to the material
4. Continuity of exposure of the material to the medium
5. Contact between the material and other materials
6. Possibility of stray currents
7. Geometry of the component

Combating corrosion by design is discussed in Section 6.7.

4.8.3 Material Parameters

The main parameters that affect the corrosion resistance of materials include chemical composition and the presence of impurities, nature and distribution of microstructural constituents, surface condition and deposits, and processing history. Generally, the corrosion resistance improves in pure metals as their purity increases. An example is the localized attack in commercially pure aluminum due to the presence of iron impurities. Table 4.9 shows the relative corrosion resistance of some commonly used metallic materials under different service conditions.

TABLE 4.9
Relative Corrosion Resistance of Some Uncoated Metallic Materials

Material	Industrial Atmosphere	Fresh-water	Sea-water	Acids H_2SO_4 5–15% Concentration	Alkalis 8%
Low-carbon steel	1	1	1	1	5
Galvanized steel	4	2	4	1	1
GCI	4	1	1	1	4
4–6% Cr steels	3	3	3	1	4
18-8 stainless steel	5	5	4	2	5
18-35 stainless steel	5	5	4	4	4
Monel (70% Ni–30% Cu)	4	5	5	4	5
Nickel	4	5	5	4	5
Copper	4	4	4	3	3
Red brass (85% Cu–15% Zn)	4	3	4	3	1
Aluminum bronze	4	4	4	3	3
Nickel silver (65% Cu–18% Ni–17% Zn)	4	4	4	4	4
Aluminum	4	2	1	3	1
Duralumin	3	1	1	2	1

Note: 1. Poor: rapid attack; 2. Fair: temporary use; 3. Good: reasonable service; 4. Very good: reliable service; 5. Excellent: unlimited service.

4.8.4 Carbon Steels and Cast Irons

Carbon steels and cast irons are used in large quantities because of their useful mechanical properties and low cost. These materials, however, are not highly corrosion resistant, with the exception of resistance to alkalis and concentrated sulfuric acid, as shown in Table 4.9. Low-carbon steels have adequate resistance to scaling in air up to about 500°C (ca. 930°F), but this temperature is reduced in the presence of sulfur in flue gases. The addition of chromium in amounts of about 3% increases the resistance to both oxidation and sulfide scaling. Chromium additions also improve resistance to atmospheric corrosion. Nickel is also added to improve the resistance to sodium hydroxide.

4.8.5 Stainless Steels

Stainless steels represent a class of highly corrosion-resistant materials and have widespread applications in engineering. It should be remembered, however, that stainless steels do not resist all corrosive environments, as shown in Table 4.9. For example, when subjected to stresses in chloride-containing environments, stainless steels are less resistant than ordinary structural steels. Stainless steels, unless correctly fabricated and heat treated, can also be more susceptible to localized corrosion such as intergranular corrosion, SCC, crevice corrosion, and pitting than ordinary structural steels.

Increasing chromium content in stainless steels increases their resistance. This is because corrosion resistance of stainless steels can be attributed to the presence of a thin film of hydrous oxide on the surface of the metal. The condition of the film depends on the composition of the stainless steel and the treatment it receives. To give the necessary protection, the film must be continuous, nonporous, self-healing, and insoluble in the corrosive medium. In the presence of such an oxide film, the stainless steels are passive and have solution potentials approaching those of noble metals. When passivity is destroyed, the potential is similar to that of iron, as shown in Table 3.1. Exposing stainless steels to mildly oxidizing corrosive agents causes them to become active, and increasing the oxygen concentration causes them to regain passivity. When the passive film is destroyed locally, stainless steels can fail catastrophically by localized mechanisms such as pitting, crevice corrosion, intergranular corrosion, or SCC, as shown in Example 3.3 and Case Study 4.5.

Chromium plays an important role in forming the passive film on the stainless steel surface. The presence of nickel in high-chromium steels greatly improves their resistance to some nonoxygenating media. It is also an austenite stabilizer. Manganese can be used as a substitute for part of the nickel as an austenite stabilizer, although it does not significantly alter the corrosion resistance of high-chromium steels. Molybdenum strengthens the passive film and improves resistance to pitting in seawater. Other elements such as copper, aluminum, and silicon also increase corrosion resistance of stainless steels.

Case Study 4.5: Corrosion of Welded 304 Stainless Steel Tank

Problem

A food processing welded 304 stainless steel tank exhibited considerable pitting corrosion near the welded joints after 6 months of service.

Analysis

Microscopic examination showed extensive precipitates in the affected areas. It is assumed that the precipitates are chromium carbides, which precipitated in the areas that were heated to the sensitizing temperature range (650–750°C). Precipitation of the carbides depleted the neighboring areas from chromium.

Solution

It is recommended that 304L stainless steel be used. With its carbon content of less than 0.03%, there is less opportunity for chromium carbides to form during welding. Other possible solutions include using stabilized stainless steels, for example, 347 or 321.

4.8.6 Nickel

Nickel has a relatively high corrosion resistance and is particularly useful for handling caustic alkalis. Nickel resists SCC in chloride environments, but may be susceptible in caustic environments if highly stressed and if it contains impurities in solution.

Inconel, 78/16/6 Ni–Cr–Fe, is resistant to many acids and has outstanding resistance to nitriding at high temperatures. Nimonic alloys, based on the 80/20 Ni–Cr basic composition, have particularly good combination of high strength and oxidation resistance at high temperatures. As shown in Table 4.9, Monel alloys, which are based on the 70/30 Ni–Cu composition, have similar resistance to pure nickel with the additional advantage of being less expensive and being able to handle seawater and brackish waters at high fluid velocities. Monel alloys present an economic means of handling hydrofluoric acid and are also resistant to other nonoxidizing acids. Monel alloys are not, however, resistant to oxidizing media such as nitric acid, ferric chloride, sulfur dioxide, and ammonia.

4.8.7 Copper

Pure copper is a noble metal and is, therefore, highly corrosion resistant. It is especially compatible with most industrial, marine, and urban atmospheres, in addition to water and seawater, as shown in Table 4.9. When copper is alloyed with zinc in concentrations more than 15%, dezincification may occur in some environments. Addition of about 1% tin can reduce this problem. Tin bronzes are resistant to a variety of atmospheres, waters, and soils. Phosphorus is added to impart oxidation resistance. Aluminum bronzes, containing about 10% Al, are resistant to corrosion from chloride–potash solutions, nonoxidizing mineral acids, and many organic acids. Cupronickels are widely used in saltwater; they have excellent resistance to biofouling and SCC.

4.8.8 Tin

Over half of the tin production is used as protective coatings of steels and other metals. In addition to its corrosion resistance, tin is nontoxic and it provides a good base for organic coatings. This explains its wide use in coating the steel cans, tin cans, used for the storage of food products and beverages. Tin is normally cathodic to iron, but the potential reverses in most sealed cans containing food products and the tin acts as a sacrificial coating, thus protecting steel. Tin is also resistant to relatively pure water and dilute mineral acids in the absence of air. This makes it suitable for coating copper pipes and sheets in contact with distilled water and medicaments. Tin is attacked by strong mineral acids and alkalies.

4.8.9 Lead

A large proportion of lead production goes into applications where corrosion resistance is important, especially those involving sulfuric acid. The corrosion resistance of lead is due to the protective sulfates, oxides, and phosphates that form on its surface as a result of reaction with corrosive environments. Lead containing about

0.06% copper is usually specified for process equipment in contact with sulfuric, chromic, hydrofluoric, and phosphoric acids. It is also used for neutral solutions, seawater, and soils. Lead is attacked by acetic, nitric, hydrochloric, and organic acids.

4.8.10 Aluminum

Aluminum is a reactive metal, but it develops an aluminum oxide film that protects it from corrosion in many environments. The film is quite stable in neutral and many acid solutions but is attacked by alkalies. The aluminum oxide film is also resistant to a variety of organic compounds, including fatty acids. This oxide film forms in many environments, but it can be artificially produced by anodization.

Pure aluminum and nonheat-treatable aluminum alloys exhibit high resistance to general corrosion but, because of their dependence on the surface oxide film, are liable to suffer local attack under deposits and in crevices. Heat-treatable alloys in the 2000 series and those in the 7000 series contain copper and exhibit lower resistance to general corrosion, as shown in Table 4.9. Such alloys are used in applications where corrosion resistance is secondary to strength.

4.8.11 Titanium

Titanium exhibits excellent corrosion resistance because of its stable, protective, strongly adherent surface oxide film. Titanium is immune to all forms of corrosive attack in seawater and chloride salt solutions at ambient temperatures, and to hot, strong oxidizing solutions. It also has very high resistance to erosion corrosion in seawater. Titanium also resists attack by moist chlorine gas, but if moisture concentration in the gas falls below 0.5%, rapid attack can result. Hydrofluoric acid is also among the substances that attack titanium by destroying the protective oxide film. Addition of alloying elements can affect corrosion resistance if they alter the properties of the oxide film.

4.8.12 Tantalum and Zirconium

Tantalum is inert to practically all organic compounds at temperatures below 150°C (300°F). Exceptions are hydrofluoric acid and fuming sulfuric acid. Zirconium is resistant to mineral acids, molten alkalies, alkaline solutions, and most organic and salt solutions. It has excellent oxidation resistance in air, steam, CO_2, SO_2, and O_2 at temperatures up to 400°C (750°F). Zirconium is attacked by corrosion in hydrofluoric acid, wet chlorine, aqua regia, ferric chloride, and cupric chloride solutions. Tantalum and zirconium will seldom be economic, but they are the only available resistant materials for a few applications.

4.8.13 Metallic Glasses

Metallic glasses, amorphous alloys, are produced by quenching from the liquid state; they have undercooled liquid structures similar to those of ceramic glasses. The compositions of these materials are adjusted to be close to low melting, stable eutectics that yield noncrystalline structures on rapid solidification. Some of the iron-based metallic glasses have corrosion resistances approaching those of tantalum or the noble metals. Typical compositions include

- 8–20% Cr, 13% P, 7% C, remainder Fe; and
- 10% Cr, 5–20% Ni, 13% P, 7% C, remainder Fe.

These materials passivate very easily, and at 8% Cr they are superior to conventional stainless steels. Their pitting resistance is equal to or greater than that of the high-nickel alloys, Hastelloy C, and titanium.

4.8.14 Plastics and Fiber-Reinforced Plastics

Because of their corrosion resistance, plastics and composites have replaced metals in many applications. Examples from the automotive industry include fenders, hoods, and other body components. However, there are several environmental effects that should be considered when selecting plastics and FRP.

Several plastics absorb moisture, which causes swelling and distortion in addition to degrading their strength and electrical resistance. Polymers can also be attacked by organic solvents, as discussed in Section 3.5 and as shown in Table 3.2. Generally, crystalline thermoplastics, such as fluorocarbons, Teflon, and nylon, have superior chemical stability than amorphous types like polycarbonate. Fluorocarbons, for example, PTFE, are among the most chemically inert materials available to the engineer. They are inert to all industrial chemicals, and they resist the attack of boiling aqua regia, fuming nitric acids, hydrofluoric acid, and most organic solvents. Other thermoplastics like polyketones and polyphenylene sulfide provide excellent chemical resistance, even at relatively elevated temperatures. Among thermosetting plastics, epoxies represent the best combination of corrosion resistance and mechanical properties.

There are several standard tests for measuring the chemical resistance of polymers and FRP. The immersion test, ASTM D 543, is used extensively as it measures the changes in weight, dimensions, and mechanical properties that result from immersion in standard reagents. Table 3.3 gives the relative chemical resistance, expressed in terms of percent retention of tensile strength, of some plastics.

4.8.15 Ceramic Materials

Most ceramic materials exhibit good resistance to chemicals, with the main exception of hydrofluoric acid. Glasses are among the most chemically stable materials, and they have exceptionally good resistance to attack by water, aqueous solutions of most acids, alkalis, and salts. However, their relative performance in various environments may vary considerably between different grades. For example, borosilicate and silica glasses show much higher resistance to boiling water and hot dilute acid solution than do soda-lime and lead-alkali glasses.

Enamels, which are made of silicate and borosilicate glass with the addition of fluxes to promote adhesion, are highly resistant to corrosion and are widely used to protect steels and cast irons.

4.8.16 Other Means of Resisting Corrosion

Occasionally, no material may offer an economical combination of corrosion resistance and other performance requirements. In such cases, a low-cost base material that satisfies the mechanical and physical requirements can be selected provided that it is adequately protected against corrosion. Protection can take the form of sacrificial coatings, passivation, corrosion inhibitors, barrier coatings, or cathodic protection, as discussed in Section 3.7. Barrier coatings are also commonly used for protection against corrosion and their selection is discussed in Section 4.9.

4.9 COATINGS FOR PROTECTION AGAINST CORROSION

Coatings are usually applied for one or more of the following purposes:

1. To modify the surface quality of color, brightness, reflectivity, or opacity
2. To provide protection against corrosion or oxidation
3. To provide protection against abrasion and wear
4. To provide electrical and thermal conductivity or insulation

The following discussion is mainly concerned with the use of coatings for protection against corrosion. In such cases, protection against corrosion can be achieved in two ways:

a. Isolation of the surface from the environment
b. Electrochemical action

Isolation of the surface is usually performed by nonmetallic coatings, and in such cases the thickness, soundness, and strength of the coating will control its effectiveness as an isolator. Nonmetallic coatings can either be inorganic, as in the case of vitreous enamels, or organic, as in the case of varnishes and lacquers. Electrochemical action is achieved with metallic coatings.

4.9.1 Metallic Coatings

The coating metal can be nobler than the base metal and thus protects it, as in the case of tin coatings on steel. However, if pores or cracks are present in the coating, more severe attack could result than if the base metal had no coating. When the coating is anodic with respect to the base metal, the coating dissolves anodically, although the base metal, which is the cathode in the galvanic cell, will not be attacked. Examples of such coatings include aluminum, cadmium, and zinc that are anodic with respect to iron. The two ways of isolation and electrochemical action can be combined by first applying an anodic coating to the base metal followed by a nonmetallic finish.

4.9.2 Organic Coatings

Organic coatings depend mainly on their chemical inertness and impermeability in providing protection against corrosion. An organic coating is made up of two principal components: a vehicle and a pigment. The vehicle contains the film-forming

TABLE 4.10
Rating of Organic Coatings

	Cost	Abrasion Resistance	Flexibility	Adhesion	Resistance to Atmosphere (Salt Spray)	Exterior Durability	Color Retention	Resistance to Chemicals (General)	Maximum Service (Temperature Rating)
Alkyd	3	2	3	3	1	3	1	1	1
Amine–alkyd	3	3	2	3	1	3	1	1	2
Acrylic	2	2	3	2	3	3	3	1	1
Cellulose (butyrate)	1	2	3	2	3	2	3	1	1
Epoxy	1	3	3	3	3	1	1	3	2
Epoxy ester	2	3	1	3	3	2	1	1	2
Fluorocarbon	0.5	1	1	2	3	3	1	3	2
Phenolic	2	3	1	3	3	3	0	2	2
Polyamide	2	3	1	2	1	0	2	1	2
Plastisol	3	3	3	2	3	2	1	3	1
Polyester (oil-free)	2	2	2	3	3	2	2	1	1
Polyvinyl fluoride (PVF)	0.5	3	3	2	3	3	2	3	1
Polyvinylidene fluoride (PVF2)	0.5	3	3	2	3	3	2	3	1
Silicone	1	2	1	1	3	3	3	1	3
Silicone alkyd	1	2	1	2	2	3	2	2	3
Silicone polyester	1	2	2	2	3	3	2	2	3
Silicone acrylic	1	2	1	2	2	3	3	2	3
Vinyl	2	2	3	1	3	3	2	1	1
Vinyl alkyd	2	2	2	2	2	1	2	1	1
Polyvinyl chloride (PVC)	1	3	3	3	3	2	1	3	1
Neoprene (rubber)	3	3	3	2	3	3	1	1	1
Urethane	0.5	3	3	3	3	3	1	1	2

Note: Properties: 3 = excellent, 2 = very good, 1 = fair, 0 = poor; cost: 3 = cheapest, 2 = moderate price, 1 = expensive, 0.5 = very expensive.

ingredients that dry to form the solid film. It also acts as a carrier for the pigment. Vehicles can be either oil or resin, but oils have limited industrial uses. Nearly all polymers can be used as film formers and frequently two or more kinds are combined to give the required properties. The properties of polymers as coatings are usually similar to those of the bulk polymers.

Pigments, which may or may not be present, give the required color, opacity, and flow characteristics. They can also contribute to the protection against corrosion of the base metal and against the destructive action of UV light on the polymeric vehicle. Table 4.10 gives the relative properties of some commonly used organic coatings.

4.9.3 Vitreous Enamels

Vitreous, or porcelain, enamels are inorganic coatings applied primarily to protect metal surfaces against corrosion. The composition should be adjusted so that the coefficient of thermal expansion closely matches with that of the base metal. Two coats are usually needed. Ground coats contain oxides that promote adhesion to the metal base, cobalt or nickel oxides for steel base and lead oxide for cast iron base, and cover coats improve the appearance and properties of the coating. The composition of porcelain enamels varies widely depending on the metal base and the application. Table 4.11 gives typical compositions of some porcelain enamels for steel and cast iron.

TABLE 4.11
Acid-Resistant Porcelain Enamels for Steel and Cast Iron

	Enamel for Steel (wt%)		Enamel for Cast Iron (wt%)	
Constituent	Ground Coat	Cover Coat	Ground Coat	Cover Coat
SiO_2	56.44	41.55	77.7	37.0
B_2O_3	14.90	12.85	6.8	4.9
Na_2O	16.59	7.18	4.3	16.8
K_2O	0.51	7.96	—	1.7
Li_2O	0.72	0.59	—	—
CaO	3.06	—	—	2.0
ZnO	—	1.13	—	5.9
$A1_2O_3$	0.27	—	7.2	1.9
TiO_2	3.10	21.30	—	7.9
CuO	0.39	—	—	—
MnO_2	1.12	—	—	—
NiO	0.03	—	—	—
CO_3O_4	1.24	—	—	—
P_2O_5	—	3.03	—	—
F_2	1.63	4.41	—	—
PbO	—	—	4.0	8.8
Sb_2O_3	—	—	—	13.1

4.10 SELECTION OF MATERIALS FOR WEAR RESISTANCE

The main factors that influence the wear behavior of a material can be grouped as

1. metallurgical variables, including hardness, toughness, chemical composition, and microstructure; and
2. service variables, including contacting materials, contact pressure, sliding speed, operating temperature, surface finish, lubrication, and corrosion.

Although the performance of a material under wear conditions is generally affected by its mechanical properties, wear resistance cannot always be related to one property. In general, wear resistance does not increase directly with tensile strength or hardness although if other factors are relatively constant, hardness values provide an approximate guide to relative wear behavior among different materials. This is particularly true for applications involving metal-to-metal sliding. In such cases, increasing the hardness increases wear resistance as a result of decreasing penetration, scratching, and deformation. Increasing toughness also increases wear resistance by making it more difficult to tear off small particles of deformed metal.

Because wear is a surface phenomenon, surface coatings and treatments play an important role in combating it. Surface coatings consist of wear-resistant materials that are applied to the surface, as discussed in Section 4.11. Surface treatment avoids having to make the entire part of a wear-resistant material, which may not provide all the other functional requirements or may be more expensive. Surface treatments include the following:

- Surface heat treatment, as in the case of flame and induction heating, allows hardening of the surface without affecting the bulk of the material (Table 4.12).
- Surface alloying, as in the case of carburizing, cyaniding, nitriding, and carbonitriding, increases the hardness of the surface by increasing its carbon and nitrogen content (Table 4.12).

In spite of the widespread use of surface treatments and surface coatings to combat wear, these solutions are not without problems. Not all materials or parts can be surface treated and surface coatings can fail by spalling. In many applications wear problems are solved, wholly or in part, by the proper selection of materials, as discussed in the following sections.

4.10.1 Wear Resistance of Steels

Mild steels, although among the cheapest and most widely used materials, have poor wear resistance and can suffer severe surface damage during dry sliding. This can be avoided by selecting compatible mating materials, such as babbitt alloys or white metals, and providing adequate lubrication. Increasing the carbon content of the steel improves the wear resistance but increases the cost.

Surface-hardenable carbon or low-alloy steels are another step higher in wear resistance. Components made from these steels can be surface hardened by

TABLE 4.12
Surface Hardening Treatments for Steels

Process	Treatment	Applications
Flame hardening	Heat the surface using torch, then quench	Hardened depth is 0.5–6 mm Used for gear teeth, crankshafts, axles
Induction hardening	Heat the surface using high frequency induction current, then quench	
Carburizing: increasing carbon content of the surface	Heat component at 850–950°C in a carbon-rich gas or solid, then quench	Hardened depth is 0.5–1.5 mm Used for gear teeth, cams, shafts, bolts, and nuts
Cyaniding: increasing carbon and nitrogen content of the surface	Heat component at 700–850°C in a cyanide-rich salt bath, e.g., sodium cyanide, then quench	Hardened depth is 0.02–0.3 mm Used for small gears, bolts, and nuts
Nitriding: increasing nitrogen content of the surface	Heat component at 500–650°C in ammonia gas	Hardened depth is 0.05–0.6 mm Used for gears, shafts, and tools
Carbonitriding: increasing carbon and nitrogen content of the surface	Heat component at 700–850°C in a mixture of carbon-rich and ammonia gases, then quench	Hardened depth is 0.05–0.6 mm Used for gears, tools, and nuts

carburizing, cyaniding, or carbonitriding (Table 4.12) to achieve better wear resistance at a still higher cost. An even higher wear resistance can be achieved either by nitriding medium-carbon–chromium or chromium–aluminum steels, or by surface hardening high-carbon–high-chromium steels. Precipitation-hardened stainless steels can be used in applications involving wear, elevated temperature, and corrosion.

Austenitic manganese steels are selected for a wide variety of applications where good abrasion resistance is important. The original austenitic manganese steel, Hadfield steel, contains 1.2% C and 12% Mn; however, several compositions are now available as covered by ASTM A128. These steels have carbon contents between 0.7% and 1.45% and manganese contents between 11% and 14%, with or without other elements such as chromium, molybdenum, nickel, vanadium, and titanium. Compared with other abrasion-resistant ferrous alloys, austenitic manganese steels have superior toughness at moderate cost. They have excellent resistance to metal-to-metal wear, as in sheave wheels, rails, and castings for railway track work. Manganese steels are also valuable in conveyors and chains subjected to abrasion and used for carrying heavy loads.

4.10.2 Wear Resistance of Cast Irons

As-cast grey cast iron has adequate wear resistance for applications such as slideways of machine tools and similar sliding members. Better wear resistance is achieved with white pearlitic and martensitic irons, which are used in chilled iron rolls and grinding balls. Alloyed white irons have even better wear resistance but are more expensive.

4.10.3 Nonferrous Alloys for Wear Applications

Aluminum bronzes (Table A.14) range from the soft and ductile alpha alloys, which are used for press guides and wear plates, to the very hard and brittle proprietary die alloys, which are used for tube bending dies and drawing die inserts. The softest alloys contain about 7% Al with some additions of Fe and Sn. Increasing the aluminum content increases the hardness. Aluminum bronzes are not self-lubricating and should only be used where adequate lubrication can be maintained. These alloys are recommended for applications involving high loads and moderate to low speeds. Increasing the hardness increases abrasion resistance but lowers conformability and embeddability of these alloys when used as sleeves for sliding bearings. This subject is discussed in more detail in Chapter 11, case study on materials selection for lubricated journal bearing.

Beryllium–copper alloys are among the hardest and strongest of all copper alloys. Properly lubricated, they have better wear resistance than other copper alloys and many ferrous alloys. An alloy containing 1.9% Be, 0.2% Co, rest Cu, is usually specified for wear applications, and it has better load-carrying capacity than all other copper-base alloys. In addition, beryllium coppers exhibit excellent corrosion resistance in industrial and marine atmospheres. Wear properties of beryllium–copper can be increased by oxidizing the surface of the alloy, by placing graphite in the surface, and by using cast parts rather than machining them from wrought alloys.

Wrought cobalt-base wear-resistant alloys have excellent resistance to most types of wear in addition to good resistance to impact and thermal shock, heat and oxidation, corrosion, and high hot hardness. The primary Co-base alloys for severe wear applications are the following:

1. *Stellite 6B*. It contains 0.9–1.4% C, 28–32% Cr, 3% Ni, 1.5% Mo, 3.5–5.5% W, 3% Fe, 2% Mn, 2% Si, rest Co.
2. *Stellite 6K*. It contains 1.4–2.2% C, 28–32% Cr, 3% Ni, 1.5% Mo, 3.5–5.5% W, 3% Fe, 2% Mn, 2% Si, rest Co.
3. *Haynes 25*. It contains 0.05–0.10% C, 19–21% Cr, 9–11% Ni, 14–16% W, 3% Fe, 1–2% Mn, 1% Si, rest Co.

Stellite has better resistance to abrasive wear, while Haynes 25 has better resistance to adhesive wear. Wrought Co-base alloys are nearly identical in chemical composition to their hard-facing alloy counterparts (Section 4.11) but with small differences in boron, silicon, or manganese levels. Another difference is the microstructure, which depends on the method of fabrication.

4.10.4 Wear Resistance of Plastics

Wear-resistant, self-lubricating plastics are favorably competing with metals in many applications including bearings, cams, and gears. In addition to ease of manufacture, these plastics have better lubricating properties and need less maintenance. Wear-resistant plastics are formulated with internal lubricating agents and are available in both unreinforced and reinforced versions. A combination of lubricating additives is usually employed to achieve optimum wear resistance. For example,

TABLE 4.13
Wear Properties of Some Lubricated Plastics on Steel

Plastic Material	Reinforcing Fibers	Wear Factor[a]	Coefficient of Friction[b]
Nylon 6/6–18% PTFE, 2% silicone		6	0.08
Nylon 6/6–13% PTFE, 2% silicone	30% carbon	6	0.11
Polyester–13% PTFE, 2% silicone	30% glass	12	0.12
Acetal–20% PTFE		13	0.13
Acetal–2% silicone		27	0.12
Polyimide–10% PTFE	15% carbon	28	0.12
Polypropylene–20% PTFE		33	0.11
Polyurethane–15% PTFE	30% glass	35	0.25
Polystyrene–2% silicone		37	0.08

[a] 10^{10} in. 3 min/ft lb h.
[b] Dynamic at 40 lb/in.^{2}, 50 ft./min.

silicone and PTFE are usually added to thermoplastics to improve their performance at high speeds and pressures. Carbon and aramid fibers, which are usually added for mechanical reinforcement, are also known to improve wear resistance. Table 4.13 lists some commonly used wear-resistant plastics in the order of decreasing resistance to wear when sliding against steel.

In spite of the advantages of plastics as wear-resistant materials, the following limitations should be kept in mind when selecting them for some applications:

1. *Plastics rubbing against plastics*. In such cases, wear is much more severe than in plastic–metal systems. The severity of wear can be reduced by adding PTFE or other internal lubricants and by similarly reinforcing the mating surfaces.
2. *Sensitivity of wear resistance to seemingly small variations in temperature*. For example, the wear rate of 15% PTFE, 30% glass-fiber nylon 6/6 at 200°C (ca. 400°F) is about 40 times its wear rate at room temperature.
3. *Sensitivity of plastics to the surface roughness of the mating metallic surface*. Finishes that are too rough or too smooth can result in excessive wear. Minimum wear of lubricated plastics is usually obtained with metallic surface roughness in the range of 12–16 μm.
4. *Type of metal can strongly affect the results*. For example, using an aluminum alloy instead of steel can dramatically increase the wear rate of plastics.

4.10.5 Wear Resistance of Ceramics

Ceramics can be used in a variety of applications where wear resistance is required. The wear behavior of ceramics is determined by the nature of the mating surfaces and the presence of surface films. In general, as the grain size and porosity of the ceramic material increases, wear increases. The presence of surface films, such as water and oils, can affect adhesion and wear. For example, wear of partially stabilized zirconia increases in aqueous environments but decreases in fatty acids such as stearic acid. For engines and similar applications, silicon carbide against lubricated steel has lower friction and less scuffing than chilled cast iron, which makes it suitable for engine valves, train components, and bearings.

4.11 WEAR-RESISTANT COATINGS

Hard-facing coatings are normally used for protection against wear. These coats may be applied to new parts made of soft materials to improve their resistance to wear, or to worn parts to restore them to serviceable condition. The selection of hard-facing alloys for a given application is guided primarily by wear and cost considerations. However, other factors, such as impact resistance, corrosion and oxidation resistance, and thermal requirements should also be considered. In general, the impact resistance of hard-facing alloys decreases as the carbide content increases. As a result, a compromise has to be made in applications where a combination of impact and abrasion resistance is required.

Most hard-facing alloys are marketed as proprietary materials and are classified as follows:

- Low-alloy steels (group 1)
- High-alloy ferrous materials (groups 2 and 3)
- Nickel-base and cobalt-base alloys (group 4)
- Carbides (group 5).

Generally, both wear resistance and cost increase as the group number increases.

The alloys in group 1 contain up to 12% Cr + Mo + Mn and have the greatest shock resistance of all hard-facing alloys, except austenitic manganese steels. They are less expensive than other hard-facing alloys and are extensively used where machinability is necessary and only moderate improvement over the wear properties of the base metal is required.

Alloys in group 2A and 2B contain up to 25% Cr + Mo and are more wear-resistant but less shock resistant than group 1 alloys. Alloys 2C and 2D contain up to 37% Mn + Ni + Cr and are highly shock resistant, but have limited wear resistance unless subjected to work hardening.

Group 3 alloys contain up to 50% Cr + Mo + Co + Ni. Their structure contains massive hypereutectic alloy carbides that improve wear resistance and give them some degree of corrosion and heat resistance.

The nonferrous Ni- and Co-base alloys of group 4A contain 50–100% Co + Cr + W, and are the most versatile hard-facing alloys. They resist heat, abrasion, corrosion, impact, galling, oxidation, thermal shock, erosion, and metal-to-metal wear. Some of these alloys retain useful hardness up to 825°C (1500°F) and resist oxidation up to 1100°C (200°F). Alloys of group 4B and 4C contain 50–100% Ni + Cr + Co + B, and are the most effective for service involving both corrosion and wear. They retain useful hardness up to about 650°C (1200°F) and resist oxidation up to 875°C (1600°F).

Group 5 materials provide maximum abrasion resistance under service conditions involving low or moderate impact. They are made of 75–96% carbides cemented by a metal matrix. Either WC or WC + TiC + TaC is used as the carbide phase, whereas Fe-, Ni-, or Co-base alloys are used as the matrix material.

4.12 SUMMARY

1. Most ductile wrought metallic materials are equally strong in tension and compression; brittle materials, however, are generally much stronger in compression than in tension.
2. The elastic modulus of a given class of materials is almost independent of chemical composition and heat treatment. The stiffness of a component may be increased by increasing the second moment of area of its cross section or by selecting a higher modulus class of materials for its manufacture.
3. When weight is an important consideration, the specific strength (strength/density) and specific stiffness (modulus of elasticity/density) may be used as the selection criteria.
4. Within a given class of materials, there is an inverse relationship between strength and toughness. Decreasing the operating temperature generally causes a decrease in toughness, particularly for bcc materials, such as carbon and low-alloy steels.
5. The fatigue strength of metallic materials generally increases with increasing tensile strength. However, the higher the strength the higher the notch sensitivity and the greater is the need to eliminate coarse second-phase particles and to produce a more refined, homogeneous structure. Some fiber-reinforced composites perform better in fatigue than some metals, especially when compared on per weight basis.
6. Many of the methods used to increase the strength at normal temperatures become ineffective at high temperatures. Fine dispersion of stable second-phase particles may be used to improve the creep strength.
7. The main parameters that affect the corrosion resistance of a metallic material are its composition and the presence of impurities, nature and distribution of microstructural constituents, surface condition and deposits, and processing history. Plastics and glasses exhibit good resistance to most chemicals with the exception of organic solvents in the case of plastics and HF acid in the case of glasses.
8. Coatings for protection against corrosion either isolate the surface from the environment, as in the case of nonmetallic coatings, or by electrochemical action, as in the case of metallic coatings.
9. Increasing the hardness and toughness increases wear resistance. Hard-surface coatings and surface hardening treatments may be used to improve wear resistance.

4.13 REVIEW QUESTIONS

4.1 a. What are the main performance requirements of the wing structure of a two-passenger training aircraft, and what are the corresponding material properties?

b. Compare the use of the following materials in making the wing structure for the two-passenger training aircraft:

	Specific Gravity	Elastic Modulus (GPa)	Tensile Strength (MPa)
Aluminum alloy	2.7	70	580
Magnesium alloy	1.7	45	280
Epoxy + 56% E glass fibers	1.97	42.8	1028

4.2 Would you use AISI 1050 steel for manufacturing a component that will serve at −50°C (−58°F)? If not suggest substitute materials.

4.3 If the available NDT equipment can detect internal cracks longer than 1 mm in length, determine the diameter of a bar that can bear a load of 150 kN without failure if it is made of AISI 4340 steel with yield strength of 1480 MPa and K_{IC} = 87.4 MPa $m^{1/2}$.

4.4 Ti–6Al–4V and aluminum 7075 T6 alloys are widely used in making lightweight structures. If the available NDT equipment can only detect flaws larger than 1 mm in length, can you safely use either of the mentioned alloys for designing a component that will be subjected to 400 MPa? Use the information in Table 4.6.

4.5 Explain why the crankshaft in a motorcar engine is only hardened on the surface and not throughout the whole cross section.

4.6 What are the material requirements for the blades of a household scissor? Suggest possible materials.

4.7 What are the material requirements for the radiator of a motorcar? Suggest possible materials.

4.8 An aluminum alloy is being considered as a replacement for steel in manufacturing a tensile member to save weight. The member has a circular cross section, a length of 1 m, and is subjected to alternating tensile load of 6000 kg. Given the following information:

a. Determine if aluminum is a viable material for saving weight in this case.
b. Compare the relative cost of the two solutions.

Characteristic	Steel	Aluminum Alloy
Ultimate tensile strength	735 MPa	315 MPa
Endurance ratio	0.43	0.31
Surface finish derating factor	0.68	0.64
Size derating factor	0.80	0.77
Reliability derating factor	0.75	0.70
Specific gravity	7.8	2.7
Relative cost	1	3.2

4.9 Why are stainless steels corrosion resistant? Explain the phenomena of passivation and sensitization.

4.10 What are the material requirements for the electric heating wires in a heat treatment furnace that is expected to operate at temperatures up to 1000°C?
4.11 What are the differences between galvanizing and tinning of steel parts? Compare the merits of using each of these methods for (a) food cans and (b) outdoor fencing.
4.12 What are the differences between organic coatings and vitreous enamels? Give examples of the uses of each type of coating in household applications.
4.13 Why is aluminum more resistant to atmospheric corrosion than plain carbon steel even though it is lower in the galvanic series?
4.14 What are the main material requirements for a kitchen knife blade? What is the type of material that you would recommend for such blades? Can ceramics be used for such an application?
4.15 What are the main material requirements for a gas turbine blade that is expected to operate at 900°C? Suggest possible materials.
4.16 What are the reasons that mechanical engineers do not always specify the strongest available material?
4.17 What are the advantages and disadvantages of low-carbon steels? How are the limitations overcome in practice?
4.18 What are the main material requirements for the following components: motorcar exhaust manifold, coil for electrical resistance heater, and railway line?

BIBLIOGRAPHY AND FURTHER READING

Ashby, M.F., *Materials Selection in Mechanical Design*, 3rd Ed., Elsevier, Amsterdam, 2005.

Ashby, M.F. and Johnson, K., *Materials and Design: The Art and Science of Materials Selection in Product Design*, Butterworth-Heinemann, Amsterdam, 2002.

Bowman, K., *Introduction to Mechanical Behavior of Materials*, Wiley, New York, 2003.

Boyer, H.E. and Gall, T.L., *Metals Handbook*, Desk Edition, ASM, Metals Park, OH, 1985.

Brooks, C.R. and Choudhury, A., *Failure Analysis of Engineering Materials*, McGraw-Hill, New York, 2001.

Collins, J.A. and Daniewicz, S.R., Failure modes: performance and service requirements for metals, in *Handbook of Materials Selection*, Kutz, M., Editor. Wiley, New York, 2002, pp. 705–773.

Collins, J.A. and Daniewicz, S.R., Failure modes: performance and service requirements for metals, in *Mechanical Engineers' Handbook: Materials and Mechanical Design*, 3rd Ed., Kutz, M., Editor. Wiley, Hoboken, NJ, 2006, pp. 860–924.

Courtney, T.H., *Mechanical Behavior of Materials*, 2nd Ed., McGraw-Hill College, Blacklick, OH, 1999.

Das, A.K., *Metallurgy of Failure Analysis*, McGraw-Hill, New York, 1997.

Dieter, G., *ASM Metals Handbook, Vol. 20, Materials Selection and Design*, ASM International, Materials Park, OH, 1997.

Dowling, N., *Mechanical Behavior of Materials*, 3rd Ed., Prentice-Hall, New York, 2006.

Flinn, R.A. and Trojan, P.K., *Engineering Materials and their Applications*, 4th Ed., Houghton Mifflin, Boston, MA, 1990.

Hosford, W.F., *Mechanical Behavior of Materials*, Cambridge University Press, London, 2005.

Jones, D.R.H., *Failure Analysis Case Studies II*, Pergamon Press, Oxford, 2001.

Kutz, M., *Handbook of Materials Selection*, Wiley, New York, NJ, 2002.

Kutz, M., *Mechanical Engineers' Handbook: Materials and Mechanical Design*, 3rd Ed., Wiley, Hoboken, NJ, 2006.

Tawancy, H.M., Ul Hamid, A., and Abbas, N.M., *Practical Engineering Failure Analysis*, Marcell Dekker, New York, 2004.

Part II

Relationship between Design, Materials, and Manufacturing Processes

THE SUCCESSFUL DESIGNER!!

The designer bent across his board,
Wonderful things in his head were stored,
And he said as he rubbed his throbbing bean,
'How can I make this hard to machine?'

'If this part here were only straight,
I'm sure the thing would work first rate,
But 't would be so easy to turn and bore,
It would never make the machinists sore.'

'I'd better put in a right angle there,
Then watch those babies tear their hair,
Now I'll put the holes that hold the cap
Way down here where they're hard to tap.'

'Now this piece won't work, I'll bet a buck,
For it can't be held in a shoe or chuck;
It can't be drilled or it can't be ground,
In fact the design is exceedingly sound.'

He looked again and cried, 'At last-
Success is mine, it can't even be cast!'

ANON.

Unlike the design made by the designer in the poem, a successful design should result in the creation of a product that satisfies a certain need and performs its function efficiently and economically within the prevailing legal, social, safety, environmental, and reliability requirements. To satisfy such requirements, the design engineer has to take into consideration a large number of diverse factors which can be grouped into three categories, as shown in Figure 5.1 and summarized as:

1. Factors related to product function and consumer requirements. These are related to human factors, ease of operation, ease of repair, aesthetics and styling, noise level, pollution, intended service environment, and possibility of reuse and recycling after retirement, design codes, capacity, size, weight, safety, expected service life, reliability, maintenance, frequency of failure, initial cost, and operating cost.
2. Material-related factors such as static strength and ductility, stiffness, toughness, fatigue resistance, creep resistance, and corrosion resistance.
3. Manufacturing-related factors such as available fabrication processes, accuracy, surface finish, shape, size, required quantity, delivery time, cost, and required quality.

Although not all of these factors are applicable or are of equal importance for all design situations, the list illustrates the multifaceted nature of design. Figure 5.1 also shows that there are other secondary relationships between material properties and manufacturing processes, between function and materials properties, and between manufacturing processes and function.

This part of the book discusses the relationship between engineering design, materials, and manufacturing processes. Chapter 5 discusses the factors related to function and consumer requirements, Chapter 6 discusses the effect of material properties on design, and Chapter 7 discusses the effect of manufacturing processes on design.

5 The Nature of Engineering Design

5.1 INTRODUCTION

Engineering design is an interdisciplinary process that transforms marketing ideas and consumer wishes into specific information and instructions that will allow successful manufacture of the product. Figure 5.1 shows the diverse factors that have to be considered when designing a component. In industry, design work is normally carried out by the design engineer in collaboration with other departments including customer service, marketing and sales; legal and patents; safety, codes and regulations; research and development; and materials and manufacturing. The design team is often required to make compromises to satisfy the conflicting requirements of the different constituencies. Engineering design work is usually performed on three different levels:

1. Development of existing products or designs, redesign, by introducing minor modifications in size, shape, or materials to improve performance or to overcome difficulties in production. This type of work represents a large proportion of the design effort in industry and may be accompanied by failure analysis to reduce the likelihood of further failures.
2. Adaptation of an existing product or design to operate in a new environment or to perform a different function. In some cases, the new design may be widely different from the starting one.
3. Creation of a totally new design that has no precedent. This type of work is most demanding in experience and creativity of the design team and is not performed as often as the other types of design work. It often requires the solution of problems that may not have been encountered before and could require a considerable effort in research and development.

The goal of this chapter is to give an overview of the parameters that influence the engineering design process in industry. The main objectives are the following:

1. To discuss the various issues that have to be considered in design
2. To review the major phases of the design process
3. To explain the use of codes and standards in design
4. To discuss the effect of component geometry on design
5. To rationalize the use of the factor of safety in design
6. To calculate the probability of failure of a component at the design stage.

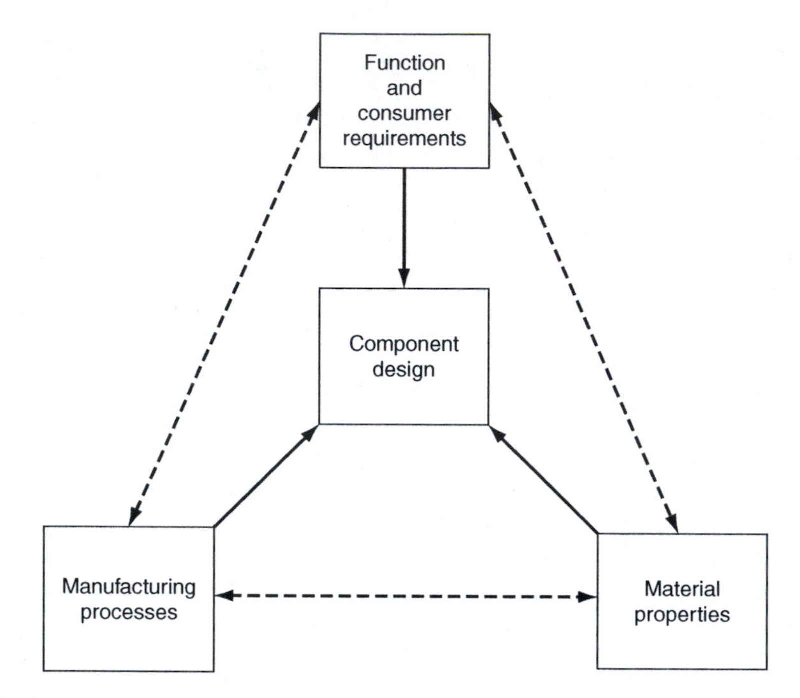

FIGURE 5.1 Factors that should be considered in component design.

5.2 GENERAL CONSIDERATIONS IN ENGINEERING DESIGN

Because of its interdisciplinary nature, an optimum engineering design involves trade-offs among the many, and often conflicting, conditions that it has to satisfy. Such conditions include human factors, marketing and esthetic considerations, functional requirements, manufacturing considerations, economic factors, as well as safety and environmental requirements.

5.2.1 Human Factors

Adaptation of the product to make it convenient for human use is an important aspect of a successful design. The design should also account for variations among human beings in terms of height, weight, physical strength, visual and hearing acuity, conceptual capacity, etc. Using the product efficiently must not require the use of excessive physical force or the performance of too many functions simultaneously. Controls must also follow natural expectations of the user. For example, moving a lever forward or the clockwise rotation of a rotary control is normally expected to increase output.

The designer must also anticipate that the product may be used in unintended ways or functions. Protective measures should also safeguard against the possibility of injuries as a result of errors in use or poor maintenance. For example, the driver cannot press the accelerator and the brake of a motor vehicle at the same time, and the blade of a shear press would not move unless the operator's hands have been completely withdrawn from the work area.

5.2.2 Industrial Design, Esthetic and Marketing Considerations

Styling of the product to reflect its function with emphasis on esthetic and visual features is the realm of industrial design. Esthetic attributes are the qualities that appeal to the senses, such as vision, touch, hearing, smell, and taste. Visual attributes include whether the surface is colorful or subdued, transparent or opaque, glossy or matte, etc. Tactile attributes include whether the surface feels soft or hard to the touch, flexible or stiff, warm or cold, etc. Sounds can be resonant or dull, muffled or sharp, low or high pitch, etc.

It is generally accepted that the success of a product is influenced by both engineering and industrial designs, both of which are emphasized when marketing the product. In more technically mature products, where differences in technical performance are slight and the prices are nearly the same, distinction from competing products can be achieved through industrial design. In such cases, a product can be distinguished by its styling, configuration, proportion, color, texture, ease of use, and character, which are all in the realm of industrial design. For example, in marketing a wristwatch, which is a mature product, the accuracy is taken for granted and more emphasis is placed on its character, for example, rugged and relatively large for a sportsperson but stylish and elegant to go with evening wear. However, the relatively new and still developing mobile (cellular) phone is marketed with more emphasis on its technological capabilities and less on styling and looks.

5.2.3 Environmental Considerations

With the increasing concern for the negative impact of many of the products that are associated with human development, environmental constraints are becoming tighter and more forcefully imposed by the many agencies concerned. Several of these agencies issue guidelines for environmentally responsible design or design for environment (DFE). The U.S. EPA and the International Organization for Standardization (ISO) are examples of agencies that issue such guidelines. The EPA provides design strategies to extend the life of products and materials, reduce material utilization, and improve process management. ISO 14000 provides the international standard for environmental management systems within a company. The ISO publications are available on the organization's website and Block (1996) describes the implementation of ISO 14001.

Following are representative examples of the guidelines for DFE:

- Design the components to be reusable or recyclable
- Minimize the number of parts for ease of disassembly
- Reduce the number of fasteners to reduce the disassembly time
- Use modular design and standardize components as much as possible
- Reduce the number of materials in the product and choose materials that are compatible and can be recycled together

- Avoid the use of materials or production aids that are toxic or harmful to the environment
- If possible, avoid the use of materials that are difficult to recycle. These include FRP, laminated materials, galvanized steels, thermosetting plastics, and ceramics
- Riveted or permanently joined assemblies that are made of different materials are difficult to recycle

Life cycle engineering, cradle to grave, is another approach in assessing the environmental impact of a product during its entire life cycle. This approach recognizes the fact that products have different environmental impact during the different stages of their life cycle. This approach is discussed in more detail in Section 8.11.

5.2.4 Functional Requirements

Functional requirements represent the minimum level of performance that any acceptable design must have. With increasing competition among different manufacturers and with more emphasis on product liability, designs are required to meet the additional requirements of reliability, safety, marketability, and cost. Manufacturing considerations affect the feasibility of making the product at a competitive price and are discussed in Chapter 7, while economic considerations are discussed in Chapter 8.

Service life represents an important design parameter as it affects both reliability and economics of the product. Service life of a component can be estimated according to safe-life or fail-safe criteria. The safe-life criterion can be applied to a component in which undetected crack or other defects could lead to catastrophic structural failure, and a life limitation must therefore be imposed on their use. The fail-safe criterion can be applied to structures in which there is sufficient tolerance of a failure to permit continuous service until discovered by routine inspection procedure, or by obvious functional deficiencies. The majority of engineering components can be designed according to the fail-safe criterion. Even a critical component can be designed according to the fail-safe criterion if failure is detectable by the maintenance program, which must define both the timing and the methods of inspection to be applied. Redistribution of the load into sufficiently robust adjacent components if failure occurs is an added safety precaution. If the use of a safe-life component is unavoidable, its safe service life must be estimated by testing and its replacement life calculated by applying an appropriate factor of safety.

5.3 MAJOR PHASES OF DESIGN

Engineering design is usually an iterative process that involves a series of decision-making steps, where each decision establishes the framework for the next one. There is no single universally recognized sequence of steps that leads to a workable design as these depend on the nature of the problem being solved as well as on the size and structure of the organization. Generally, however, a design usually passes through most of the phases that are shown in Figure 5.2 and grouped into the following three categories:

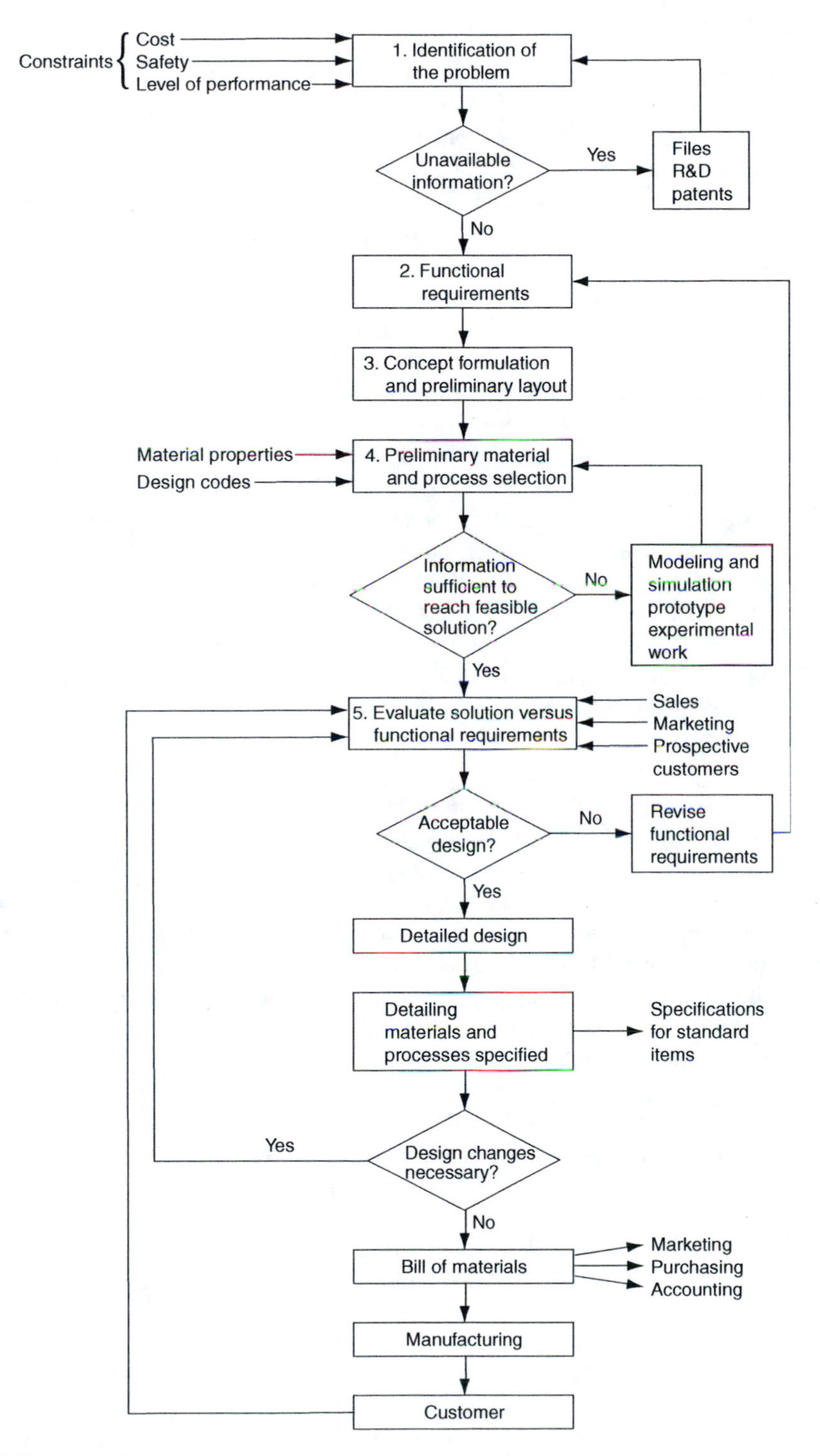

FIGURE 5.2 Major phases of design.

Preliminary and conceptual design

1. Identifying the need, evaluating the product feasibility, selecting the most promising concept, and defining the objective of the design represent the first phase of product development in most cases. The major constraints such as cost, safety and level of performance, and the overall specifications are also defined at this stage. Effective communication with other departments in the organization, such as marketing, legal, R&D, and manufacturing, is essential at this stage. Unavailable information is identified at this stage and strategy for obtaining it is outlined.
2. Functional requirements and operational limitations are directly related to the required characteristics of the product and are specified as a result of the activities of phase 1. Although it is not always possible to assign quantitative values to these product characteristics, they must be related to measurable quantities that will allow future evaluation of the product performance. This is sometimes called the conceptual design stage.
3. System definition, concept formulation, and preliminary layout are usually completed, in this order, before evaluating the operating loads and determining the form of the different components or structural members. Allowances must be made for uncertainties of loading and approximations in calculations. The consequences of component failure must also be considered at this stage. Whether the component can be easily and cheaply replaced or whether large costs will be incurred, will significantly influence the design.

Configuration (embodiment) design

4. Consulting design codes and collecting information on material properties will allow the designer to perform preliminary materials selection, preliminary design calculations, and rough estimation of manufacturing requirements. Preliminary design begins by expanding the conceptual design into a detailed structure of subsystems and sub-subsystems. In many cases, several solutions to the design problem can be proposed at this stage.
5. The evaluation phase involves a comparison of the expected performance of the design with the performance requirements established in phase 2. Evaluation of the different solutions and selection of the optimum alternative can be performed using decision-making techniques, modeling techniques, experimental work, and prototypes. Having arrived at an optimum solution, it is often necessary to revise the design and to make more precise design calculations as well as to specify materials and manufacturing processes in more detail. Questions related to design for manufacture and assembly are also considered at this stage. Iteration steps of this type are easier to perform if computer techniques are used in design and materials selection.

In some cases, it is not possible to arrive at a design that fulfills all the requirements and complies with all the limitations established in phase 2. This means that these requirements and limitations have to be reconsidered and the phases 3–5 have to be repeated until an acceptable design is arrived at.

Detail (parametric) design

6. Having arrived at a final design, the project then enters the detailed design stage where it is converted into a detailed and finished form suitable for use in manufacturing. The preliminary design layout, any available detail drawings, models and prototypes, and access to the developer of the preliminary design usually form the basis of the detailed design.
7. The next step in the detailed design phase is detailing, which involves the creation of detailed drawings for every part. All the information that is necessary to unambiguously define the part should be recorded in the detail drawing. The material of the part should also be selected and specified by reference to standard codes. The temper condition of the stock material, the necessary heat treatment, and the expected hardness may also be specified for quality control purposes. In the course of detailing it may become clear that the manufacture, assembly, or disassembly of some parts of the layout could be improved if they are changed. In this case, communication should take place between the detail designer and the preliminary designer to agree on the proposed changes.
8. An important part of the detail design phase is the preparation of the bill of materials, sometimes called parts list. The bill of materials is a hierarchical listing of everything that goes into the final product including fasteners and purchased parts. The bill of materials is used by a variety of departments including purchasing, marketing, and accounting. When the detailed design is released for manufacturing, a working bill of materials should go with it. The manufacturing plan, production planning, and assembly will all be based on the bill of materials. An appropriate version of the bill of materials is also shipped along with the finished product for guidance in operating and maintenance. Close interaction between design, manufacturing, and materials engineers is important at this stage.
9. The relationship between the designer and the product does not usually end at the manufacturing or even delivery stages. The manufacturing engineer may ask the detail designer for a change in some parts to make fabrication easier or cheaper. Finally when the product gets into use, the reaction of the consumer and the performance of the product in service are of concern to the designer as the feedback represents an important source of information for future design modifications.

5.4 DESIGN CODES AND STANDARDS

A design code is a set of standards or specifications for the analysis, design, manufacture, and construction of a structure or a component. Codes of practice are set by professional groups and government bodies to achieve a specified degree of safety, efficiency, performance or quality, as well as a common standard of good design practice. Codes serve to disseminate proved data and research results to the average designer who is not expected to have the expertise to appreciate and critically examine all the specialized information associated with the part being designed. Codes are often legal requirements that are adopted and enforced by a government.

A standard specification is a published document that describes the characteristics of a part, material, or process, which is acceptable by an authority or by general consent as a basis of comparison or for approval. Standards can vary from those developed by a company for use in-house to those that represent industry consensus such as those published by American National Standards Institute (ANSI) and ISO. Standards can also be issued by governments to regulate their own purchases and operations. When cited by the purchaser and accepted by a supplier, a standard becomes part of the purchase agreement. The widespread use of standards has benefited companies by reducing the number of products, materials, or components that need to be manufactured or held in stock. Specification of dimensions, shapes, and sizes also helps in achieving interchangeability of components.

As a standard is a document that can be used to control procurement, it should contain both technical and commercial requirements. Specifications normally cover the following information:

1. Product classification, scope of application, size range, condition, and processing details that could help either the supplier or the user.
2. Allowable ranges of chemical composition.
3. All physical and mechanical properties necessary to characterize the product are given in addition to the test methods used to determine these properties.
4. If applicable, other requirements such as special tolerances, surface preparation, loading instructions, packaging, etc., are included.

As shown in Figure 5.2, the designer should consult the relevant design codes soon after formulating the design concept to make certain that the design meets the users' expectations, which normally include the intended function and safety requirements.

5.5 EFFECT OF COMPONENT GEOMETRY

In almost all cases, engineering components and machine elements have to incorporate design features that introduce changes in their cross section. For example, shafts must have shoulders to take thrust loads at the bearings and must have keyways or splines to transmit torques to or from pulleys and gears mounted on them. Other features that introduce changes in cross section include oil holes, fillets, undercuts, bolt heads, screw threads, and gear teeth. These changes cause localized stress concentrations that are higher than those based upon the nominal cross section of the part.

5.5.1 Stress Concentration Factor

The severity of the stress concentration depends on the geometry of discontinuity and nature of the material. A geometric, or theoretical, stress concentration factor,

K_t, is usually used to relate the maximum stress, S_{max}, at the discontinuity to the nominal stress, S_{av}, according to the relationship:

$$K_t = \frac{S_{max}}{S_{av}} \tag{5.1}$$

The value of K_t depends only on the geometry of the part and for the simple case of an elliptical hole in an infinitely large plate, it is given by

$$K_t = 1 + \frac{2b}{a} \tag{5.2}$$

where $2b$ is the dimension of the hole perpendicular to the stress direction and $2a$ the dimension of the hole parallel to the stress direction.

In the case of a circular hole in an infinite plate, a is equal to b and $K_t = 3$. The value of K_t for other geometries can be determined from stress concentration charts, such as those given by Peterson (1974) and Shigley and Mitchell (1983). Other methods of estimating K_t for a certain geometry include photoelasticity, brittle coatings, and finite element techniques. Table 5.1 gives some typical values of K_t.

Experience shows that, under static loading, K_t gives an upper limit to the stress concentration value and applies only to brittle and notch sensitive materials. With more ductile materials, local yielding in the very small area of maximum stress causes considerable relief in the stress concentration. Consequently, for ductile materials under static loading, it is not usually necessary to consider the stress concentration factor. However, due consideration should be given to the stress concentration when designing with high-strength, low-ductility, case-hardened, or heavily cold-worked materials.

5.5.2 Stress Concentration in Fatigue

Stress concentration should also be considered in designing components that are subject to fatigue loading. Under such conditions, a fatigue stress-concentration factor, or fatigue-strength reduction factor, K_f, is usually defined as

$$K_f = \frac{\text{endurance limit of notch} - \text{free part}}{\text{endurance limit of notched part}} \tag{5.3}$$

The relationship between K_f and K_t is discussed in Section 4.6 and a notch sensitivity factor, q, was defined in Equation 4.9. The value of q was shown to vary between 1 and 0. When $K_f = K_t$, the value of q is 1 and the material is fully sensitive to notches. However, when the material is not at all sensitive to notches, $K_f = 1$ and $q = 0$.

In making a design, K_t is usually determined from the geometry of the part. Then, when the material is selected, q can be specified, and Equation 4.9 is solved for K_f. Generally, the value of q approaches unity as the material strength increases, for example, ultimate tensile strength (UTS) is more than 1400 MPa (200 ksi) for steels. Whenever in doubt, the designer can take $K_f = K_t$ and err on the safe side.

5.5.3 Guidelines for Design

Stress concentration can be a source of failure in many cases, especially when designing with high-strength materials and under fatigue loading. In such cases, the

TABLE 5.1
Approximate Values of Stress Concentration Factor (K_t)

Component Shape	Value of Critical Parameter	K_t
Round shaft with transverse hole		
Bending	$d/D = 0.025$	2.65
	$= 0.05$	2.50
	$= 0.10$	2.25
	$= 0.20$	2.00
Torsion	$d/D = 0.025$	3.7
	$= 0.05$	0.36
	$= 0.10$	3.3
	$= 0.20$	3.0
Round shaft with shoulder		
Tension 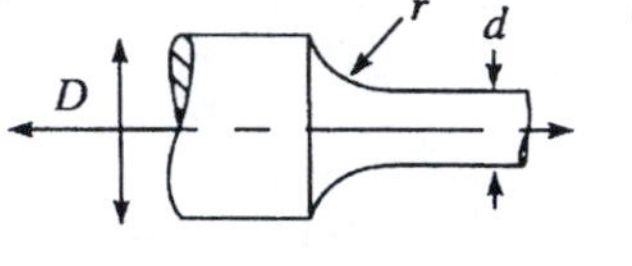	$d/D = 1.5,\ r/d = 0.05$	2.4
	$r/d = 0.10$	1.9
	$r/d = 0.20$	1.55
	$d/D = 1.1,\ r/d = 0.05$	1.9
	$= 0.10$	1.6
	$= 0.20$	1.35
Bending	$d/D = 1.5,\ r/d = 0.05$	2.05
	$r/d = 0.10$	1.7
	$r/d = 0.20$	1.4
	$d/D = 1.1,\ r/d = 0.05$	1.9
	$r/d = 0.10$	1.6
	$r/d = 0.20$	1.35
Torsion 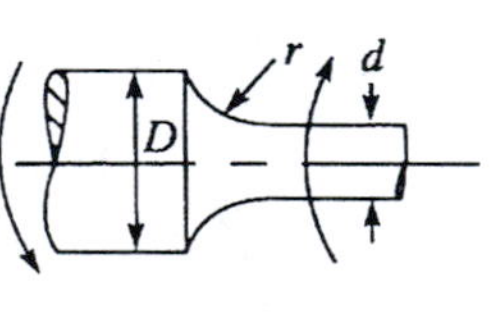	$d/D = 1.5,\ r/d = 0.05$	1.7
	$r/d = 0.10$	1.45
	$r/d = 0.20$	1.25
	$d/D = 1.1,\ r/d = 0.05$	1.25
	$r/d = 0.10$	1.15
	$r/d = 0.20$	1.1
Grooved round bar 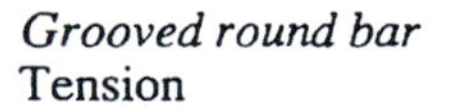		
Tension 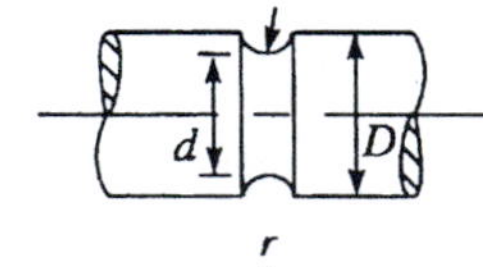	$d/D = 1.1,\ r/d = 0.05$	2.35
	$r/d = 0.10$	2.0
	$r/d = 0.20$	1.6
Bending 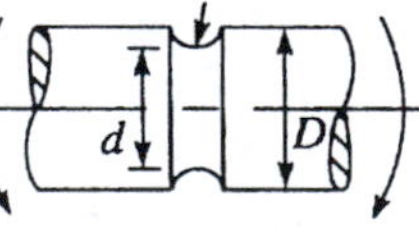	$d/D = 1.1,\ r/d = 0.05$	2.35
	$r/d = 0.10$	1.9
	$r/d = 0.20$	1.5
Torsion	$d/D = 1.1,\ r/d = 0.05$	1.65
	$r/d = 0.10$	1.4
	$r/d = 0.20$	1.25

following design guidelines should be observed if the deleterious effects of stress concentration are to be kept to a minimum:

1. Abrupt changes in cross section should be avoided. If they are necessary, generous fillet radii or stress-relieving grooves should be provided (Figure 5.3a).
2. Slots and grooves should be provided with generous run-out radii and with fillet radii in all corners (Figure 5.3b).

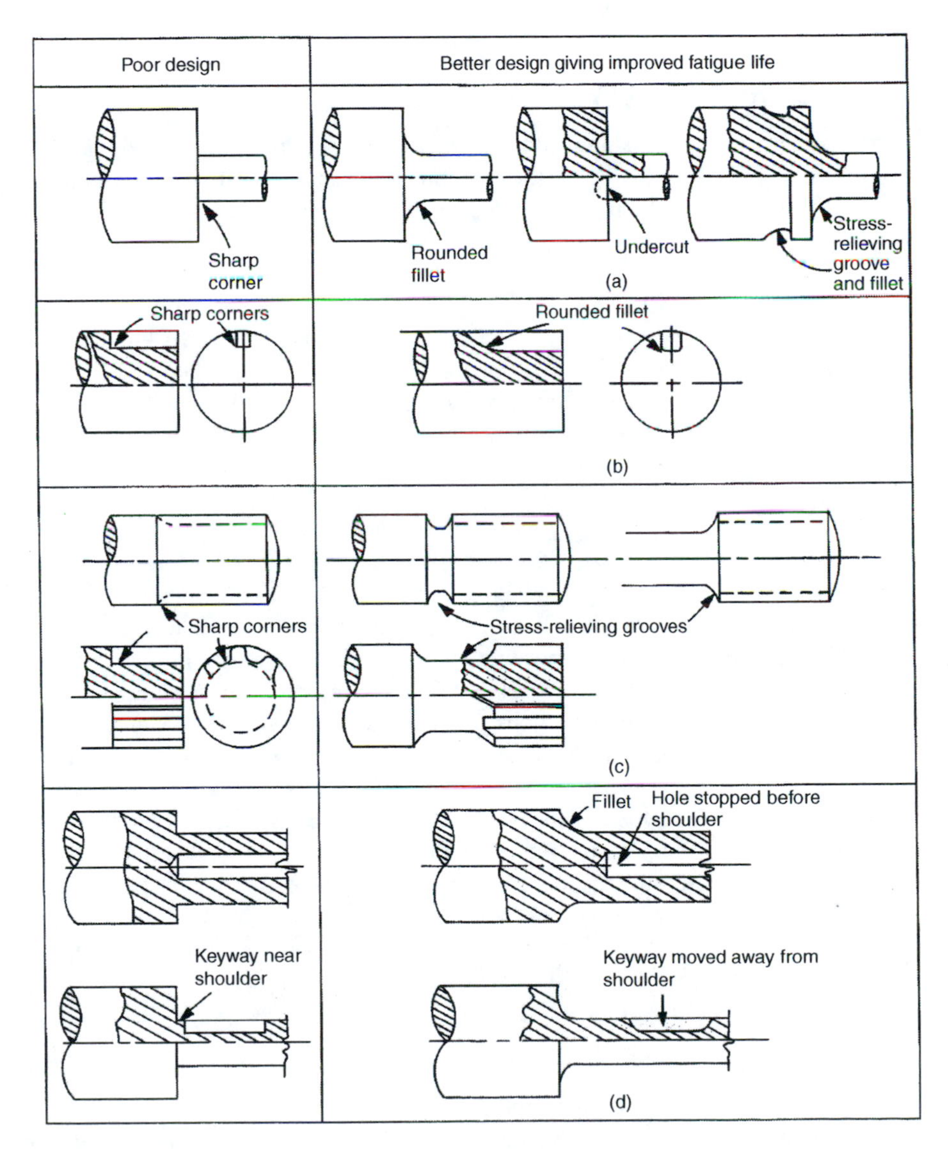

FIGURE 5.3 Design guidelines for shafts subjected to fatigue loading.

3. Stress-relieving grooves or undercuts should be provided at the end of threads and splines (Figure 5.3c).
4. Sharp internal corners and external edges should be avoided.
5. Oil holes and similar features should be chamfered and the bore should be smooth.
6. Weakening features like bolt and oil holes, identification marks, and part numbers should not be located in highly stressed areas.
7. Weakening features should be staggered to avoid the addition of their stress concentration effects (Figure 5.3d).

5.6 FACTOR OF SAFETY

The term factor of safety is applied to the factor used in designing a component to ensure that it will satisfactorily perform its intended function. The main parameters that affect the value of the factor of safety, which is always greater than unity, can be grouped into

1. uncertainties associated with material properties due to variations in composition, heat treatment and processing conditions as well as environmental variables such as temperature, time, humidity, and ambient chemicals;
2. parameters related to manufacturing processes also contribute to the uncertainties of component performance. These include variations in surface roughness, internal stresses, sharp corners, identifying marks, and other stress raisers; and
3. uncertainties in loading and service conditions.

Generally, ductile materials that are produced in large quantities generally show fewer property variations than less ductile and advanced materials that are produced by small batch processes. In composite materials, small variations in fiber orientation or volume fraction can have considerable effect on properties. Manufacturing processes can also add to the variations in component behavior. For example, parts manufactured by processes such as casting, forging, and cold forming are known to have variations in properties from point to point. Dimensional and geometrical variations resulting from manufacturing process tolerances can also affect the load-carrying capacity of components. Improved quality control techniques should result in more uniform material properties and more consistent component behavior in service and therefore lower values for the factor of safety. To account for these uncertainties, the factor of safety is used to divide into the nominal strength (S) of the material to obtain the working or allowable stress (S_a) as follows:

$$n_s = \frac{S}{S_a} \tag{5.4}$$

where n_s is the material factor of safety.

In simple components, S_a in Equation 5.4 can be viewed as the minimum allowable strength of the material. However, there is some danger involved in this use especially in the cases where the load-carrying capacity of a component is not directly related to the strength of the material used in making it. Examples include long compression members, which could fail as a result of buckling; and components of complex shapes, which could fail as a result of stress concentration. Under such conditions, it is better to consider S_a as the load-carrying capacity, which is a function of both material properties and geometry of the component.

In assessing the uncertainties in loading, two types of service conditions have to be considered:

1. Normal working conditions, which the component has to endure during its intended service life
2. Limit working conditions, such as overloading, which the component is only intended to endure on exceptional occasions, and which if repeated frequently could cause premature failure of the component.

In a mechanically loaded component, the stress levels corresponding to both normal and limit working conditions can be determined from a duty cycle. The normal duty cycle for an airframe, for example, includes towing and ground handling, engine run, take-off, climb, normal gust loadings at different altitudes, kinetic and solar heating, descent, and normal landing. Limit conditions can be encountered in abnormally high gust loadings or emergency landings. Analyses of the different loading conditions in the duty cycle lead to determination of the maximum load that will act on the component. This maximum load can be used to determine the maximum stress, or damaging stress, which if exceeded would render the component unfit for service before the end of its normal expected life. The load factor of safety (n_l) in this case can be taken as

$$n_l = \frac{L}{L_a} \tag{5.5}$$

where L is the maximum load and L_a the normal load.

The total or overall factor of safety (n), which combines the uncertainties in material properties and manufacturing processes as well as the uncertainties in external loading conditions, can be calculated as

$$n = n_s n_l \tag{5.6}$$

Factors of safety ranging from 1.1 to 20 are known, but common values range from 1.5 to 10.

In some applications, a designer is required to follow established codes when designing certain components, such as pressure vessels and piping systems. Under these conditions, the factors of safety used by the writers of the codes may not be specifically stated but an allowable working stress is given instead.

Derating factors are numbers less than unity and are used to reduce material strength values to take into account manufacturing imperfections and the expected severity of service conditions. When a component is subjected to fatigue loading, for example, several derating factors can be used to account for imperfections in surface finish, size of the component, stress concentration, etc., as discussed in Section 6.5.

5.7 RELIABILITY OF COMPONENTS

As discussed earlier, the actual behavior of the material in a component could vary from one point to another and from one component to another. In addition, it is usually difficult to precisely predict the external loads acting on the component under actual service conditions. To account for these variations and uncertainties, both the load carrying capacity S and the externally applied load L can be expressed in statistical terms. As both S and L depend upon many independent factors, it would be reasonable to assume that they can be described by normal distribution curves. Consider that the load-carrying capacity of the population of components has an average of $\bar{S}$ and a standard deviation σ_S, whereas the externally applied load has an average of $\bar{L}$ and a standard deviation σ_L. The relationship between the two distribution curves is important in determining the factor of safety and reliability of a given design. Figure 5.4 shows that failure takes place in all the components that fall in the area of overlap of the two curves, that is, when the load-carrying capacity is less than the external load. This is described by the negative part of the $(\bar{S} - \bar{L})$ curve of Figure 5.4. Transforming the distribution $(\bar{S} - \bar{L})$ to the standard normal deviate z, the following equation is obtained:

$$z = \frac{[(S - L) - (\bar{S} - \bar{L})]}{[(\sigma_S)^2 + (\sigma_L)^2]^{1/2}} \tag{5.7}$$

From Figure 5.4, the value of z at which failure occurs is

$$z = \frac{[0 - (\bar{S} - \bar{L})]}{[(\sigma_S)^2 + (\sigma_L)^2]^{1/2}} = \frac{-(\bar{S} - \bar{L})}{[(\sigma_S)^2 + (\sigma_L)^2]^{1/2}} \tag{5.8}$$

For a given reliability, or allowable probability of failure, the value of z can be determined from the cumulative distribution function for the standard normal distribution. Table 5.2 gives some selected values of z that will result in different values of probabilities of failure.

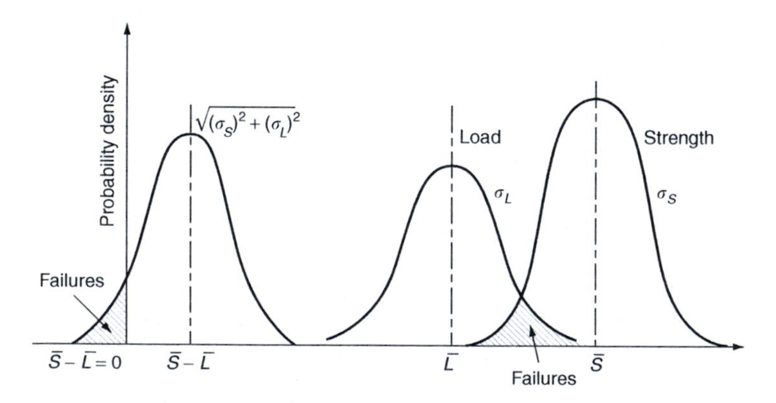

FIGURE 5.4 Effect of variation in load and strength on the failure of components.

TABLE 5.2
Values of z and Corresponding Levels of Reliability and Probability of Failure

z	Reliability	Probability of Failure
−1.00	0.8413	.1587
−1.28	0.9000	.1000
−2.33	0.9900	.0100
−3.09	0.9990	.0010
−3.72	0.9999	.0001
−4.26	0.99999	.00001
−4.75	0.999999	.000001

Knowing σ_S, σ_L and the expected $\overline{L}$, the value of $\overline{S}$ can be determined for a given reliability level. As defined earlier, the factor of safety in the present case is simply $\overline{L}/\overline{S}$. Example 5.1 illustrates the use of the above concepts in design.

Design Example 5.1: Estimating the Probability of Failure of a Structural Member

Problem

A structural element is made of a material with an average tensile strength of 2100 MPa. The element is subjected to a static tensile stress of an average value of 1600 MPa. If the variations in material quality and load cause the strength and stress to vary according to normal distributions with standard deviations of $\sigma_S = 400$ and $\sigma_L = 300$, respectively, what is the probability of failure of the structural element?

Solution

From Figure 5.4,

$$\overline{S} - \overline{L} = 2100 - 1600 = 500 \text{ MPa}$$

Standard deviation of the curve

$$(S - L) = [(\sigma_S)^2 + (\sigma_L)^2]^{1/2} = [(400)^2 + (300)^2]^{1/2} = 500$$

From Equation 5.8,

$$z = \frac{-500}{500} = -1$$

From Table 5.2, the probability of failure of the structural element is 0.1587, that is, 15.87%, which is too high for many practical applications.

One solution to reduce the probability of failure is to impose better quality measures on the production of the material and, thus, reduce the standard deviation of the strength. Another solution is to increase the cross-sectional area of the element to reduce the stress. For example, if the standard deviation of the strength is reduced to $\sigma_S = 200$, the standard deviation of the curve $(\bar{L} - \bar{S})$ will be $[(200)^2 + (300)^2]^{1/2} = 360$,

$$z = \frac{-500}{360} = -1.4$$

which according to Table 5.2 gives a more acceptable probability of failure value of 0.08, that is, 8%.

Alternatively, if the average stress is reduced to 1400 MPa,

$$(\bar{S} - \bar{L}) = 700 \text{ MPa}$$

$$z = \frac{-700}{500} = -1.4$$

with a similar probability of failure as the first solution.

As the discussion shows, statistical analysis allows the generation of data on the probability of failure and reliability, which is not possible when a deterministic safety factor is used. One of the difficulties with this statistical approach, however, is that material properties are not usually available as statistical quantities. In such cases, the following approximate method can be used.

In the case where the experimental data are obtained from a reasonably large number of samples, more than 100, it is possible to estimate statistical data from non-statistical sources that only give ranges or tolerance limits. In this case, the standard deviation σ_S is approximately given by

$$\sigma_S = \frac{(\text{maximum value of property} - \text{minimum value})}{6} \tag{5.9}$$

This procedure is based on the assumption that the given limits are bounded between plus and minus three standard deviations. Example 5.2 illustrates this point.

Design Example 5.2: Estimating the Coefficient of Variation in Material Strength

Problem

If the range of strength of an alloy is given as 800–1200 MPa, what is the mean strength, the standard deviation, and coefficient of variation?

Solution

The mean strength can be taken as 1000 MPa.

The standard deviation σ can be estimated as

$$\sigma = \frac{(1200 - 800)}{6} = 66.67 \text{ MPa}$$

The coefficient of variation ν' is then:

$$\nu' = \frac{66.67}{1000} = 0.0667$$

If the results are obtained from a sample of about 25 tests, it may be better to divide by 4 in Equation 5.9 instead of 6. With a sample of about 5 it is better to divide by 2.

In the cases where only the average value of strength is given, the following values of coefficient of variation, which is defined as $\nu' = \sigma_S/S$, can be taken as typical for metallic materials:

$\nu' = 0.05$ for ultimate tensile strength

$\nu' = 0.07$ for yield strength

$\nu' = 0.08$ for endurance limit for steel

$\nu' = 0.07$ for fracture toughness

5.8 SUMMARY

1. Engineering design is an interdisciplinary process that transforms consumer needs into instructions that allow successful manufacture of the product.
2. A good design should result in an attractive and user-friendly product that performs its function efficiently and economically within the prevailing legal, social, safety, and reliability requirements.
3. Major phases of design can be grouped into three categories: preliminary and conceptual design, configuration (embodiment) design, and detail (parametric) design. Materials and manufacturing processes are better defined as the design progresses.
4. A design code is a set of specifications for the analysis, design, manufacture, and construction of a structure or a product. A standard specification is a published document that describes the characteristics of a part, material, or process and should contain both technical and commercial requirements.
5. The factor of safety is used in design to ensure satisfactory performance. This factor is normally in the range of 1.5–10 and is used to divide into the strength of the material to obtain the allowable stress or the load to obtain the allowable load.
6. The lack of homogeneity of a material property, or variations in the externally applied load can be statistically described by a mean value, a standard deviation, and a coefficient of variation. These parameters can be used to estimate a factor of safety and to calculate the probability of failure of a component and its reliability in service. When the available material properties are not available in a statistical form, approximate methods may be used.

5.9 REVIEW QUESTIONS

5.1 An aluminum 2014 T6 tube of 75 mm (3 in.) outer diameter and 1 mm (0.04 in.) thickness is subjected to internal pressure of 8.4 MPa (1200 lb/in.^{2}). What is the factor of safety that was taken against failure by yielding when the pipe was designed? (Answer: 1.35)

5.2 Distinguish between the factor of safety and the derating factor. What are the main factors that affect the value of the factor of safety?

5.3 A structural member is made of steel of mean yield strength of 200 MPa (28.6 ksi) and a standard deviation on strength of 30 MPa (4.3 ksi). The applied stress has a mean value of 150 MPa (14.3 ksi) and a standard deviation of 5 MPa (700 lb/in.^{2}). (a) What is the probability of failure? (Answer: 6.9%) (b) What factor of safety is required if the allowable failure rate is 1%? (Answer: 1.55)

5.4 What are the steps required for a manufacturer of domestic water heaters to convert gas into solar energy heating.

5.5 A structural element is made of FRP with an average tensile strength of 2400 MPa. The element is subjected to a static tensile stress of an average value of 1800 MPa. If the variations in material quality and load cause the strength and stress to vary according to normal distributions with standard deviations of $\sigma_S = 380$ and $\sigma_L = 330$, respectively, what is the probability of failure of the structural element?

5.6 Select an everyday product such as a bicycle, child toy, and can opener. List the different areas in the product in order of importance to the function and safety. Assign factors of safety that you would use in designing each area.

5.7 For the product selected in question 5.6, which features would you emphasize when you plan the marketing campaign.

5.8 A manufacturer of sports equipment is considering the possibility of using FRP in making racing bicycle frames. It was suggested that fatigue failures of the joints could be a problem in this case. Describe a testing program that can address this problem.

5.9 Two batches of steel components were heat treated in two different shops. The RC hardness results were as follows:
Shop 1: 48, 51, 52, 49, 50, 50, 47, 50, 51, and 47
Shop 2: 50, 49, 47, 48, 50, 48, 49, 52, 51, and 48
Did treating the steel in the different shops make a significant difference? (Answer: no)

5.10 Study a car jack. Suggest changes in the design to make it easier to be used by a handicapped driver.

BIBLIOGRAPHY AND FURTHER READING

Ashby, M.F. and Johnson, K., *Materials and Design: The Art and Science of Material Selection in Product Design*, Butterworth-Heinemann, Amsterdam, 2002.

Block, M.R., *Implementing ISO 14001*, American Society for Quality, Milwaukee, WI, 1996.

Dieter, G.E., *Engineering Design, A Materials and Processing Approach*, McGraw-Hill, New York, 1983.

Dieter, G.E., Editor, *Materials Selection and Design, ASM Handbook*, Vol. 20, ASM International, Materials Park, OH, 1997.

Dixon, J.R., Overview of the design process, in *ASM Handbook*, Vol. 20, Dieter, G.E., Editor. ASM International, Materials Park, OH, 1997, pp. 7–14.

Farag, M.M., *Materials Selection for Engineering Design*, Prentice-Hall, London, 1997.

Fleischmann, S.T., Environmental aspects of design, in *Materials Selection and Design, ASM Handbook*, Vol. 20, Dieter, G.E., Editor. ASM International, Materials Park, OH, 1997, pp. 131–138.

Hubka, V., *Principles of Engineering Design*, Butterworth Scientific, London, 1982.

Hunter, T.A., Designing to codes and standards, in *ASM Handbook*, Vol. 20, Dieter, G.E., Editor. ASM International, Materials Park, OH, 1997, pp. 66–71.

Peterson, R.E., *Stress-concentration design factors*, John Wiley, New York, 1974.

Ray, M.S., *Elements of Engineering Design*, Prentice-Hall, Englewood Cliffs, NJ, 1985.

Shigley, J.E. and Mitchell, L.D., *Mechanical Engineering Design*, 4th Ed., McGraw-Hill, New York, 1983.

6 Effect of Material Properties on Design

6.1 INTRODUCTION

Figure 5.1 illustrates the direct relation between material properties and component design. This relation, however, is complex because the behavior of the material in the finished product can be quite different from that of the stock material used in making it. This point is illustrated in Figure 6.1, which shows the direct influence of stock material properties, production method, component geometry, and external forces on the behavior of materials in the finished component. The figure also shows that secondary relationships exist between geometry and production method, between stock material and production method, and between stock material and component geometry. The effect of stock material properties, component geometry, and applied forces on the behavior of materials is discussed in this chapter and the effect of production method on the design and component geometry is discussed in Chapter 7.

The goal of this chapter is to illustrate how the material-related factors affect the design. The main objectives are to get a better understanding of the material properties that need to be considered when designing for

1. static strength, stiffness, and toughness;
2. fatigue resistance;
3. high-temperature conditions;
4. hostile environments; and
5. wear resistance.

6.2 DESIGNING FOR STATIC STRENGTH

Designs based on the static strength usually aim at avoiding yielding of the component in the case of soft, ductile materials and at avoiding fracture in the case of strong, low-toughness materials. Designs based on soft, ductile materials is discussed in this section, whereas those based on strong, low-toughness materials is discussed in Section 6.4.

6.2.1 Designing for Simple Axial Loading

Components and structures made from ductile materials are usually designed so that no yield will take place under the expected static loading conditions. When the component is subjected to uniaxial stress, yielding will take place when the local stress

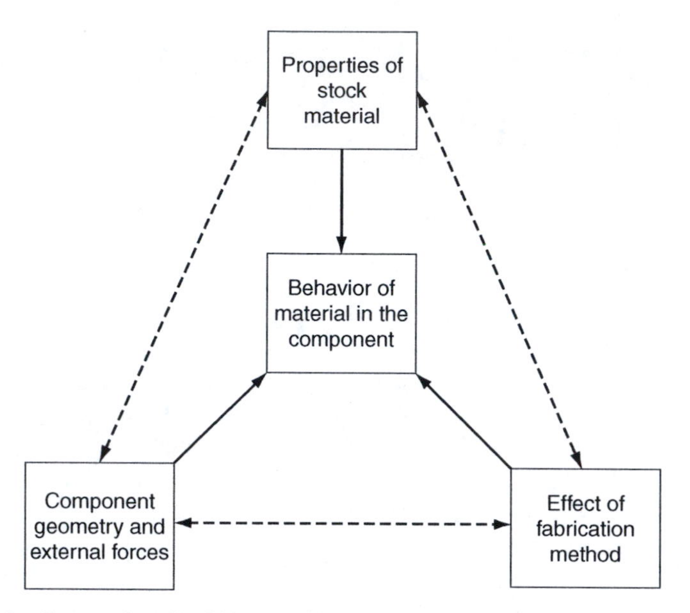

FIGURE 6.1 Factors that should be considered in anticipating the behavior of a material in the component.

reaches the yield strength of the material. The critical cross-sectional area, A, of such a component can be estimated as

$$A = \frac{K_t nL}{\text{YS}} \tag{6.1}$$

where

K_t = Stress concentration factor
L = Applied load
n = Factor of safety
YS = Yield strength of the material

6.2.2 Designing for Torsional Loading

The critical cross-sectional area of a circular shaft subjected to torsional loading can be determined from the relationship

$$\frac{2I_p}{d} = \frac{K_t nT}{\tau_{max}} \tag{6.2}$$

where

d = Shaft diameter at the critical cross section
τ_{max} = Shear strength of the material
T = Transmitted torque
I_p = Polar moment of inertia of the cross section
= $\pi d^4/32$ for a solid circular shaft
= $\pi(d_o^4 - d_i^4)/32$ for a hollow circular shaft of inner diameter d_i and outer diameter d_o

Although Equation 6.2 gives a single value for the diameter of a solid shaft, a large combination of inner and outer diameters can satisfy the relationship in the case of a hollow shaft. Under such conditions, either one of the diameters or the required thickness has to be specified to calculate the other dimensions.

The ASTM code of recommended practice for transmission shafting gives an allowable value of shear stress of 0.3 of the yield or 0.18 of the ultimate tensile strength, whichever is smaller. With shafts containing keyways, ASTM recommends a reduction of 25% of the allowable shear strength to compensate for stress concentration and reduction in cross-sectional area.

6.2.3 Designing for Bending

When a relatively long beam is subjected to bending, the bending moment, the maximum allowable stress, and dimensions of the cross section are related by the equation

$$Z = \frac{(nM)}{\text{YS}} \tag{6.3}$$

where

M = Bending moment
Z = Section modulus = I/c
I = Moment of inertia of the cross section with respect to the neutral axis normal to the direction of the load. (Figure 4.3 gives the formulas for calculating the value of I for some commonly used cross sections)
c = Distance from center of gravity of the cross section to the outermost fiber

Example 6.1 illustrates the use of Equation 6.3 in design.

Design Example 6.1: Designing a Cantilever Beam

Problem

Determine the dimensions of a cantilever beam of length 1 m and rectangular cross section of depth-to-width ratio 2:1. The cantilever is expected not to deflect more than 50 mm for every 1000 N increment of load at its tip. The material used in making the beam is steel AISI 4340 with a yield strength of 1420 MPa and UTS 1800 MPa. What is the maximum permissible load? Assume a suitable factor of safety.

Answer

Deflection (y) of a cantilever beam under a load (L) acting on its tip is given by the relationship

$$y = \frac{(Ll^3)}{(3EI)}$$

where

l = Length of the cantilever
E = Elastic modulus of the cantilever material = 210 GPa
I = Second moment of area of the cross section

From Figure 4.3,

$$I = \frac{b(2b)^3}{12} = \frac{Ll^3}{3yE} = \frac{1000 \times (1000)^3}{3 \times 50 \times 210 \times 10^3}$$

The preceding equation gives the width of the beam, b = 14.77 mm.

Taking a factor of safety n = 1.5 and using Equation 6.3,

$$Z = \frac{14.77(2 \times 14.77)^3}{12 \times 14.77} = 2148 \text{ mm}^3 = \frac{nM}{\text{YS}} = \frac{1.5 \times L \times 1000}{1420}$$

The safe value of L = 2033 N.

6.3 DESIGNING FOR STIFFNESS

In addition to being strong enough to resist the expected service loads without yielding, there may also be the added requirement of stiffness to ensure that deflections do not exceed certain limits. Stiffness is important in applications such as machine elements to avoid misalignment and to maintain dimensional accuracy of machine parts. In such cases, the dimensions of a component are calculated once for resistance to yielding and another for resistance to deflection and the larger dimensions are selected. This point is illustrated in Example 6.2.

Design Example 6.2: Selection of a Material for a Tie Rod

Problem

It is required to select a structural material for the manufacture of the tie-rods of a suspension bridge. A representative rod is 10 m long and should carry a tensile load of 50 kN without yielding. The maximum extension should not exceed 18 mm. Which one of the steels listed in Table 6.1 will give the lightest tie-rod?

Solution

For the present case, calculations of the area will be carried out twice:

1. Area based on yield strength = load/YS
2. Area based on deflection = (load × length)/(E × deflection)

The larger of the two areas will be taken as the design area and will be used to calculate the mass.

TABLE 6.1
Candidate Materials for Suspension Bridge Tie-Rods

Material	YS (MPa)	E (GPa)	Specific Gravity	Area Based on Yield Strength (mm^2)	Area Based on Deflection (mm^2)	Mass (kg)
ASTM A675 Grade 60	205	212	7.8	244	131	19
ASTM A572 Grade 50	345	211	7.8	145	131	11.3
ASTM A717 Grade 70	485	211	7.8	103	131	10.2
Maraging steel Grade 200	1400	211	7.8	36	131	10.2
Al 5052-H38	259	70.8	2.7	193	392	10.6
Cartridge brass 70% hard temper	441	100.6	8.0	113	276	22.1

The results of the calculations, given in Table 6.1, show that steel A717 grade 70 and maraging steel grade 200 give the least mass. As the former steel is more ductile and less expensive it will be selected.

6.3.1 Design of Beams

When an initially straight beam is loaded, it becomes curved as a result of its deflection. As the deflection at a given point increases, the radius of curvature at this point decreases. The radius of curvature, r, at any point on the curve is given by the relationship

$$r = \frac{EI}{M} \tag{6.4}$$

Equation 6.4 shows that the stiffness of a beam under bending is proportional to the elastic constant of the material, E and the moment of inertia of the cross section, I. Selecting materials with higher elastic constant and efficient disposition of material in the cross section are essential in designing of beams for stiffness. Placing the material as far as possible from the neutral axis of bending is generally an effective means of increasing I for a given area of cross section.

When designing with plastics, whose elastic modulus is 10–100 times less than that of metals, stiffness must be given special consideration. This drawback can usually be overcome by making some design adjustments. Figure 6.2 shows examples of how the low stiffness of plastics is overcome by increasing the second moment of area of the critical cross section. Example 6.3 illustrates this point.

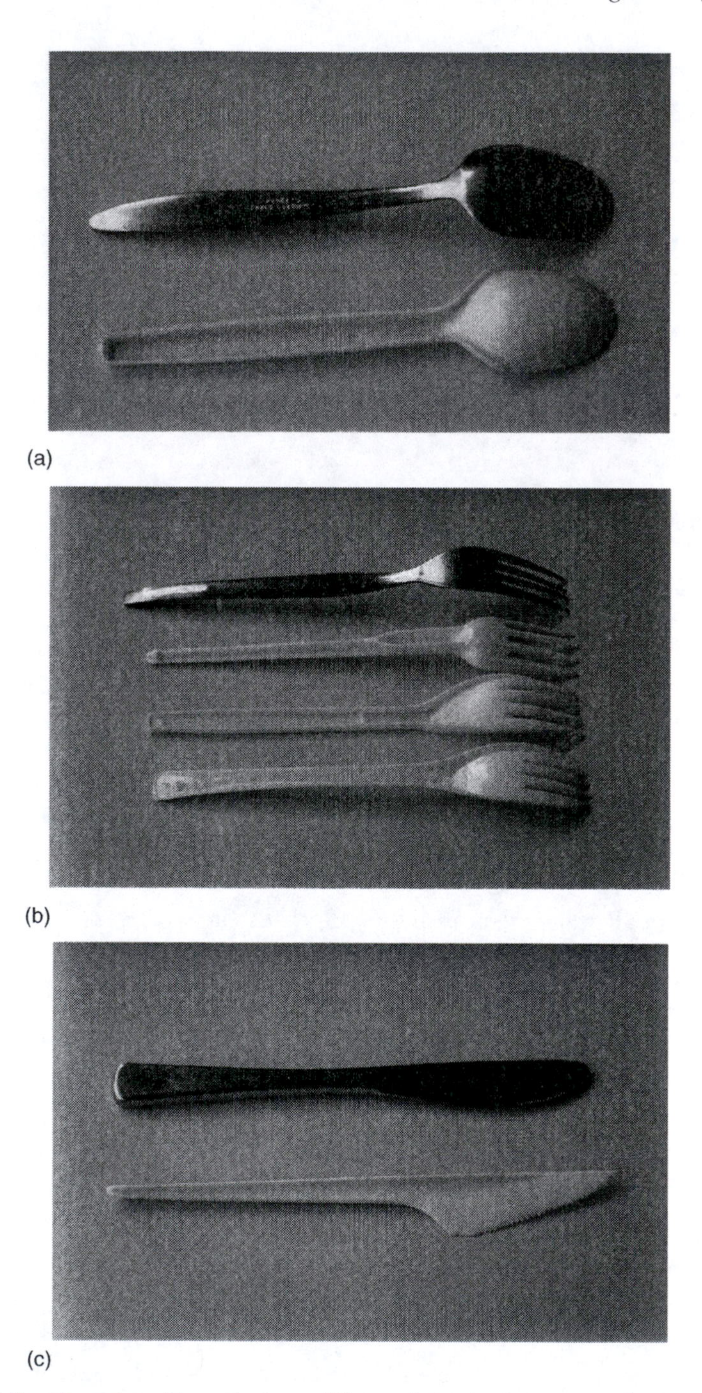

(a)

(b)

(c)

FIGURE 6.2 Examples of how the low stiffness of plastics is overcome by increasing the second moment of area of the critical cross section.

Design Example 6.3: Substitution of HDPE for Stainless Steel

Problem
What design changes are required when substituting HDPE for stainless steel in making a fork for a picnic set while maintaining similar stiffness.

Solution
E for stainless steel = 210 GPa
E for HDPE = 1.1 GPa

The narrowest cross section of the original stainless steel fork is rectangular of 0.6 mm × 5 mm.

I for the stainless steel section $= \dfrac{5 \times (0.6)^3}{12} = 0.09\ \text{mm}^4$

From Equation 6.4, EI should be kept constant for equal deflection under load.

EI for stainless steel = 210 × 0.09 = 18.9
EI for HDPE design = 1.1 × I
I for HDPE design = 17.2 mm^4.

Taking a channel section of thickness 0.5 mm, web height 4 mm, and width 8 mm, from Figure 4.3,

$$I = \frac{[8 \times (4)^3 - 7 \times (3.5)^3]}{12} = 17.7\ \text{mm}^4$$

which meets the required value.

Area of the stainless steel section = 3 mm^2
Area of the HDPE section = 7.5 mm^2

The specific gravity of stainless steel is 7.8 and that of HDPE is 0.96.

$$\frac{\text{Relative weight of HDPE}}{\text{Stainless steel}} = \frac{(7.5 \times 0.96)}{(3 \times 7.8)} = 0.3$$

6.3.2 Design of Shafts

The torsional rigidity of a component is usually measured by the angle of twist, θ, per unit length.

For a circular shaft, θ is given in radians by

$$\theta = \frac{T}{GI_p} \tag{6.5}$$

where

T = Torque
I_p = Polar moment of area
G = Modulus of elasticity in shear

$$G = \frac{E}{2(1 + \nu)} \tag{6.6}$$

where

v = Poisson's ratio
E = Elastic modulus

The usual practice is to limit the angular deflection in shafts to about 1°, that is, $\pi/180$ radians, at a length of 20 times the diameter.

6.3.3 Designing of Columns

Elastic instability becomes an important design criterion in the case of columns, struts, and thin-wall cylinders subjected to compressive axial loading where failure can take place by buckling. Buckling takes place if the applied axial compressive load exceeds a certain critical value, P_{cr}. The Euler column formula is usually used to calculate the value of P_{cr}, which is a function of the elastic modulus of the material, geometry of the column, and restraint at the ends. For the fundamental case of a pin-ended column, that is, ends are free to rotate around frictionless pins, P_{cr} is given as

$$P_{cr} = \frac{\pi^2 EI}{L^2} \quad (6.7)$$

where

I = Least moment of inertia of the cross-sectional area of the column
L = Length of the column

Equation 6.7 can be modified to allow for end conditions other than the pinned ends. The value of P_{cr} for a column with both ends fixed, that is, built-in as part of the structure, is four times the value given by Equation 6.7. However, the critical load for a free-standing column, that is, one end is fixed and the other free as in a cantilever, is only one quarter of the value given by Equation 6.7.

The Euler column formula shows that the critical load for a given column is only a function of E and I and is independent of the compressive strength of the material. This means that the resistance to buckling of a column of a given material and a given cross-sectional area can be increased by distributing the material as far as possible from the principal axes of the cross section. Hence, tubular sections are preferable to solid sections in carrying loads. Reducing the wall thickness of such sections and increasing the transverse dimensions increase the stability of the column. However, there is a lower limit for the wall thickness below which the wall itself becomes unstable and causes local buckling.

Experience shows that the values of P_{cr} calculated according to Equation 6.7 are higher than the buckling loads observed in practice. The discrepancy is usually attributed to manufacturing imperfections, such as lack of straightness of the column and lack of alignment between the direction of the compressive load and the axis of the column. This discrepancy can be accounted for by using an appropriate imperfection parameter or a factor of safety. For normal structural work, a factor of safety of 2.5 is usually used. As the extent of the imperfections mentioned is expected to increase with increasing slenderness of the column, it is suggested that the factor of safety be increased accordingly. A factor of safety of 3.5 is recommended for columns with $[L(A/I)^{1/2}] > 100$, where A is the cross-sectional area.

Equation 6.7 shows that the value of P_{cr} increases rapidly as the length of the column, L, decreases. For a short enough column, P_{cr} becomes equal to the load required for yielding or crushing of the material in simple compression. Such a case represents the limit of applicability of the Euler formula as failure takes place by yielding or fracture rather than elastic instability. Such short columns are designed according to the procedure described for simple axial loading. This design procedure is illustrated in Example 4.1 in Chapter 4.

6.4 DESIGNING WITH HIGH-STRENGTH LOW-TOUGHNESS MATERIALS

High-strength materials are being increasingly used in designing critical components to save weight or to meet difficult service conditions. Unfortunately, these materials tend to be less tolerant of defects than the traditional lower-strength, tougher materials. Although a crack-like defect can safely exist in a component made of low-strength ductile material, it can cause catastrophic failure if the same part is made of a high-strength low-toughness material. This has led to more demand for accurate calculation of acceptable defect levels, and to an increased use of NDT in manufacture. These defects can be the result of

1. initial flaws in the material, for example, inclusions and cavities;
2. production deficiencies, for example, welding defects; and
3. service conditions, for example, fatigue cracks or stress corrosion cracks.

6.4.1 Fail-Safe Design

Fail safety requires a structure to be sufficiently damage-tolerant to allow defects to be detected before they develop to a dangerous size. This means that inspection has to be conducted before the structure is put into service to ensure that none of the existing defects exceed the critical size. In addition, the structure has to be inspected periodically during its service life to ensure that none of the subcritical defects grow to a dangerous size, as illustrated in Figure 6.3. The figure shows that it is not strictly necessary to select a material with a low crack propagation rate. In principle, the structure can be made fail-safe when cracks propagate fast if the inspection interval is short enough. However, short inspection periods are not always possible or cost-effective. A better alternative is to use a more sensitive inspection method to reduce the minimum detectable defect size.

6.4.2 Guidelines for Design

In designing with high-strength low-toughness materials, the interaction between fracture toughness of the material, the allowable crack size, and the design stress should be considered. An analogy can be drawn between these parameters and the yield strength, and the nominal stress, which are considered in designing with ductile unflawed part. In the latter case, as the load increases the nominal stress increases until it reaches the yield stress and plastic deformation occurs. In the case of

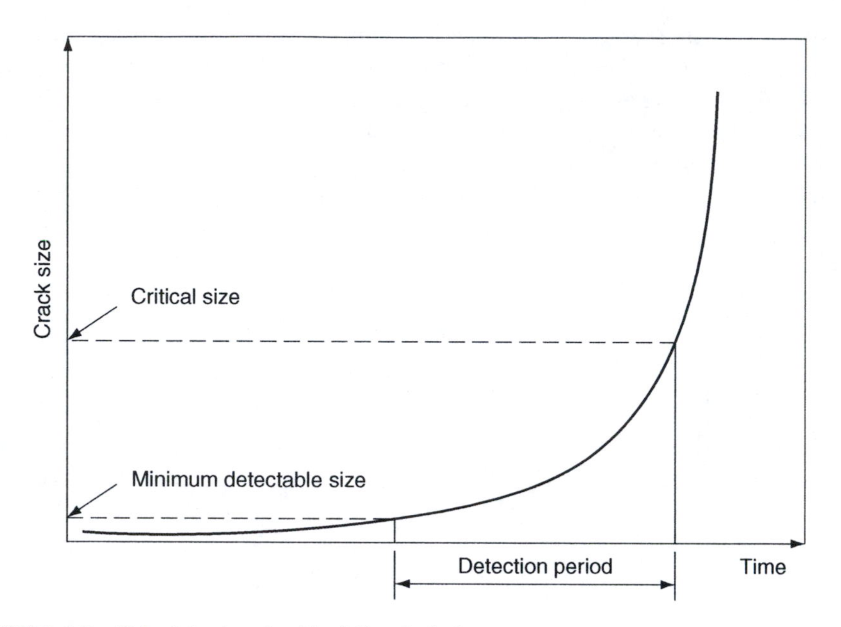

FIGURE 6.3 Principles involved in fail-safe designs.

high-strength low-toughness material, as the design stress increases (or as the size of the flaw increases), the stress concentration at the edge of the crack, stress intensity K_I, increases until it reaches K_{IC} and fracture occurs. Thus, the value of K_I in a structure should always be kept below K_{IC} value in the same manner that the nominal design stress is kept below the yield strength. It was shown in Section 2.3 that the condition for failure under plane strain conditions, where a crack of length $2a$ exists in a thick infinitely large plate, is given by

$$K_I = K_{IC} = Y\sigma_f (\pi a)^{1/2} \tag{6.8}$$

where σ_f is the fracture stress and is controlled by the applied load and shape of the part and a a quality control parameter, which is controlled by the manufacturing method and NDT technique used. Y is a dimensionless shape factor, which is a function of crack geometry. The value of Y can be estimated experimentally, analytically, or numerically. For simplicity, Y is taken as 1 for infinite plate with center through thickness crack or surface crack and as 1.12 for edge cracks. More accurate estimations of Y can be found in ASTM STP 380 and 410 standards.

Equation 6.8 may be used in several ways to design against failure. For example, selecting a material to resist other service requirements automatically fixes K_{IC}. In addition, if the minimum crack size that can be detected by the available NDT methods is known, Equation 6.8 is used to calculate the allowable design stress, which must be less than $K_{IC}/Y(\pi a)^{1/2}$. Alternatively, if the space and weight limitations necessitate a given material and operating stress, the maximum allowable crack size can be calculated to check whether it can be detected using routine inspection methods. Example 6.4 illustrates the use of K_{IC} in design.

Design Example 6.4: Designing with K_{IC}

Problem
If the minimum detectable crack in the beam described in Example 6.1 is 3 mm, what is the maximum permissible load? K_{IC} of the beam material is 87.4 MPa $(m)^{1/2}$.

Answer
From Equation 6.8,

$$K_{IC} = 87.4 = Y\sigma_f\,(\pi a)^{1/2}$$

Taking $Y = 1$ and $a = 1.5$ mm, then $\sigma_f = 1273.2$ MPa
From Equation 6.3,

$$L = \frac{(\sigma_f \times Z)}{(n \times l)} = \frac{(1273.2 \times 2148)}{(1.5 \times 1000)} = 1824 \text{ N}$$

Conclusion
This load is less than the "safe load" calculated in Example 6.1, which means that, had the presence of cracks been ignored, failure would have taken place.

6.4.3 Leak-before-Burst

Another design approach that utilizes the fracture mechanics approach is the leak-before-burst concept, which can be used in designing pressure vessels and similar structures. This approach is based on the concept that if a vessel containing pressurized gas or liquid contains a growing crack, the toughness should be sufficiently high to tolerate a defect size that will allow the contents to leak out before it grows catastrophically. For leakage to occur, the crack must grow through the vessel wall thickness, t. This means that the crack length $2a$ is about $2t$. From Equation 6.8, it can be seen that this condition is satisfied when $(K_{IC}/Y\sigma)^2$ is larger than πt. Example 6.5 illustrates the use of the leak-before-burst approach in design.

Design Example 6.5: Designing of a Pressure Vessel for Leak-before-Burst Conditions

Problem
Design a pressure vessel of the following specifications:

- Internal pressure, $p = 16$ MN/m^2 (2320 psi).
- Internal diameter, $D = 400$ mm (15.75 in.).

- The pressure vessel will be manufactured by welding of sheets, and the welded joints will then be inspected using an NDT technique, which is capable of detecting surface cracks of sizes greater than 3 mm (ca. 0.118 in.).

Solution

As a preliminary step, consider the use of AISI 4340 in the manufacture of the pressure vessel (see Table 4.6). When tempered at 260°C (500°F), this steel has a yield strength of 1640 MPa (238 ksi) and $K_{IC} = 50.0$ MPa $(m)^{1/2}$ (45.8 ksi $(in)^{1/2}$).

Treating the pressure vessel as a thin-wall cylinder, the wall thickness, t, can be calculated as

$$t = \frac{pD}{2S_w} \tag{6.9}$$

where S_w is the working stress.

Taking a factor of safety of 2, the working stress is 820 MPa (119 ksi) and the wall thickness is 3.9 mm (0.15 in.).

The critical surface crack size can then be calculated from Equation 6.8 as

$$K_{IC} = 50 = Y\sigma_f (\pi a)^{1/2} = Y \times 820(\pi a)^{1/2}$$

Taking Y as 1, for $\sigma_f/YS = 0.5$, the critical crack length is found to be 2.37 mm (0.093 in.). This length is too small to be detected by the available NDT technique and does not satisfy the condition for leak before burst, which makes the selected steel unsuitable.

Taking the same steel, AISI 4340, but tempering at 425°C (800°F) gives a yield strength of 1420 MPa (206 ksi) and K_{IC} of 87.4 MPa $(m)^{1/2}$ (80 ksi in.$^{1/2}$).

Following the same procedure as before, $t = 4.5$ mm (0.177 in.) and the critical crack length is 9.66 mm (0.38 in.), which can be detected by the available NDT technique and also satisfies the condition for leak before burst. This means that the steel in this tempered condition is acceptable for making the pressure vessel from the fracture mechanics point of view. Other factors such as availability of material, weldability, weight of vessel, and cost have to be taken into consideration before this steel is finally selected.

6.5 DESIGNING AGAINST FATIGUE

The fatigue behavior of materials is usually described by means of the *S–N* diagram that gives the number of cycles to failure, *N*, as a function of the maximum applied alternating stress, S_a, as shown in Figure 2.9.

6.5.1 Factors Affecting Fatigue Behavior

In the majority of cases, the reported fatigue strengths or endurance limits of materials are based on tests of carefully prepared small samples under laboratory conditions. Such values cannot be directly used for design purposes because the behavior of a component or structure under fatigue loading depends not only on the fatigue or endurance limit of the material used in making it, but also on several other factors including

- size and shape of the component or structure,
- type of loading and state of stress,
- stress concentration,
- surface finish,
- operating temperature,
- service environment, and
- method of fabrication.

The effect of some of these parameters on the fatigue strength of some steels is shown in Table 6.2. The influence of the abovementioned factors on fatigue behavior of a component can be accounted for by modifying the endurance limit of the material using a number of factors, as discussed in the following paragraphs.

6.5.1.1 Endurance-Limit Modifying Factors

A variety of modifying factors, or derating factors, are usually used to account for the main parameters that affect the behavior of components or structures in service. The numerical value of each of the modifying factors is less than unity and each one is intended to account for a single effect. This approach is expressed as follows:

$$S_e = k_a\, k_b\, k_c\, k_d\, k_e\, k_f\, k_g\, k_h\, S'_e \tag{6.10}$$

where

S_e = Endurance limit of the material in the component
S'_e = Endurance limit of the material as determined by laboratory fatigue test
k_a = Surface finish factor
k_b = Size factor
k_c = Reliability factor

TABLE 6.2
Effect of Surface Condition and Environment on the Fatigue Strength of Steels

	Fatigue Strength as Percentage of Maximum Endurance Limit						
UTS (MPa) (ksi)	Mirror Polish	Polished	Machined	0.1 mm Notch	Hot-Worked Surface	Under Fresh Water	Under Salt Water
280 (40)	100	95	93	87	82	72	52
560 (80)	100	92	88	77	63	53	36
840 (122)	100	90	84	66	47	37	25
1120 (162)	100	88	78	55	37	25	17
1400 (203)	100	88	72	44	30	19	14
1540 (223)	100	88	69	39	30	19	12

k_d = Operating temperature factor
k_e = Loading factor
k_f = Stress concentration factor
k_g = Service environment factor
k_h = Manufacturing processes factor

Equation 6.10 can be used to predict the behavior of a component or a structure under fatigue conditions, provided that the values of the different modifying factors are known. The effect of some of the abovementioned factors on fatigue behavior is known and can be estimated accurately, others are more difficult to quantify, as the following discussion shows.

6.5.1.1.1 Surface Finish Factor

Surface finish factor, k_a, is introduced to account for the fact that most machine elements and structures are not manufactured with the same high-quality finish that is normally given to laboratory fatigue test specimens. The value of k_a can vary between unity and 0.2 depending on the surface finish and the strength of the material. As shown in Tables 6.2 and 6.3, stronger materials are more sensitive to surface roughness variations.

6.5.1.1.2 Size Factor

Size factor, k_b, accounts for the fact that large engineering parts have lower fatigue strengths than smaller test specimens. In general, the larger the volume of the material under stress, the greater is the probability of finding metallurgical flaws that could cause fatigue crack initiation. Although there is no quantitative agreement on the precise effect of size, the following values can be taken as rough guidelines:

k_b = 1.0 for component diameters less than 10 mm (0.4 in.)
k_b = 0.9 for diameters in the range of 10–50 mm (0.4–2.0 in.)
$k_b = 1 - [(D - 0.03)]/15$

where D is the diameter expressed in inches, for sizes 50–225 mm (2–9 in.).

TABLE 6.3
Effect of Surface Finish and UTS on Surface Finish Factor (k_a) for Steels

UTS (MPa) (ksi)	Forged R_a=500−125	Hot Rolled R_a=250−63	Machined or Cold Drawn R_a=125−32	Ground R_a=63−4	Polished R_a<16
420 (60)	0.54	0.70	0.84	0.90	1.00
700 (100)	0.40	0.55	0.74	0.90	1.00
1000 (143)	0.32	0.45	0.68	0.90	1.00
1400 (200)	0.25	0.36	0.64	0.90	1.00
1700 (243)	0.20	0.30	0.60	0.90	1.00

6.5.1.1.3 Reliability Factor

Reliability factor, k_c, accounts for the random variations in fatigue strength. The published data on endurance limit usually represent average values representing 50% survival in fatigue tests. Since most designs require higher reliability, the published values of endurance limit must be reduced by the reliability factor, k_c. The following values can be taken as guidelines:

$$k_c = 0.900 \text{ for } 90\% \text{ reliability}$$
$$k_c = 0.814 \text{ for } 99\% \text{ reliability}$$
$$k_c = 0.752 \text{ for } 99.9\% \text{ reliability}$$

6.5.1.1.4 Operating Temperature Factor

Operating temperature factor, k_d, accounts for the difference between the test temperature, which is normally room temperature, and the operating temperature of the component or structure. For carbon and alloy steels, the fatigue strength is not greatly affected by operating temperature in the range from 45°C to 450°C (−50 to 840°F) and, therefore, k_d can be taken as 1.0 in this temperature range. At higher operating temperatures, k_d can be calculated according to the following relationships:

$$k_d = 1 - 5800(T - 450) \text{ for } T \text{ between } 450°\text{C and } 550°\text{C}$$

or

$$k_d = 1 - 3200(T - 840) \text{ for } T \text{ between } 840°\text{F and } 1020°\text{F}$$

6.5.1.1.5 Loading Factor

Loading factor, k_e, can be used to account for the differences in loading between laboratory tests and service. Transient overloads, vibrations, shocks, and changes in load spectrum that may be encountered during service can greatly affect the fatigue life of a component or structure. Experience shows that repeated overstressing, that is, stressing above the fatigue limit, can reduce the fatigue life. Under such conditions, k_e should be given a value less than unity. The type of loading also affects the fatigue life. Most published fatigue data are based on reversed bending test. Other types of loading, for example, axial or torsional, generate different stress distributions in the material, which could affect the fatigue results. The factor k_e can be used as a correction factor to allow the use of reversed bending data in a different loading mode. Thus,

$$k_e = 1 \text{ for applications involving bending}$$
$$k_e = 0.9 \text{ for axial loading}$$
$$k_e = 0.58 \text{ for torsional loading}$$

6.5.1.1.6 Stress Concentration Factor

Stress concentration factor, k_f, accounts for the stress concentrations that may arise due to changes in cross section or similar design features, as discussed in Section 5.5. Experience shows that low-strength, ductile steels are less sensitive to notches than high-strength steels.

TABLE 6.4
Effect of Environment and UTS on the Service Environment Factor (k_g) for Steels

UTS (MPa) (ksi)	k_g under Freshwater	k_g under Saltwater
280 (40)	0.72	0.52
560 (80)	0.52	0.36
840 (122)	0.37	0.25
1120 (162)	0.25	0.17
1400 (203)	0.19	0.14
1540 (223)	0.19	0.12

6.5.1.1.7 Service Environment Factor

Service environment factor, k_g, accounts for the reduced fatigue strength due to the action of hostile environment. The sensitivity of the fatigue strength of steels to corrosive environments is also affected by their strength, as shown in Table 6.4.

6.5.1.1.8 Manufacturing Process Factor

Manufacturing process factor, k_h, accounts for the influence of fabrication parameters such as heat treatment, cold working, residual stresses, and protective coatings on the fatigue strength of the material. Although the factor, k_h, is difficult to quantify, it is included here as a reminder that the abovementioned parameters should be taken into account.

Example 6.6 illustrates the use of the mentioned parameters in design.

Design Example 6.6: Design of an Axle for Fatigue Resistance

Problem

Calculate the diameter of the two rear axle shafts for a truck of gross mass of 6000 kg when fully loaded. Assume that two-thirds of the mass of the truck is supported by the rear axles, and the construction is such that the axles can be treated as cantilever beams each of 1 m length with the load acting on its end. The shaft material is heat-treated 4340 steel of tensile strength 952 MPa. The shaft construction requires a change in diameter to allow the fitting of bearings.

Solution

From Table 4.7, the endurance ratio of the steel is 0.56 and the endurance limit is 532 MPa. Assuming that the shaft will be finished by machining, the surface finish factor, k_a, can be taken as 0.68 (Table 6.3). As the shaft diameter is expected to be in the range of 50–225 mm, the size factor, k_b, can be taken as 0.8, assuming a 3 in. diameter. The reliability of the shaft should be high and the reliability factor is taken as k_c = 0.752. The loading factor can be taken as k_e = 1, as the shaft is loaded in bending. Stress concentration due to change in diameter can be reduced by taking a relatively large fillet radius at the change in diameter, and in this case $k_f = 0.7$. From Equation 6.10, the endurance limit of the shaft material S_e is

$$S_e = 532 \times 0.68 \times 0.8 \times 0.752 \times 1.0 \times 0.7 = 152.3 \text{ MPa}$$

The load acting on each shaft $= 6000 \times \frac{2}{3} \times \frac{1}{2} \times 9.8 = 19{,}600$ N

Bending moment $M = 19{,}600 \times 1 = 19{,}600$ N m

Stress acting on the surface of the shaft $= S = M/Z = 32\,M/\pi D^3$

$$D^3 = \frac{(32 \times 19{,}600 \times 1000)}{(\pi \times 152.3)} = 1{,}310{,}860 \text{ mm}^3$$

$$D = 109.5 \text{ mm}$$

6.5.2 Effect of Mean Stress

In the majority of cases, the fatigue behavior is determined using the rotating bending test, which applies alternating tension–compression and a stress ratio $R = -1$, as shown in Figure 2.8. In practice, however, conditions are often met where a static mean stress, S_m, is also present, as shown in Figure 2.8. Several methods are available for describing the fatigue behavior of materials under such conditions, as shown in Figure 6.4. The point S_e represents the fatigue or endurance limit under a stress ratio $R = -1$, whereas the points UTS and YS represent the ultimate tensile strength and yield strength under static loading, that is, $S_a = 0$. In general, experimental results for ductile materials fall between the Goodman line and the Gerber parabola, but because of scatter in the results, the Goodman line is usually preferred for design. The Goodman line can be represented by the relationship

$$\left(\frac{S_m}{\text{UTS}}\right) + \left(\frac{S_a}{S_e}\right) \tag{6.11}$$

The Soderberg line is more conservative and uses the yield strength as the limiting mean stress instead of the ultimate tensile strength, as shown in Figure 6.4. The Soderberg line can be represented by Equation 6.11 by substituting YS for UTS.

A criterion that follows the line *abc* of Figure 6.4 is not as conservative as Soderberg criterion but avoids gross yielding at high mean stresses. Along the line *bc*, which is drawn from the yield stress at 45°, the sum of the mean and alternating stresses equals the yield stress. Operating below this line avoids gross yielding of smooth components under combined alternating and static stresses.

A factor of safety is usually applied when using Equation 6.11 for design. It is generally preferable to employ separate factors of safety for the static and alternating strengths of the material, n_m and n_a, respectively. When the component contains a stress concentration, the maximum allowable static stress is reduced by an amount proportional to the stress concentration factor, K_t, and the fatigue strength is reduced by an amount proportional to the fatigue stress concentration factor, K_f.

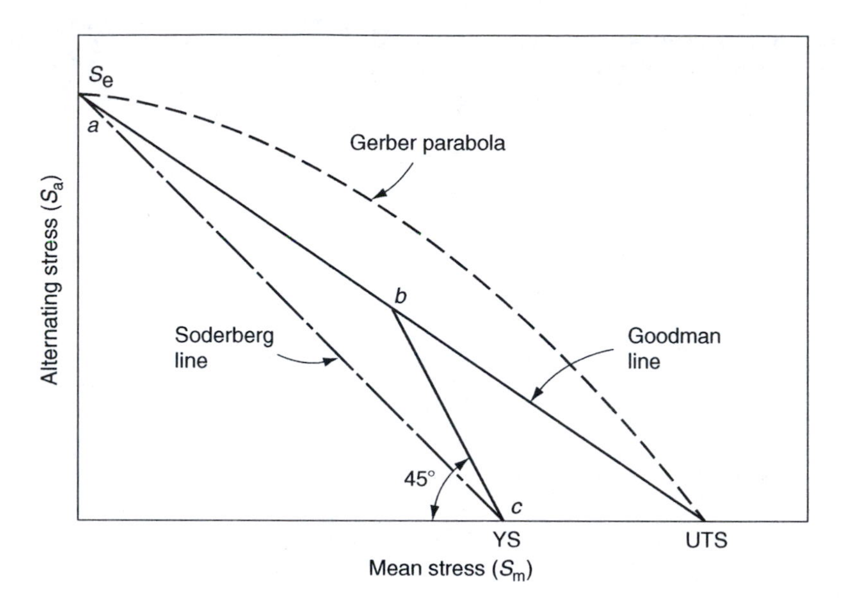

FIGURE 6.4 Fatigue behavior under combined static and alternating stresses.

The factors of safety and stress concentration factors can be incorporated in Equation 6.11, which will allow it to be used for design. Thus,

$$\left(\frac{n_m K_t S_m}{\text{UTS}}\right) + \left(\frac{n_a K_f S_a}{S_e}\right) = 1 \tag{6.12}$$

The use of the abovementioned parameters in design is illustrated in Example 6.7.

Design Example 6.7: Design of a Cantilever Beam for Fatigue Resistance

Problem

If the load acting on the cantilever beam described in Example 6.1 is fluctuating around a mean value of 1000 N, what is the maximum alternating load that can be applied without causing fatigue failure? The endurance ratio of the AISI 4340 steel is 0.56. Use suitable factors of safety.

Answer

From Example 6.1, $Z = 2148\ \text{mm}^3$,

$$\text{Mean stress} = S_m = \frac{M}{Z} = \frac{1000 \times 1000}{2148} = 465.3\ \text{MPa}$$

Endurance limit $= S_e = 1800 \times 0.56 = 1008$ MPa.

Taking a factor of safety of 1.5 for the static load and 3 for the dynamic load, from Equation 6.12,

$$\left(\frac{1.5 \times 465.3}{1800}\right) + \left(\frac{3 \times S_a}{1008}\right) = 1$$

$S_a = 205.7$ MPa

$$\text{Alternating load} = \frac{(205.7 \times 2148)}{1000} = 440 \text{ N}$$

6.5.3 Cumulative Fatigue Damage

Engineering components and structures are often subjected to different fatigue stresses in service. Estimation of the fatigue life under variable loading conditions is normally based on the concept of cumulative fatigue damage, which assumes that successive stress cycles cause a progressive deterioration in the component.

The Palmgren–Miner rule, also called Miner's rule, proposes that if cyclic stressing occurs at a series of stress levels, $S_1, S_2, S_3, \ldots, S_i$, each of which would correspond to a failure life of $N_1, N_2, N_3, \ldots, N_i$, if applied singly. Then, the fraction of total life used at each stress level is the actual number of cycles applied at this level, $n_1, n_2, n_3, \ldots, n_i$, divided by the corresponding life. The part is expected to fail when the cumulative damage satisfies the relationship

$$\frac{n_1}{N_1} + \frac{n_2}{N_2} + \frac{n_3}{N_3} + \cdots + \frac{n_i}{N_i} = C \tag{6.13}$$

The constant C can be determined experimentally and is usually found to be in the range of 0.7–2.2. When such experimental information is lacking, C can be taken as unity.

The Palmgren–Miner rule takes into account neither the sequence of loading nor the effect of mean stress, and it should only be taken as a rough guide to design.

6.5.4 Other Fatigue-Design Criteria

Components made of steel or other materials that have well-defined endurance limits can be designed for an indefinite fatigue life, provided that working stresses do not exceed this critical value. According to the discussion in Section 6.5.1, the endurance limit has to be reduced to account for adverse service environment as well as material, manufacturing, and design inaccuracies. In some cases, components and structures whose design is based on the resulting working fatigue strength would be too heavy or too bulky for the intended application. In such cases, other design criteria such as safe life, fail safe, or damage tolerance may be employed.

Safe-life, or finite-life, design is based on the assumption that the component or structure is free from flaws but the stress level in certain areas is higher than the endurance limit of the material. This means that fatigue-crack initiation is inevitable and the life of the component is estimated on the basis of the number of stress cycles that is necessary to initiate such a crack. Fail-safe design is based on the philosophy that cracks that form in service will be detected and repaired before they can lead to failure. Materials with high-fracture toughness, crack-stopping features, and a reliable NDT program should be employed when the fail-safe criterion is adopted. Damage-tolerant design is an extension of the fail-safe criterion; it assumes that flaws exist in engineering components and structures before they are put in service. Fracture mechanics techniques as discussed in Section 2.3 and this section are used to determine whether such cracks will grow large enough to cause failure before they are detected during a periodic inspection.

6.6 DESIGNING UNDER HIGH-TEMPERATURE CONDITIONS

From discussions in Sections 2.6 and 4.7, it becomes clear that the service temperature has a considerable influence on the strength of materials and, consequently, on the working stress used in design. Depending on the temperature range, the design can be based on the following:

1. Short-time properties of the material, as described by the yield strength and ultimate tensile strength, for moderate temperatures
2. Both short-time and creep properties for intermediate temperature range
3. Creep properties of the materials for high temperatures

For example, in the case of carbon steels, 300°C (575°F) and 400°C (750°F) can be taken as the upper and lower temperature limits of ranges 1 and 3, respectively. Adding alloying elements to the steel generally increases the limiting temperatures of the three ranges. For example, short-time properties of 18-8 stainless steel can be used at temperatures up to 425°C (800°F).

In addition to creep, the other factors that must be taken into consideration when designing for elevated temperatures include

1. metallurgical and microstructural changes that occur in the material due to long-time exposure to elevated temperature,
2. influence of method of fabrication, especially welding, on creep behavior, and
3. oxidation and hot corrosion that may take place during service and shutdown periods.

6.6.1 Design Guidelines

For design purposes, creep properties are usually presented on plots that yield reasonable straight lines. Common methods of presentation include log–log plots of stress versus steady-state creep rate and stress versus time to produce different amounts of total strain (instantaneous strain on load application plus creep strain), as shown schematically in Figure 6.5. A change in the microstructure of the material is usually accompanied by a change in creep properties, and consequently a change in the slope of the line.

Generally, designing under high-temperature conditions is carried out according to well-established codes. Most countries have one or more design codes that cover plants and structures operating at high temperatures. Examples of such codes include ASME Boiler and Pressure Vessel Code: section I (power boilers) and section VIII (pressure vessels), BS 806:1975 (piping for land boilers), and BS 5500:1976 (unfired fusion-welded pressure vessels). The common feature of such codes is that calculations and stress analysis are kept to a minimum. Design of local areas, such as branch connections and supports, is provided by simple formulas and by reference to charts.

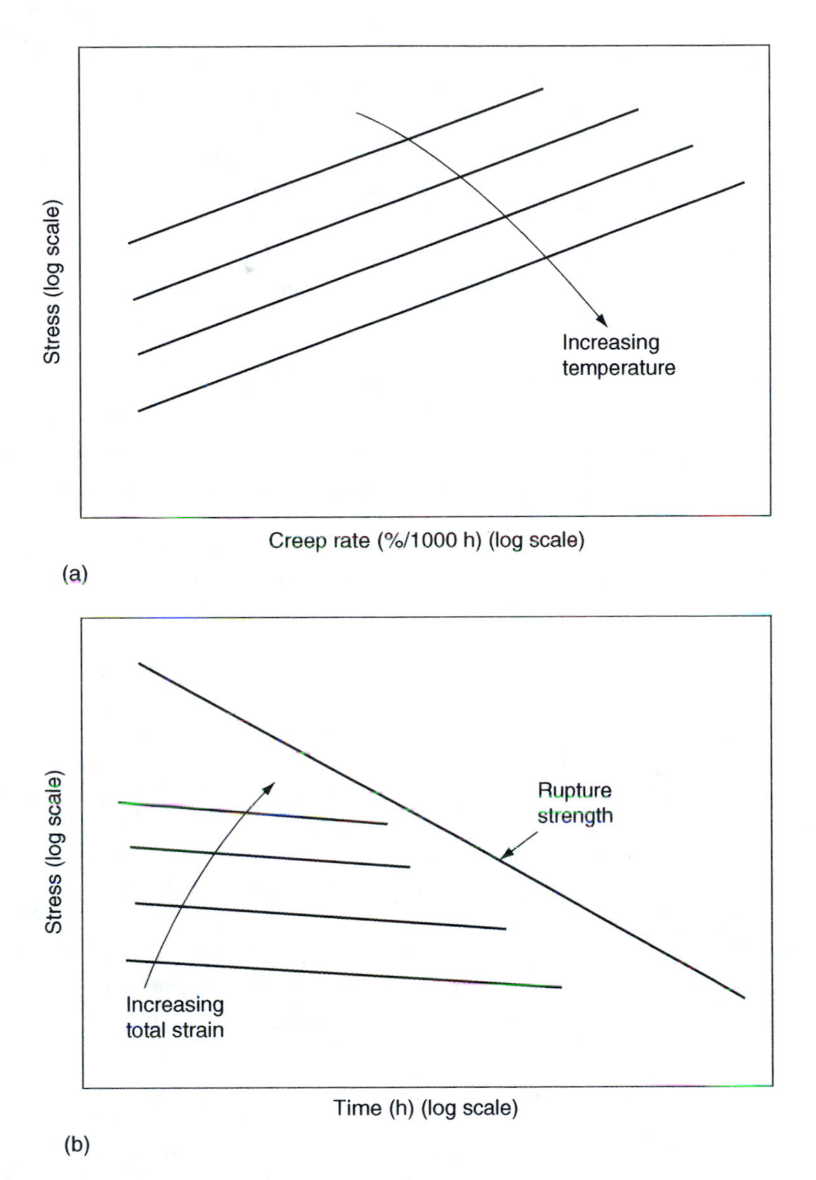

FIGURE 6.5 Presentation of creep data for design purposes. (a) Variation of stress with steady-state creep rate at various temperatures; (b) variation of stress with time to produce different amounts of total strain at a given temperature. The uppermost curve is the stress rupture, which occurs at different total strains to failure.

For moderate temperatures, range 1, both the ASTM Pressure Vessel Code and Boiler Code specify the allowable stresses as the lowest obtained from

1. 25% of the tensile strength at room temperature,
2. 25% of the tensile strength at service temperature, and
3. 62.5% of the yield strength (or 0.2% offset) at service temperature.

For high temperatures, range 3, the Boiler Code specifies that the stress values are based on 60% of the stress to produce a creep rate of 1/100%/1000 h. In addition, the stress values are also limited to 80% of the stress to produce rupture at the end of 100,000 h. Generally, service temperatures and pressures are lower than design values, wall thicknesses are often increased by a corrosion allowance, and material properties are usually higher than those specified. All these factors result in an increased factor of safety.

At intermediate temperatures, range 2, the code limits the stress to values obtained from a smooth curve joining the values for the low- and high-temperature ranges.

Example 6.8 illustrates the use of the concepts in design.

Design Example 6.8: Designing of a Cantilever Beam to Resist Creep

Problem

If the cantilever described in Example 6.1 is made of 5% Cr–0.5% Mo steel, and is to serve at 300°C, what is the maximum allowable load that can be endured at least 100,000 h?

Analysis

The allowable design stress for 5% Cr–0.5% Mo steel according to the ASME Boiler Code is shown in Figure 6.6, which is based on data given by Clark.

From the figure, the allowable stress at 300°C is about 90 MPa.

$$\text{Maximum allowable load} = \frac{(90 \times 2148)}{1000} = 193.3\ \text{N}$$

Conclusion

Comparing the preceding result with Example 6.1 shows that the maximum allowable load at 300°C is less than 10% of the safe load at room temperature.

6.6.2 Larson–Miller Parameter

In many cases, creep data are incomplete and have to be supplemented or extended by interpolation or, more hazardously, extrapolation. This is particularly true of long-time creep and stress-rupture data where the 100,000 h (11.4 years) creep resistance of newly developed materials is required. Reliable extrapolation of creep and stress-rupture curves to longer times can be made only when no structural changes occur in the region of extrapolation. Such changes can affect the creep resistance, which would result in considerable errors in the extrapolated values.

As an aid in extrapolation creep data, several time–temperature parameters have been developed for trading off temperature for time. The basic idea of these parameters is that they permit the prediction of long-time creep behavior from the results of shorter time tests at higher temperatures at the same stress. A widely used

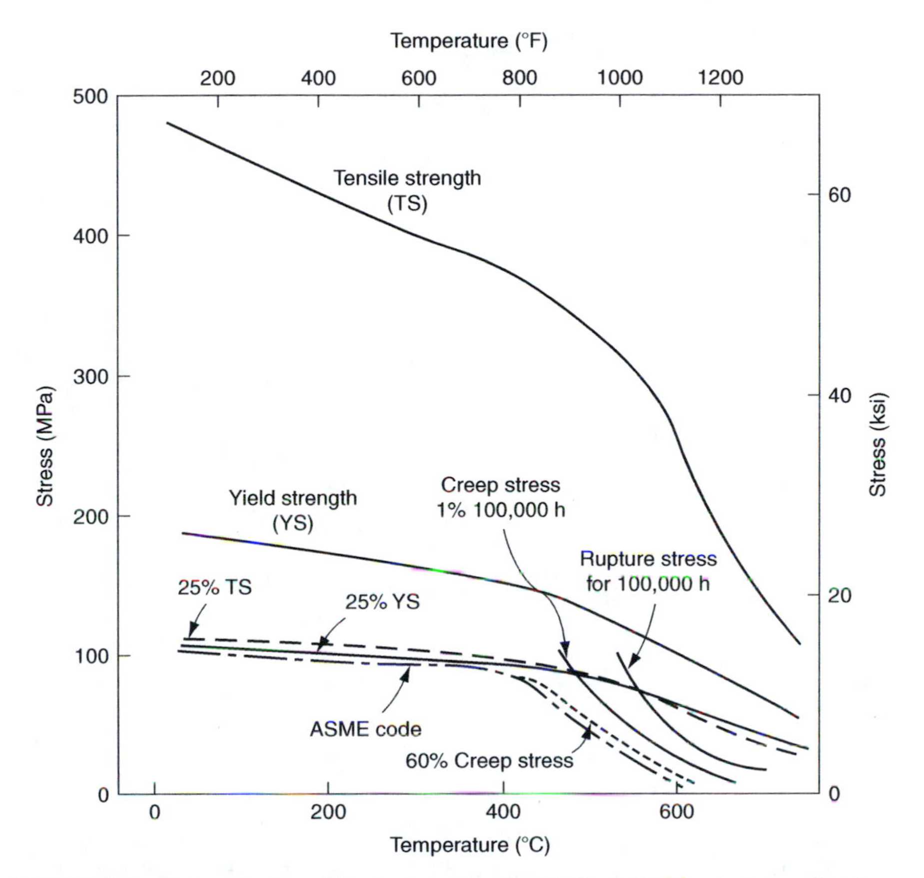

FIGURE 6.6 Determination of the design stress for 5% Cr–0.5% Mo steel according to the ASTM Boiler Code.

parameter for correlating the stress-rupture data is the Larson–Miller parameter (LMP), which is described in Figure 6.7. LMP is described by the relationship

$$\text{LMP} = T\,(C + \log t_{\text{r}}) \tag{6.14}$$

where

T = Test temperature in degrees Kelvin (°C + 273) or degrees Rankin (°F + 460)

t_{r} = Time to rupture in hours, the log is to the base 10

C = Larson–Miller constant, generally falls between 17 and 23, but is often taken to be 20

Example 6.9 illustrates the use of the LMP in design.

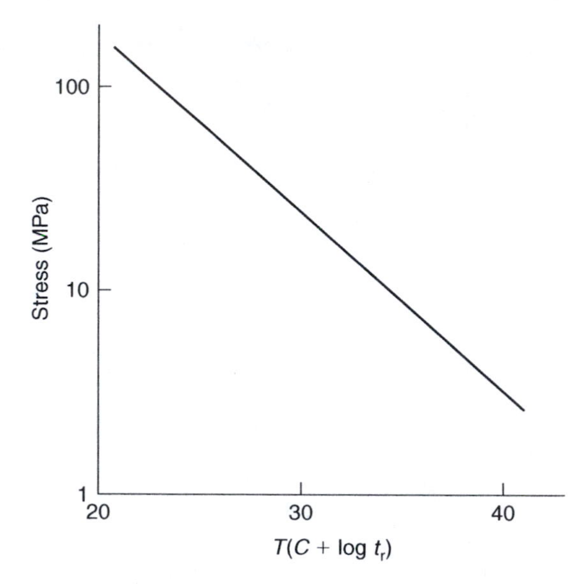

FIGURE 6.7 The Larson–Miller plot.

Design Example 6.9: Designing a Turbine Blade to Resist Creep Using LMP

Problem

Assuming that a turbine blade made of Nimonic 105 alloy had a life of 10,000 h at 150 MPa with a service temperature of 810°C, what is the expected life at the same stress but with a service temperature of 750°C?

Solution

From Equation 6.14,

$$\text{LMP} = (810 + 273)(20 + 4) = (750 + 273)(20 + \log t_r)$$

The expected life at the new service temperature is about 255,637 h.

The LMP can also be expressed in terms of time to give a specified strain, t_s, thus,

$$\text{LMP} = T(C + \log t_s)$$

6.6.3 Life under Variable Loading

The stress-rupture life of a component or a structure that is subjected to variable loading can be roughly estimated if the expected life at each stress level is known. Under such conditions, the life fraction rule assumes that rupture occurs when

$$\frac{t_1}{t_{r1}} + \frac{t_2}{t_{r2}} + \frac{t_3}{t_{r3}} + \cdots = 1 \tag{6.15}$$

where

$t_1, t_2, t_3, \ldots$ = Time spent by the part under stress levels 1, 2, 3, ..., respectively

$t_{r1}, t_{r2}, t_{r3}, \ldots$ = Rupture lives of the part under stress levels 1, 2, 3, ..., respectively

6.6.4 Life under Combined Fatigue and Creep Loading

Similar reasoning can also be applied to predict the life of a component or a structure when subjected to combined creep and fatigue loading. Cumulative fatigue damage laws, for example, Palmgren–Miner law, can be combined with the life fraction rule, given in Equation 6.15, to give a rough estimate of expected life under combined creep-fatigue loading. Thus,

$$\frac{t_1}{t_{r1}} + \frac{t_2}{t_{r2}} + \frac{t_3}{t_{r3}} + \cdots \frac{n_1}{N_1} + \frac{n_2}{N_2} + \frac{n_3}{N_3} + \cdots = 1 \quad (6.16)$$

where

$n_1, n_2, n_3, \ldots$ = Number of cycles at stress levels 1, 2, 3, . . ., respectively

$N_1, N_2, N_3, \ldots$ = Fatigue lives at stress levels 1, 2, 3, . . ., respectively

6.7 DESIGNING FOR HOSTILE ENVIRONMENTS

Service in a hostile environment is a major source of failure in many areas of engineering. Such failures can be prevented or at least reduced by selecting the appropriate material and by observing certain design rules, as will be discussed in this section.

6.7.1 Design Guidelines

Galvanic corrosion usually takes place as a result of design errors where dissimilar metals are placed in electrical contact. Under such conditions, corrosion occurs in the anodic material whereas the cathodic material is protected. The rate of corrosion and damage caused depends on the difference between the two materials in the galvanic series (Table 3.1) and on the relative areas of the exposed parts. A small anode and a large cathode will result in intensive corrosion of the anode, whereas a large anode and a small cathode is not as serious.

The safest way of avoiding galvanic corrosion is to ensure that dissimilar metals are not in electrical contact by using insulating washers, sleeves, or gaskets. When protective paints are used, both metals or only the cathodic metal should be painted. Painting only the anode will concentrate the attack at the breaks or defects in the coating.

Severe corrosion can take place in crevices formed by the geometry of the structure. Common sites for crevice corrosion include riveted and welded joints, areas of contact between metals and nonmetals, and areas of metal under deposits or dirt, as discussed in Section 3.3. Crevices can also be created between flanged pipes as a result of incorrect usage of gaskets, as shown in Figure 6.8a. Fibrous materials that can draw the corrosive medium into the crevice by capillary action should not be

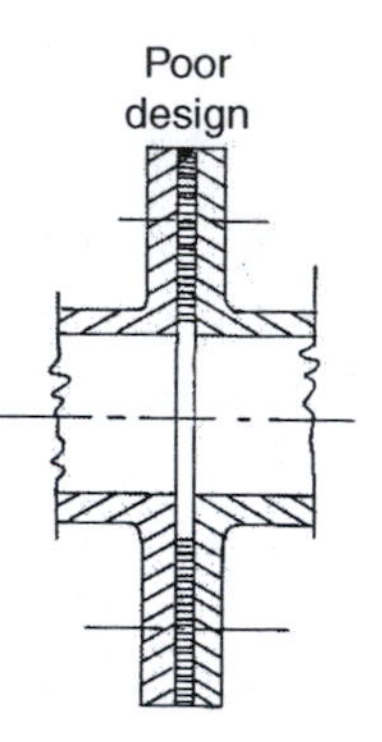

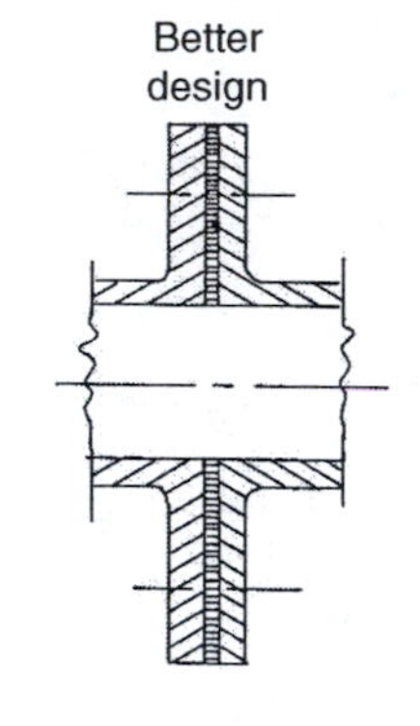

(a)

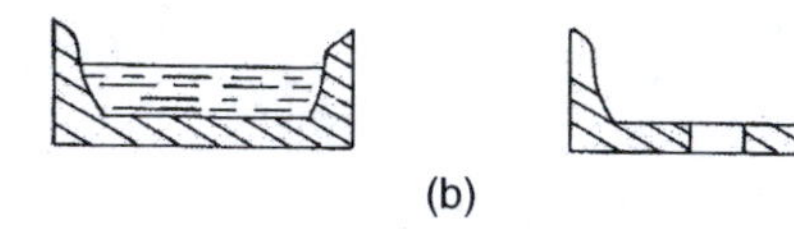

(b)

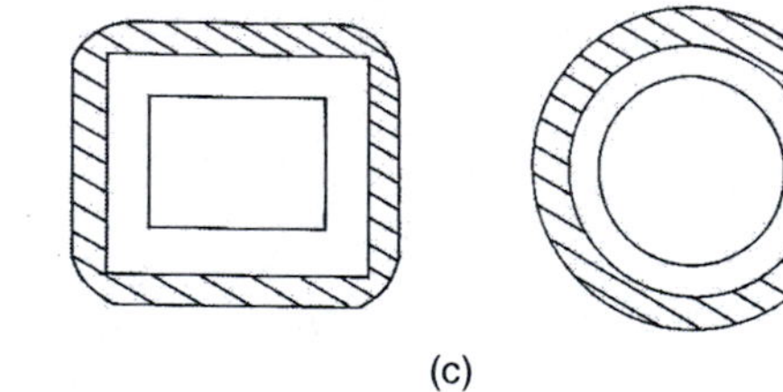

(c)

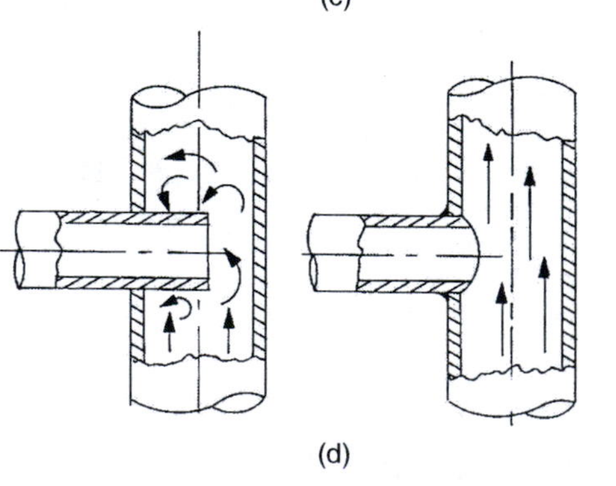

(d)

FIGURE 6.8 Examples of design features that should be considered in designing for a corrosive environment.

used as gaskets, washers, or similar applications. If crevices cannot be avoided in design, they should be sealed by welding, soldering, or brazing with a more noble alloy, adhesives, or caulking compounds.

Design features that retain undrained liquids in reservoirs or that collect rainwater should be avoided as they cause accelerated corrosion rates. Figure 6.8b shows examples of such features and how to avoid them. A similar problem is faced in closed tanks and sections where inadequate ventilation can cause condensation, or sweating, and accelerated corrosion rate. Closed sections are also difficult to paint and maintain. Avoiding closed sections and providing adequate ventilation can overcome this problem.

Sharp corners and convex surfaces that tend to have thinner coatings or are subject to coating cracks should be avoided. Similarly, coated surfaces that are exposed to direct impingement of airborne abrasive particles should be reduced. Where practicable, rounded contours and corners are preferable to angles, as illustrated in Figure 6.8c.

Design features that cause turbulence and rotary motion in moving liquids or gases should be avoided as this can cause impingement attack and other forms of accelerated corrosion near the obstruction. Figure 6.8d shows an example of such a design feature.

Fretting corrosion can take place at the interface of two closely fitting surfaces when they are subject to slight oscillatory motion. Assemblies like shrink-and-press fits, bolted joints, splined couplings, and keyed wheels are vulnerable to such attacks. Fretting corrosion can be prevented by preventing slippage or relative motion between surfaces that are not meant to move. This can be done by roughening the surface to increase friction or by eliminating the source of vibrations. Other solutions include using a soft metal surface in contact with a hard one, using low-viscosity lubricants, and phosphating.

6.8 DESIGNING WITH SPECIFIC MATERIALS (MATERIAL-SPECIFIC DESIGN FEATURES)

6.8.1 Designing with Metallic Materials

The load-carrying capacity of a component can be related to the yield strength, fatigue strength, or creep strength of the material depending on loading and service conditions. These properties are structure-sensitive and can be considerably changed by changing the chemical composition of the alloy, method, and conditions of manufacture as well as heat treatment. It should be noted, however, that increasing the strength of most metallic materials causes their ductility and toughness to decrease, which could adversely affect the performance of the component in service. This subject is discussed in more detail in Section 4.5.

Electrical and thermal conductivities of metallic materials can greatly affect the design and impose severe limitations on material selection in many applications. For example, electrical quality copper and aluminum may represent the only possible materials for the manufacture of a component where electrical conductivity is a primary requirement. Corrosion resistance and specific gravity requirements could impose similar limitations. However, judicious design and careful selection can, in many cases, overcome these limitations.

The design and material selection for a given component are also influenced by manufacturing considerations. The majority of metallic components have either cast or wrought microstructures. Wrought microstructures are usually stronger and more ductile than cast microstructures and this is one of the reasons why about 80% of metallic materials are produced in the wrought form.

6.8.2 Designing with Polymers

From the performance point of view, there are many applications where plastics outperform other materials. Their lightweight, corrosion resistance, low coefficient of friction, low thermal and electrical conductivities, optical properties, and decorative appeal are among the factors that explain the increasing use of plastics. Specific examples where plastics successfully replaced metals in the motorcar include bumpers, front and rear ends, radiator end tanks, and door handles. In most cases, plastics are not direct substitution for metals. This is because plastics have widely different properties and employ widely different processing techniques. The design should be changed to make use of the advantages of plastics and to avoid their limitations.

When the part is to carry loads, that is, a structural part, it should be remembered that the strength and stiffness of plastics vary significantly with temperature. Room temperature data cannot be used in design calculations if the part is going to be used at any other temperature. Long-term properties, that is, creep behavior, cannot be predicted from short-term properties. Many of the commonly used engineering plastics have a notched impact strength less than 5.4 J/cm (10 ft lb/in.). From the design point of view, this means that they are brittle and the effect of stress raisers must be considered. For example, one type of acetal has an unnotched Izod impact strength of greater than 54 J/cm (100 ft lb/in.) and a notched Izod impact of only about 1 J/cm (ca. 2 ft lb/in.). A complicating factor is that many plastics achieve their impact resistance through the addition of plasticizers. With time, these additives may vaporize, leaving the aged plastic relatively brittle.

6.8.3 Designing with Ceramics

Designing ceramic products needs special considerations in view of their brittleness and relatively low mechanical and thermal shock resistance. If the same configuration is used when a ceramic material is substituted for a metallic alloy, the ceramic part could fail in service or even during assembly. Special designs should be developed to make use of the advantages of ceramics and to avoid their limitations. As the ratios between tensile strength, modulus of rupture, and compressive strength are usually in the range 1:2:10, every effort should be made to load ceramic parts in compression and to avoid tensile loading.

As ceramics are brittle, they are sensitive to stress concentration. Features such as sharp corners, notches, and unstrengthened holes should, therefore, be avoided. Press fits and shrink fits permit successful attachment of ceramics to steel and allow prestressing the ceramic part in compression, which increases its load-carrying capacity. Variability of properties in addition to lower toughness and thermal shock resistance result in lower reliability of ceramic components in comparison with metallic ones.

6.8.4 Designing with Composites

The characteristics of composite materials are different in many respects from those of common metallic materials. This means that major design and manufacturing changes have to be made when replacing a metallic material by a composite material. Efficient use of composite materials can be achieved by tailoring the material for the application. For example, to achieve maximum strength in one direction in a fibrous composite material, the fibers should be well aligned in that direction.

The desirable feature of tailorability is illustrated by the use of carbon fibers, which exhibit a negative coefficient of thermal expansion, in building dimensionally stable structures. The negative coefficient of thermal expansion of carbon fibers is balanced by the positive coefficient of thermal expansion of the epoxy matrix and the resulting composite has a zero coefficient of thermal expansion. Space structures and finely tuned optical equipment are typical applications for such composites.

An important factor that should be borne in mind when designing with composite materials is that their high strengths are obtained only as a result of large elastic strains in the fibers. In some structures, these strains can cause unacceptable deflections. An example is the case of an aircraft wing where a strain of about 1.7% could cause the wing tips to become vertical. This indicates that composite materials should not simply be substituted for other materials without making the necessary modifications in design and construction.

Differences between the fatigue behavior of metals and composite materials should also be taken into account when designing with composites. Unlike steels, which show an endurance limit or a stress below which fatigue failure does not occur, composite materials suffer fatigue damage even at relatively low stress levels. Although fatigue crack propagation in metals is usually limited to a single crack, which progressively grows with each cycle, fibrous composites may have many cracks, which can be growing simultaneously. Cracks may propagate through the matrix, be arrested by a fiber, or move along a fiber–matrix interface. This makes fatigue failure of fibrous composites a more gradual, noncatastrophic event than for metals. Fatigue loading of fibrous composites is also accompanied by a progressive reduction in modulus with increasing number of cycles. In addition, cyclic creep, which is the increase in strain at the minimum fatigue stress, can also occur in chopped fiber-reinforced plastics. This latter weakness can be overcome by using continuous fibers that are oriented in the principal stress direction.

In spite of their high strength, fibrous composites can exhibit low fracture toughness. This is because the conditions required for high strength, for example, short-critical aspect ratio and strong interfacial bond between fibers and matrix, are in conflict with the conditions required for high fracture toughness, for example, long-critical aspect ratio and weak interfaces parallel to the fibers. In general, a compromise between strength and toughness will be required for a given application. Several hybrid composites have been developed to achieve such compromise. For example, Kevlar fibers, which have good impact resistance but relatively low compressive strength, complement graphite fibers, which have roughly four times the compressive strength but are less tough. Because each fiber retains its own identity, hybrid technology is becoming increasingly important as a way of selectively utilizing the dominant properties of each fiber and reducing the overall cost of the composite. For example, mixing carbon and glass fibers in a composite and placing the carbon fibers at critical locations, high-performance lightweight structures with minimal amount.

6.9 SUMMARY

1. A successful design should take into account the function, material properties, and manufacturing processes. The relationship between material properties and design is complex because the behavior of the material in the finished product can be quite different from that of the stock material.
2. The stiffness of components under bending is proportional to the elastic constant of the material (E) and the moment of inertia of the cross section (I). Selecting materials with higher E and efficient disposition of material in the cross section are essential in designing such components for stiffness. Placing material as far as possible from the neutral axis of bending is generally an effective means of increasing I for a given area of cross section.
3. Fail-safe design requires a structure to be sufficiently damage tolerant to allow defects to be detected before they develop to a dangerous size. In the case of high-strength materials, the interaction between fracture toughness, crack size, and the design stress should be considered.
4. Leak-before-burst is a useful design criterion in the case of pressure vessels and similar products. The toughness of the material used in such applications should be sufficiently high to tolerate a defect size that will allow the contents to leak out before it grows catastrophically.
5. In designing for fatigue resistance, the designer should consider the fatigue strength of the material as well as size and shape of the component or structure, type of loading and state of stress, stress concentration, surface finish, operating temperature, service environment, and method of fabrication.
6. Safe-life, or finite-life, design is based on the assumption that the component or structure is free from flaws but the stress level in certain areas is higher than the endurance limit of the material. This means that fatigue-crack initiation is inevitable and the life of the component is estimated on the basis of the number of stress cycles that are necessary to initiate such a crack.
7. The LMP can be used when creep data are incomplete and have to be supplemented or extended by interpolation or extrapolation.
8. Many of the causes of failure in hostile environment can be avoided by observing appropriate design rules such as avoiding electrical contact between dissimilar metals, sealing crevices, providing appropriate ventilation and drainage of liquids in closed containers, avoiding sharp corners in coated components, preventing turbulence in moving liquids, and preventing slippage between surfaces that are not meant to move.
9. A successful design with a specific material must use its points of strength and avoid its points of weakness. With metallic materials, the ductility and toughness generally decrease as the strength increases. Polymers are light but their stiffness and impact strength are much lower than metals. Ceramics are corrosion resistant and strong in compression but are very brittle and sensitive to mechanical and thermal shocks. Composite materials generally have high strength/weight ratio and their properties can be tailored to suit the type of loading.

6.10 REVIEW QUESTIONS

6.1 Determine the dimensions of a cantilever beam of length 2 m (80 in.) and rectangular section of depth-to-width ratio 3:1. The cantilever is expected not to deflect more than 25.4 mm (1 in.) for every 1000 N (220 lb) increment of load at its tip. The material used in making the beam is steel AISI 4340 with a yield strength of 1420 MPa (206 ksi) and UTS 1800 MPa (257 ksi). What is the maximum permissible load? Assume a suitable factor of safety.

6.2 For the cantilever beam in question 6.1, if the minimum detectable crack is 3 mm (0.118 in.), what is the maximum permissible load? K_{IC} of the beam material is 87.4 MPa $(m)^{1/2}$ (80 ksi $(in.)^{1/2}$).

6.3 If the load of the cantilever beam in question 6.1 is fluctuating around a mean value of 1000 N (220 lb), what is the maximum alternating load that can be applied without causing fatigue failure? The endurance ratio for this steel is 0.56. Use a suitable factor of safety and modifying factors for the endurance limit.

6.4 A component is made of steel with a static tensile strength of 600 MPa and endurance ratio of 0.4 is to be subjected to a combination of static and alternating stresses. If the average static stress is 150 MPa, what is the maximum alternating stress that can be safely applied to the component?

6.5 Distinguish between the stiffness of a component and the modulus of elasticity of the material used in making that component.

6.6 Low elastic modulus is one of the main limitations in using plastics for load-bearing components. Using actual products, select three design features that were used to overcome this limitation.

6.7 Compare the uses of titanium and aluminum alloys as supersonic aircraft wing tip materials.

6.8 What are the main material requirements for the following components: motorcar exhaust manifold; coil for electrical resistance heater; kitchen knife blade; and railway line? Suggest suitable materials for these components.

6.9 What are the main requirements for materials used for gas turbine blades? Suggest suitable materials for the turbine blades.

6.10 What are the main requirements for materials used for the sleeves of lubricated journal bearings? Suggest suitable materials for the sleeves.

6.11 What are the main requirements for materials used for making motorcar bodies? Suggest suitable materials for such bodies.

6.12 What would be the advantages and disadvantages of substituting fiber-reinforced plastics for steel in making motorcar bumpers?

6.13 (a) What are the main material requirements for the paper clip? (b) What is the type of material that you would recommend for such application? (c) Can plastics be used for such application? If so, what are the design changes that would be necessary?

6.14 What are the main materials requirements for the wing of a small aircraft wing structure? Compare the uses of the following materials in making the wing structure:

	Specific Gravity	Elastic Modulus (GPa)	Tensile Strength (MPa)
Aluminum alloy	2.7	70	580
Magnesium alloy	1.7	45	280
Polyester + 65% glass fibers	1.8	19.6	340

6.15 Calculate the diameter of a steel shaft that is subjected to a cyclic tensile load of 5000 kg. The shaft material has a tensile strength of 950 MPa and its endurance ratio is 0.5. Assume that the surface finish factor $k_a = 0.68$, the reliability factor $k_c = 0.75$, the size factor $k_b = 0.8$, and the stress concentration factor $k_f = 0.7$. Take a factor of safety of 2 to account for possible overloads and material defects.

6.16 The deflection (y) of a simply supported beam of length (l) under a concentrated load (L) is given by the formula

$$y = \frac{(Ll^3)}{(48EI)}$$

where E is the elastic constant of the material and I the second moment of area.

a. Discuss the possible alternatives that are open to the design engineer in maximizing the stiffness of the beam.
b. Compare the use of steels and carbon fiber–reinforced plastics for such applications ($E_{steel} = 210$ GPa, $E_{cfrp} = 200$ GPa).

6.17 You have used the tensile-testing machine in the materials testing laboratory to test the mechanical properties of different materials: (a) What are the main material properties that are usually measured in a tensile test? (b) How can you compute the toughness of materials from the tensile test? (c) What are the main design requirements for the structural members used in making tensile-testing machines? (d) What are the important properties of materials used in making the structural members of tensile-testing machines?

6.18 What are the main material requirements for airplanes? Compare the uses of aluminum alloys, titanium alloys, magnesium alloys, and fiber-reinforced plastics in the various parts of a civilian aircraft.

BIBLIOGRAPHY AND FURTHER READING

Collins, J.A. and Daniewicz, S.R., Failure modes: performance and service requirements for metals, in *Handbook of Materials Selection*, Kutz, M., Editor. Wiley, New York, 2002, pp. 705–773.

Collins, J.A. and Daniewicz, S.R., Failure modes: performance and service requirements for metals, in *Mechanical Engineers' Handbook: Materials and Mechanical Design*, 3rd Ed., Kutz, M., Editor. Wiley, Hoboken, NJ, 2006, pp. 860–924.

Cook, N.H., *Mechanics and Materials for Design*, McGraw-Hill, New York, 1985.

Dieter, G.E., *Engineering Design: A Materials and Processing Approach*, 2nd Ed., McGraw-Hill, New York, 1991.

Farag, M.M., *Selection of Materials and Manufacturing Processes for Engineering Design*, Prentice-Hall, New York, 1989.

Farag, M.M., *Materials Selection for Engineering Design*, Prentice-Hall, Materials Park, London, 1997.

Farag, M.M., Properties needed for the design of static structures, in *Materials Selection and Design, ASM Metals Handbook,* Vol. 20, Dieter, G., Volume Chair. ASM International, Materials Park, OH, 1997, pp. 509–515.

Kutz, M., *Handbook of Materials Selection*, Wiley, New York, NJ, 2002.

Kutz, M. Ed., *Mechanical Engineers Handbook: Materials and Mechanical Design*, 3rd Ed., Wiley, Hoboken, NJ, 2006.

Parker, A.P., *The Mechanics of Fracture and Fatigue*, E. and F.N. Spon, Ltd., London, 1981.

Peterson, R.E., *Stress-Concentration Design Factors*, Wiley, New York, 1974.

Shigley, J.E. and Mitchell, L.D., *Mechanical Engineering Design*, 4th Ed., McGraw-Hill, New York, 1983.

Wilshire, B. and Owen, D.R.J., Editor, *Engineering Approaches to High Temperature Design*, Pineridge Press, Swansea, U.K., 1983.

Woodford, D. (Editor of Section 6), Properties versus performance of materials, in *ASM Handbook*, Vol. 20, Dieter, G.E., Editor. ASM International, Materials Park, OH, 1997, pp. 507–666.

7 Effect of Manufacturing Processes on Design

7.1 INTRODUCTION

Manufacturing can be defined as the act of transforming materials into usable and saleable end products. This means that the products must function satisfactorily and provide value for money. Cost, rate of production, availability, and quality of product are important criteria that should be considered when selecting a manufacturing process. With the increasing sophistication of many products and the increasing range of available materials, new manufacturing technologies have been developed to produce more sound and accurate components economically.

It is now widely recognized that design, materials selection, and manufacturing are intimately related activities that cannot be performed in isolation of one another. It is uneconomical to make the design first and then modify it later to accommodate manufacturing requirements. This is because the later the design changes are made in the product development cycle, the more expensive they are. It is also recognized that over 70% of the final product costs are determined during the design phase, which means that it is more effective to consider materials and manufacturing costs while developing the design. Recognition of the capabilities and limitations of the available processes would make it possible to specify the correct manufacturing instructions and to select the processing route that will yield the required product quality at the optimum cost.

The goal of this chapter is to illustrate how the manufacturing processes influence the design of components, with emphasis on the following topics:

1. Types of available manufacturing processes and their selection
2. Design for manufacture and assembly
3. Design considerations for cast components
4. Design considerations for molded plastic components
5. Design considerations for forged components
6. Design considerations for powder metallurgy (P/M) parts
7. Designs involving welding processes
8. Designs involving machining processes

7.2 CLASSIFICATION OF MANUFACTURING PROCESSES

7.2.1 Processing of Metallic Materials

The relationship between the different manufacturing processes that are normally used in processing metallic materials can generally be presented as shown in Figure 7.1.

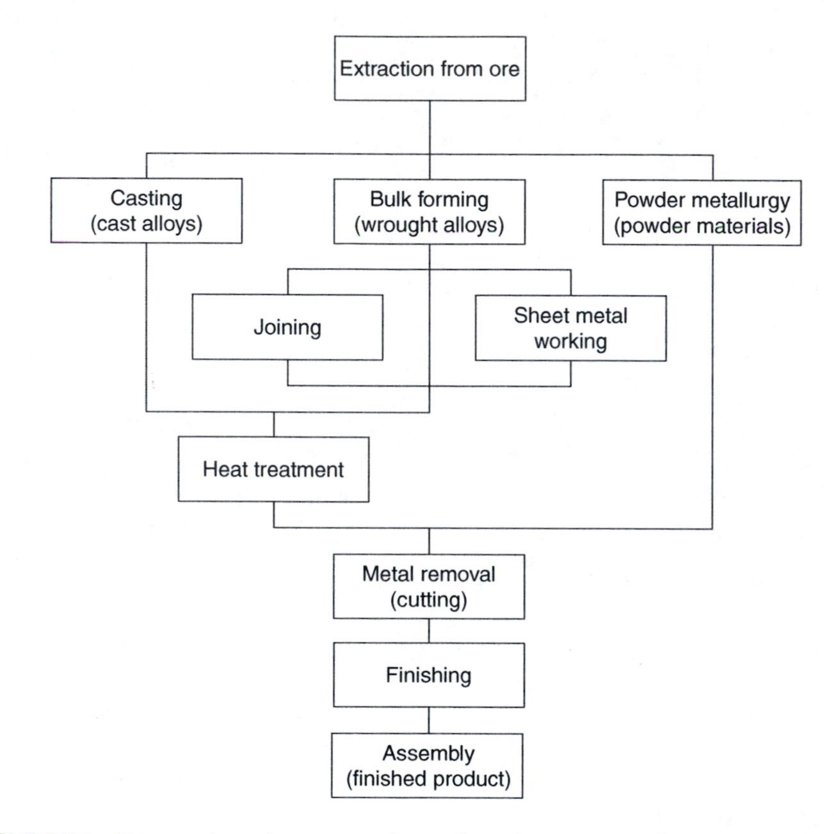

FIGURE 7.1 Types and usual sequence of manufacturing processes that are normally used in processing metallic materials.

Most metallic materials are found in nature in the form of ores, which have to be reduced or refined and then cast as ingots of convenient size and shape for further processing. Casting processes involve pouring the liquid metal, or alloy, into a mold cavity and allowing it to solidify. Powdered materials can be shaped by pressing in suitably shaped dies and then sintering, heating, to achieve the required properties. An important advantage of casting and P/M processes is that parts of complex shapes can be obtained in one step.

Bulk forming processes are generally used to change the shape of metallic materials by plastic deformation. The deformation can be carried out at relatively high temperatures, as in the case of hot working processes, or at relatively low temperatures, as in the case of cold working processes. The basic bulk deformation processes are forging; rolling; extrusion; swaging; and drawing of rod, wire, and tube. These processes are called primary working when applied to ingots to break down their cast structure into wrought structure and to change their shape to slabs, plates, or billets. Secondary working involves further processing of the products from primary working into final or semifinal shapes.

Sheet metal working processes are normally carried out at room temperature and usually involve the change of sheet form without greatly affecting its thickness. The basic sheet metal working processes include shearing, bending, stretch forming, bulging, deep drawing, spinning, and press forming. Other sheet metal forming operations have been developed for the manufacture of certain components and special materials.

Heat treatment is normally carried out to control the structure and properties of the material. By proper control of temperature and cooling rate, the material can be softened to permit further processing or hardened to increase its mechanical strength. Heat treatment can also be used to remove internal stresses, control grain size, or produce a hard surface on a ductile interior. With the proper heat treatment, a less-expensive material could replace a more-expensive material or a less-costly processing could be employed.

Material removal or cutting processes are normally used to remove unwanted material in the form of chips by using cutting tools, which are mounted in machine tools. The traditional, basic machine tools are lathes, boring machines, shapers and planers, milling machines, drill presses, saws, broaches, and grinding machines. The productivity in cutting processes can be improved by using machining centers, which are single machines that can perform the functions of several basic machine tools. When cutting very hard metals or when machining intricate shapes and delicate parts, nontraditional or chipless processes can be used. These cutting methods include ultrasonic, electrical discharge, electrochemical, chemical milling, abrasive jet, electro-arc, plasma-arc, electron beam, and laser cutting.

In many cases, products are manufactured as separate units and then assembled either by fastening, as in the case of temporary or semipermanent joints, or by joining, as in the case of permanent joints. Examples of fastening methods include screws, bolts, and rivets, whereas joining methods include welding, brazing, soldering, and adhesive bonding. Assembling by press and shrink fitting is also used in some applications. Finishing processes are normally used to control the quality of the surface and to make it ready for service. Cleaning, deburring and polishing, anodizing, tinning, galvanizing, plating, and painting are among the frequently used finishing processes. In addition to controlling the appearance of the surface, many of the finishing processes provide some protection against corrosion.

7.2.2 Processing of Plastic Parts

There are many methods of manufacturing plastic parts, which can be considered as molding processes. These processes usually employ the following sequence of steps:

1. Plastics in the form of powder, pellets, or granules are usually heated above the softening point.
2. The molten plastic is forced or placed into a mold that determines the dimensions of the molded part.

3. The material is then allowed to harden, by curing or freezing, and is then ejected from the mold.
4. In some molding processes, the ejected part is ready for use. With others, trimming and other finishing processes are necessary to make it ready for use.

Pressure-molding processes cover compression molding, transfer molding, and injection molding. These processes are similar in some ways to die-casting processes used for metals. It should be noted, however, that molten plastics are more viscous and much less conducting to heat than metals. These differences are reflected in the design and operation of the plastics-molding machines.

7.2.3 Processing of Ceramic Products

The raw materials used for making ceramic parts are usually in the form of particles or powder. After mixing and blending the appropriate ingredients, processing is carried out either in a dry, semidry, or liquid state, and either cold or hot condition. Forming processes include slip casting, molding, jiggering, extrusion, and pressing. After forming the plastic ceramic mass into the required shape, it is dried to remove the water and then fired to sinter the ceramic powder into a final product.

7.2.4 Manufacture of Reinforced Plastic Components

Thermoplastics that are reinforced with short-chopped, randomly oriented fibers are easily fabricated using conventional techniques of injection molding and extrusion. Composites based on thermosetting plastics are processed using specially developed methods like

1. contact molding, which employs single-surface molds as in the case of hand layup, spray-up, and filament winding;
2. compression-type molding, as in the case of sheet molding, bulk molding, preform molding, and cold molding;
3. resin-injection molding, which is similar to the process used for nonreinforced materials and reinforced thermoplastics; and
4. pultrusion, which is a modification of the extrusion process.

7.2.5 Manufacture of Reinforced Metal Components

Metal–matrix composites can be prepared in a variety of ways, which can be broadly classified into the following:

1. Processes based on solid-state diffusion
2. Liquid-phase infiltration
3. *In situ* processes

Selection of the appropriate process is not only a function of the required shape and properties of the composite, but also of the fiber–matrix combination.

7.3 SELECTION OF MANUFACTURING PROCESSES

Figure 5.1 shows that in addition to having a direct influence on the design of components, manufacturing processes are also closely related to the material out of which the component is made and to its function and shape. The importance of manufacturing processes is further emphasized in Figure 5.2, where it is shown that the behavior of the material in the component is not only a function of the stock material properties but is also strongly influenced by the fabrication method. Such intimate relations between manufacturing processes, material properties, and design have led to widely used practice of concurrent, or simultaneous, engineering in product development. As the design progresses from concept to configuration and the material choices get narrower, manufacturing processes, which have initially been broadly defined, also need to be better identified. Successful product development requires a good match between the capabilities and limitation of the manufacturing process and the component attributes, which include its size, geometrical features, number required, and the type of material used.

As most components require a sequence of processes for their manufacture, there will be hundreds of possible process combinations, which can be used to make a given part. The choices can be narrowed down based on the following:

1. Not all processes are suitable for all materials. For example, cast iron cannot be forged and P/M is uneconomical for a limited production run. Table 7.1 outlines the compatibility between some widely used metallic materials and processes; further details appear throughout this chapter.
2. Many processes form a natural sequence for shape generation. For example, casting and forging are normally followed by machining and then surface finishing, if needed, as shown in Figure 7.1. From this point of view, processes can be grouped as follows:
   a. Primary processes, including casting, bulk forming (forging, rolling, extrusion), etc.
   b. Primary/secondary processes, including joining and welding, sheet metal work, heat treatment, metal cutting, etc.
   c. Tertiary processes or finishing processes, including surface treatment, grinding, coating, etc.

7.4 DESIGN FOR MANUFACTURE AND ASSEMBLY

The intimate relation between design and manufacturing has resulted in a variety of methods to improve the design of components and products with respect to specific manufacturing processes and activities. Examples include design for manufacture, design for assembly, design for disassembly, design for casting, design

TABLE 7.1
Compatibility between Some Widely Used Metallic Materials and Processes

	Carbon Steel	Stainless Steel	Cast Iron	Aluminum Alloys	Copper and Alloys	Magnesium and Alloys	Zinc and Alloys	Titanium and Alloys	Superalloys
Sand casting	XX	XX	XX	XX	XX	XX	X	Nr	XX
Investment casting	XX	XX	Nr	XX	XX	X	Nr	X	XX
Die casting	Nr	Nr	Nr	XX	X	XX	XX	Nr	Nr
Powder metallurgy	XX	XX	Nr	XX	XX	Nr	Nr	X	XX
Forging	XX	XX	Nr	XX	XX	XX	Nr	X	X
Rolling	XX	XX	Nr	XX	XX	XX	X	X	XX
Extrusion	X	X	Nr	XX	XX	XX	X	X	X
Sheet metal work	XX	XX	Nr	XX	XX	X	X	X	X
Cold heading	XX	XX	Nr	XX	XX	Nr	Nr	Nr	X
Metal cutting	XX	XX	XX	XX	XX	XX	XX	X	X
Fusion welding	XX	XX	X	XX	XX	XX	Nr	XX	XX

Note: XX = common practice, X = less common or performed with difficulty, and Nr = not recommended.

for injection molding, and design for machining. Designing for "X" (DFX) is the collective name of this methodology, which seeks to lower costs and cycle times by ensuring that components are designed to be compatible with the activity of process concerned.

Design for manufacture and assembly seeks to minimize the cost of a product by designing components that are easier to manufacture (DFM) and designing components that are easier to assemble (DFA). Although DFM and DFA are separate design activities, they are often interrelated and are best addressed at the configuration stage of the design, as discussed in Section 5.3. Changes during the detail design stage can also be introduced to further improve ease of manufacture and assembly. DFMA provides a structured procedure for analyzing a design from the point of view of manufacture and assembly. Such procedure normally results in simpler designs with fewer components in an assembly. In addition to reduction in assembly time and cost, having fewer components means fewer detail drawings, smaller number of materials and specifications, less inventory, and lower overheads. Boothroyd and Dewhurst (1994) developed a DFA software to provide guidance in reducing the number of parts in an assembly. The software asks questions such as: (a) Is the part or subassembly used only for fastening or securing other items? (b) Is the part or subassembly used only for connecting other items? If the answer to either of these questions is "yes," then the part or subassembly is considered theoretically unnecessary. If the answer to both questions is "no," the following questions are then asked: Does the part move relative to all other parts? Must the part be made of different material or isolated from other parts? If the answer is "no," the part is considered theoretically unnecessary. The design team attempts to integrate the parts that are considered theoretically unnecessary with other parts, thus reducing the total part count and assembly time of the product.

One of the methods reported by Boothroyd (1997) for DFM uses processability evaluation method. Processability is defined as being proportional to cost of the component, which is determined by its shape, material, and the processing method. Various designs, processing methods, and materials are considered and the cost is determined for the different combinations. For example, the cost of an injection-molded plastic component is compared with an equivalent die-casting or sheet metal stamping.

Example 7.1 is based on that given by Boothroyd and the figures are reprinted with permission of ASM International.

Design Example 7.1: Application of DFMA Principles to the Design of a Motor-Drive Assembly (Based on Boothroyd, 1997)

Problem

It is required to design a motor-drive assembly that senses and controls its position on two guide rails. The main requirements are that the motor must have

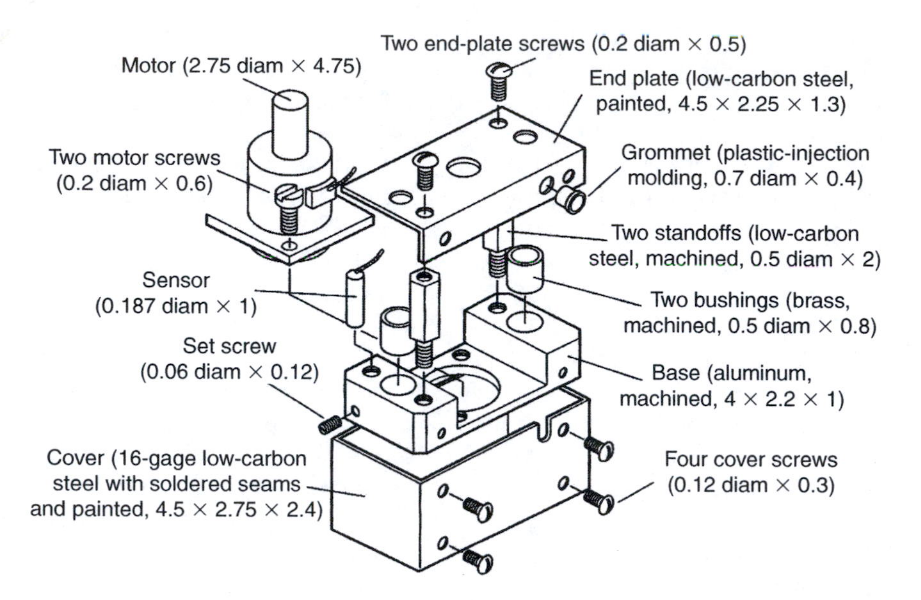

FIGURE 7.2 Initial design of motor-drive assembly. Dimensions in inches. (Reprinted from Dieter, G.E., in *ASM Handbook Vol. 20 Materials Selection and Design, Materials Seletion and Design, Performance Indices pp. 281–290, Manufacture and Assembly Design pp. 676–686*, ASM International, Materials Park, OH, 1997. With permission.)

a removable cover, and a rigid base that supports both the motor and the sensor, in addition to sliding up and down the guide rails. Figure 7.2 shows a proposed design, which requires two subassemblies for the motor and sensor in addition to eight additional parts and nine screws, making a total of 19 items to be assembled.

Analysis

Using the DFMA software and asking the kind of questions mentioned earlier, it is possible to eliminate the parts that do not meet the minimum part-count criteria and Figure 7.3 shows the resulting design. The number of items to be assembled has been reduced to six and the assembly time has been reduced from 160 to 46 s, thus reducing the assembly cost from $1.33 to $0.38.

Although the shape of the base remains essentially the same, it is machined out of nylon in the new design instead of aluminum to eliminate the bushings and reduce the cost of material (aluminum, $2.34 and nylon, $0.49). The new base also has less drilled and tapped holes, with further reduction of cost of manufacturing. The total cost of components in the motor-drive assembly has been reduced from $35.08 for the original design to $22.00 for the new design.

These costs show that the saving in the cost of components was much more than the saving in assembly cost in this case.

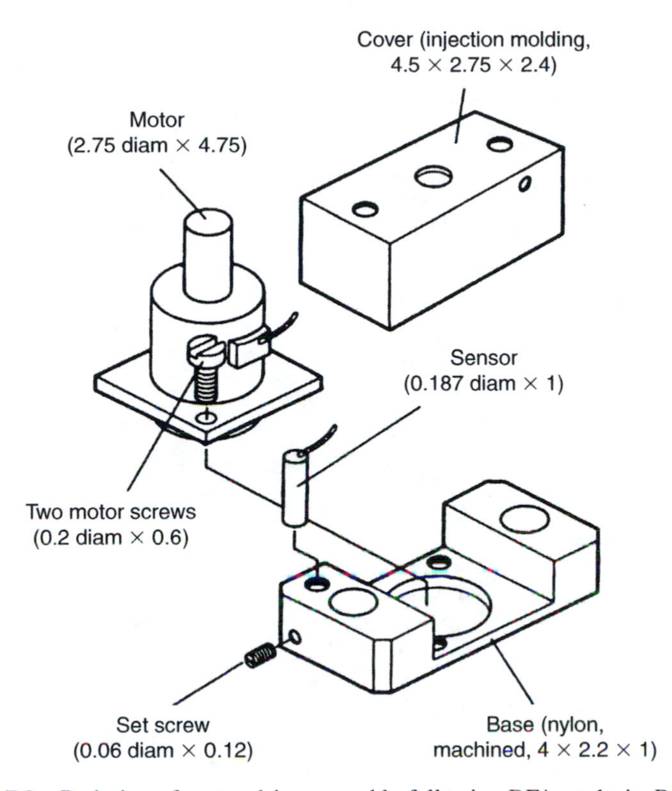

FIGURE 7.3 Redesign of motor-drive assembly following DFA analysis. Dimensions in inches. (Reprinted from Dieter, G.E., in *ASM Handbook Vol. 20 Materials Selection and Design, Materials Seletion and Design, Performance Indices pp. 281–290, Manufacture and Assembly Design pp. 676–686*, ASM International, Materials Park, OH, 1997. With permission.)

7.5 DESIGN CONSIDERATIONS FOR CAST COMPONENTS

Casting covers a wide range of processes that can be used to shape almost any metallic material and some plastics in a variety of shapes, sizes, accuracy, and surface finishes. The tolerances and surface finishes that can be achieved by standard commercial foundry practice are given in Tables 7.2 and 7.3. The number of castings can vary from very few to several thousands. In some cases, casting represents the obvious and only way of manufacturing, as in the case of components made of different types of cast iron or cast alloys. In many other applications, however, decision has to be made whether it is advantageous to cast a product or to use another method of manufacture. In such cases, the following factors should be considered:

1. Casting is particularly suited for parts that contain internal cavities that are inaccessible, too complex, or too large to be easily produced by machining.
2. It is advantageous to cast complex parts when required in large numbers, especially if they are to be made of aluminum or zinc alloys.

TABLE 7.2
Approximate Values of Surface Roughness and Tolerance That Are Normally Obtained with Different Manufacturing Processes

	Typical Tolerance (±)		Typical Surface Roughness (R_a)	
Process	**(mm)**	**(in. × 10³)**	**(μm)**	**(μin.)**
Sand casting	0.5–2.0	20–80	12.5–2.5	500–1000
Investment casting	0.2–0.8	8–30	1.6–3.2	63–125
Die casting	0.1–0.5	4–20	0.4–1.6	16–63
Powder metallurgy	0.2–0.4	8–16	0.8–3.2	32–125
Forging	0.2–1.0	8–40	3.2–12.5	125–500
Hot rolling	0.2–0.8	8–30	6.3–25	250–1000
Hot extrusion	0.2–0.8	8–30	6.3–25	250–1000
Cold rolling	0.05–0.2	2–8	0.4–1.6	16–32
Cold drawing	0.05–0.2	2–8	0.4–1.6	16–32
Cold extrusion	0.05–0.2	2–8	0.8–3.2	32–125
Flame cutting	1.0–5.0	40–200	12.5–25	500–1000
Sawing	0.4–0.8	15–30	3.2–25	125–1000
Turning and boring	0.025–0.05	1–2	0.4–6.3	16–250
Drilling	0.05–0.25	2–10	1.6–6.3	63–250
Shaping and planning	0.025–0.125	1–5	1.6–12.5	63–500
Milling	0.01–0.02	0.5–1	0.8–6.3	32–250
Chemical machining	0.02–0.10	0.8–4	1.6–6.3	63–250
Electro-discharge machining (EDM) and electro-chemical machining (ECM)	0.02–0.10	0.8–4	1.6–6.3	63–250
Reaming	0.02–0.05	0.4–2	0.8–3.2	32–125
Broaching	0.01–0.05	0.4–2	0.8–3.2	32–125
Grinding	0.01–0.02	0.4–0.8	0.1–1.6	4–63
Honing	0.005–0.01	0.2–0.4	0.1–0.8	4–32
Polishing	0.005–0.01	0.2–0.4	0.1–0.4	4–16
Lapping and surface finishing	0.004–0.01	0.16–0.4	0.05–0.4	2–16

3. Casting techniques can be used to produce a part, which is one of a kind in a variety of materials, especially when it is not feasible to make it by machining.
4. Precious metals are usually shaped by casting since there is little or no loss of material.
5. Parts produced by casting have isotropic properties, which could be an important requirement in some applications.
6. Casting is not competitive when the parts can be produced by punching from sheet or by deep drawing.
7. Extrusion can be preferable to casting in some cases, especially in the case of lower melting nonferrous alloys.
8. Casting is not usually a viable solution when the material is not easily melted, as in the case of metals with very high melting points such as tungsten.

TABLE 7.3
Characteristics of Different Casting Processes and Powder Metallurgy

Process	Alloy	Weight kg (lb)	Surface Finish μm (μin.)	Tolerance (m/m[in./in.])	Minimum Section Thickness (mm [in.])	Porosity Rating	Least Economy Quantity	Relative Production Rate
Sand casting	Most	0.2 and up (0.4 and up)	12.5–25 (500–100)	0.03–0.2	3–5 (0.12–0.20)	Fair	1	1
Shell molding	Most	0.2–10 (0.44–22)	1.6–12.5 (63–500)	0.01–0.03	2–5 (0.08–0.20)	Good	500	4
Gravity die casting	Nonferrous	0.2–10 (0.44–22)	1.6–12.5 (63–500)	0.02–0.05	3–5 (0.12–0.2)	Very good	500	4–5
Pressure die casting	Al, Zn, Mg, Cu alloys	0.2–10 (0.44–22)	0.4–1.6 (16–63)	0.001–0.05	1–2 (0.04–0.08)	Excellent	10,000	10
Investment casting	Most	0.1–10 (0.22–22)	1.6–3.2 (63–125)	0.002–0.005	0.5–1 (0.02–0.04)	Very good	50	6
Powder metallurgy	Most	0.01–5 (0.022–11)	0.8–3.2 (32–125)	0.002–0.005	0.8 (0.03)	Variable	5000	8

7.5.1 Guidelines for Design

When casting is selected as the manufacturing process, it is important for the designer to observe the general rules that are related to solidification, material of the casting, position of the casting in the mold, and required accuracy of the part. Designs that violate the rules of sound casting can make production impossible or possible only at higher expense and large rejection rates.

A general rule of solidification is that the shape of the casting should allow the solidification front to move uniformly from one end toward the feeding end, that is, directional solidification. This can most easily be achieved when the casting has virtually uniform thickness in all sections. In most cases this is not possible. However, when section thickness must change, such change should be gradual, to avoid feeding and shrinkage problems. Sudden changes in section thickness give rise to stress

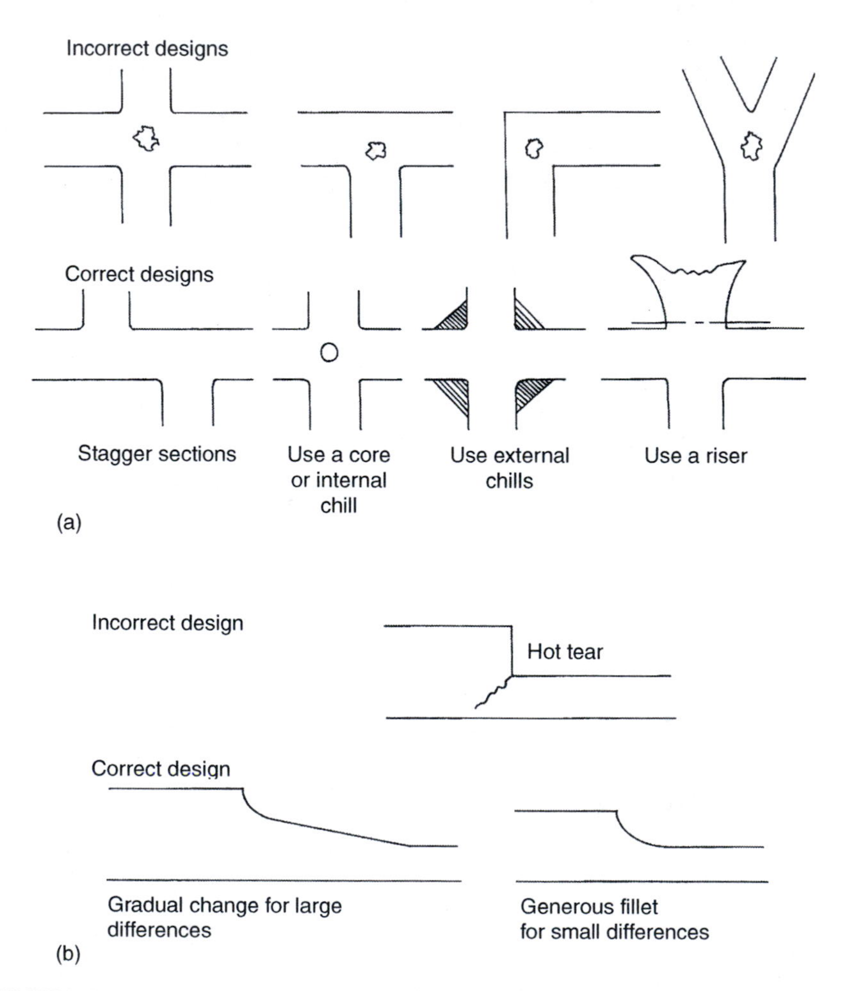

FIGURE 7.4 Design considerations for case components. (a) Solidification of intersecting sections results in hot spots and shrinkage cavities and (b) large changes in section thickness cause hot tears.

concentration and possible hot tears in the casting. Figure 7.4 gives some guidelines to avoid these defects.

Another problem that arises in solidification is caused by sharp corners; these also give rise to stress concentration and should be replaced by larger radii. When two sections cross or join, the solidification process is interrupted and a hot spot results. Hot spots retard solidification and usually cause porosity and shrinkage cavities. Some solutions to this problem are given in Figure 7.4. Large unsupported flat areas should also be avoided, as they tend to warp during cooling.

7.5.2 Effect of Material Properties

The type and composition of the material play an important part in determining the shape, minimum section thickness, and strength of the casting. Materials that have large solidification shrinkage and contain low-melting phases are susceptible to hot tears. Collapsible mold materials and a casting shape that allows shrinkage with least stresses in the casting are required in such cases.

Another material variable is castability, which can be related to the minimum section thickness that can be easily achieved. Table 7.3 provides some guidelines to minimum section thickness, which can be economically obtained for different casting processes. It should be noted that the shape and size of the casting as well as the casting process and foundry practice can affect the minimum section thickness. As cooling rates are directly related to section thickness, sections of different thickness can develop different structures and mechanical properties. In cast iron, the grain size as well as the graphite size and the amount of combined carbon are affected. Thinner sections result in faster cooling rates and higher strength and hardness in the different classes of gray cast iron.

7.6 DESIGN CONSIDERATIONS FOR MOLDED PLASTIC COMPONENTS

Compression, transfer, and injection molding processes are the commonly used methods of molding plastic components. These processes involve the introduction of a fluid or a semifluid material into a mold cavity and permitting it to solidify into a desired shape. Although section thickness, dimensional accuracy, or the incorporation of inserts would make it desirable to use one molding technique in preference to another, there are general design rules that should be observed to ensure the quality of the product.

7.6.1 Guidelines for Design

Experience shows that the mechanical, electrical, and chemical properties of molded components are influenced by the flow of molten plastic as it fills the mold cavity. Streamlined flow will avoid gas pockets in heavy-sectioned areas.

An important common feature in molding processes is draft, which is required for easy ejection of molded parts from the mold cavity. A taper of 1–4° is usually used for polymers, but tapers of less than 1° can be used for deep articles. Another common feature is the uniformity of wall thickness. In general, molded parts should be designed to have uniform wall thickness. Nonuniformity of thickness in a molded piece tends to produce nonuniform cooling and unbalanced shrinkage—leading to internal stresses and warpage.

If thickness variations are necessary, generous fillets should be used to allow gradual change in thickness. The effect of junctions and corners can also be reduced by using a radius instead, as shown in Figure 7.5a. The nominal wall thickness must obviously be such that the part is sufficiently strong to carry the expected service loads. However, it is better to adjust the shape of the part to cope with the applied load than to increase the wall thickness. This is because thick sections retard the molding cycle and require more material. Ribs, beads, bosses, edge stiffeners, and flanges should be used instead. However, shrinkage dimples, sink marks, may appear opposite to ribbed surfaces if they are not proportioned correctly. Adopting the proportions shown in Figure 7.5b may eliminate sink marks.

Large, plain, flat surfaces should be avoided, as they are prone to warping and lack rigidity. Such surfaces should be strengthened by ribbing or doming. The presence of holes disturbs the flow of the material during molding and a weld line occurs on the side of the hole away from the direction of flow. This results in a potentially weak point and some form of strengthening, such as bosses, may be necessary, as shown in Figure 7.5c. Through holes are preferred to blind holes from a manufacturing standpoint. This is because core prints can often be supported in both halves of the mold in the case of through holes, but can only be supported from one end in the case of blind holes.

Undercuts are undesirable features in molded parts as they cause difficulties in ejection from the mold. Examples of external and internal undercuts are shown in Figure 7.5d. Some parts with minor undercuts may be flexible enough to be stripped from the mold without damage. Many thermoplastics can tolerate about 10% strain during ejection from the mold.

Parts with external or internal threads can be made by molding and are usually removed from the mold by unscrewing. The mold is usually costly, since unscrewing devices may need to be incorporated. External threads may be produced without special devices if they are located on the parting plane of the mold. The thread may need secondary finishing in this case. Threads can also be formed by tapping, especially in the case of diameters less than 8 mm (5/16 in.).

Plastic parts can be given a wide variety of surface finishes including mirror-like finish, dull-satin, wood grain, leather grain, and other decorative textures. A highly smooth surface is usually required for surfaces that are to be painted or vacuum-metallized. Decorative textures are often helpful in hiding any possible surface imperfections, such as flow lines or sink marks. Raised letters on a molded part are easier and cheaper to produce than depressed letters because the lettering is machined into the mold cavity. The position of the parting plane of the mold should be carefully considered as it is normally accompanied by an unsightly flash line.

7.6.2 Accuracy of Molded Parts

Dimensional tolerances in molded plastic parts are affected by the type and constitution of the material, shrinkage of the material, heat and pressure variables in the

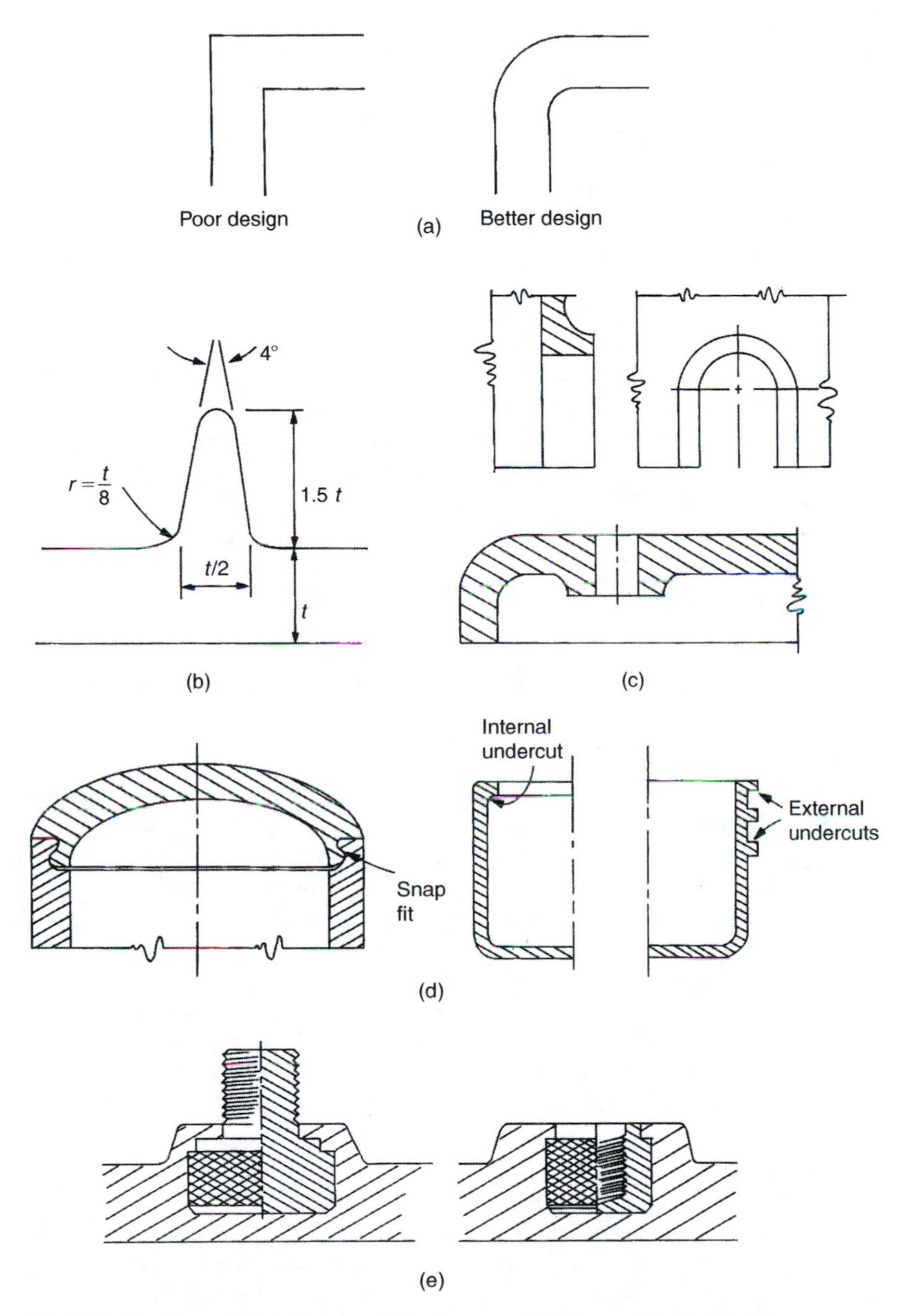

FIGURE 7.5 Some design features of plastic parts. (a) Using radii instead of sharp corners, (b) recommended rib proportions to avoid sink marks, (c) use of bosses to strengthen areas round holes and slots, (d) examples of undercuts, and (e) accommodating metal inserts.

molding process, and the toolmaker's tolerances on the mold manufacture. Generally, shrinkage has two components:

- Mold shrinkage, which occurs upon solidification
- After-shrinkage, which occurs in some materials after 24 h

For example, a thermosetting plastic like melamine has a mold shrinkage of about 0.7–0.9% and an after-shrinkage of 0.6–0.8%. Thus a total shrinkage of about 1.3–1.7% should be considered. However, a thermoplastic like polyethylene may shrink as much as 5% and nylon as much as 4%. In addition, the value of the tolerance depends on the size of the part. Larger dimensions are normally accompanied by larger tolerances. For example, dimensions less than 25 mm (1 in.) can be held within ±50 μm (±0.002 in.). Larger dimensions are usually given tolerances of ±10 to 20 μm/cm (±0.001 to 0.002 in./in.). The value of tolerance also depends on the direction in relation to the parting plane. Generally, dimensions at right angles to the parting plane should be given higher tolerance than dimensions parallel to it. As a rough guide, if a tolerance of ±50 μm is allowed in the direction parallel to the parting plane of the mold, a tolerance of ±200 μm should be allowed at right angles. Generous dimensional tolerances make economical production possible.

7.7 DESIGN CONSIDERATIONS FOR FORGED COMPONENTS

Forging processes represent an important means of producing relatively complex parts for high-performance applications. In many cases forging represents a serious competitor to casting, especially for solid parts that have no internal cavities. Forged parts have wrought structures that are usually stronger, more ductile, contain less segregation, and are likely to have less internal defects than cast parts. This is because the extensive hot working, which is usually involved in forging, closes existing porosity, refines the grains, and homogenizes the structure. However, cast parts are more isotropic than forged parts, which usually have directional properties. This directionality is due to the fiber structure that results from grain flow and elongation of second phases in the direction of deformation. Forged components are generally stronger and more ductile in the direction of fibers than across the fibers. This directionality can be exploited in some cases to enhance the mechanical performance of the forged part, as shown in Figure 7.6.

7.7.1 Guidelines for Design

When forging is selected as the manufacturing process, it is important for the designer to observe the general rules that are related to the flow of material in the die cavity. As with casting, it is better to maintain uniform thickness in all sections. Rapid changes in thickness should be avoided because these could result in laps and cracks in the forged metal as it flows in the die cavity. To prevent these defects, generous radii must be provided at the locations of large changes in thickness. Another similarity with casting is that vertical surfaces of a forging must be tapered to permit removal from the die cavity. A draft of 5–10° is usually provided.

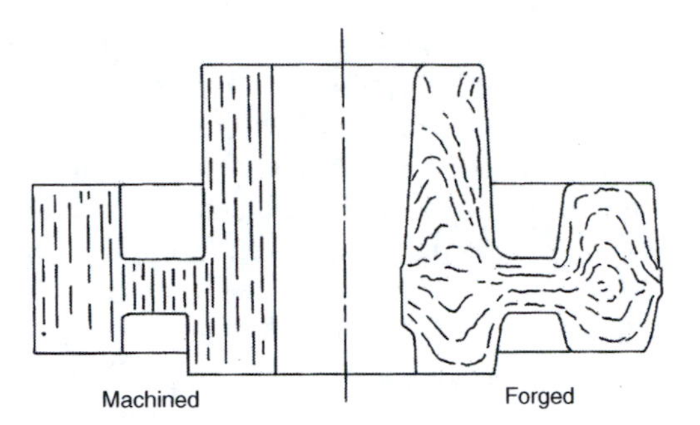

FIGURE 7.6 Schematic comparison of the grain flow in forged and machined parts.

It is better to locate the parting line near the middle of the part. This avoids deep impressions in either of the two halves of the die and allows easier filling of the die cavity. Inaccuracies in die forging result from mismatch between the die halves, due to the lateral forces that occur during forging, and from incomplete die closure, which is usually introduced to avoid die-to-die contact. A design would be more economically produced by forging if dimensions across the parting line are given appropriate mismatch allowance, and parallel dimensions are given a reasonable die closure allowance. Specifying close tolerances to these dimensions could require extensive machining, which would be expensive. Allowance must also be made for surface scale and warpage in hot forged parts. The dimensional tolerances and surface roughness that are commercially achieved in forging processes are shown in Table 7.2.

7.8 DESIGN CONSIDERATIONS FOR POWDER METALLURGY PARTS

P/M techniques can be used to produce a large number of small parts to the final shape in few steps, with little or no machining, and at high rates. Many metallic alloys, ceramic materials, and particulate reinforced composites can be processed by P/M techniques. Generally, parts prepared by the traditional P/M techniques, which involve mechanical pressing followed by sintering, contain from 4 to 10 vol% porosity. The amount of porosity depends on part shape, type and size of powder, lubricant used, pressing pressure, sintering temperature and time, and finishing treatments.

The distribution and volume fraction of porosity greatly affect the mechanical, chemical, and physical properties of parts prepared by P/M techniques. Using higher compaction pressures or employing techniques like P/M forging and hot isostatic pressing (HIP) can greatly decrease porosity and provide strength properties close to those of wrought materials. The HIP process is particularly suited for producing parts from high-temperature alloys that are difficult to forge and machine.

An added advantage of P/M is versatility. Materials that can be combined in no other way can be produced by P/M. Examples include aluminum–graphite bearings, copper–graphite electrical brushes, cobalt–tungsten carbide cutting tools (cermets), and porous bearings and filters. P/M is also the only practical way of processing tungsten and other materials with very high melting points.

The final tolerances in mechanically pressed and sintered components are comparable to those achievable by machining on production machine tools (Tables 7.2 and 7.3). Tolerances in the axial direction, die-fill direction, are usually about ±2% of the dimension. Closer tolerances in the diameter of circular sections are usually possible and can be less than ±0.5% of the diameter.

On a unit weight basis, powdered metals are considerably more expensive than bulk wrought or cast materials. However, the absence of scrap, elimination of machining, the fewer production steps, and the higher rates of production often offset the higher material cost. Dies needed for mechanical pressing are also an expensive item in P/M techniques. Production volumes of less than 10,000 parts are usually not practical for mechanically pressed parts. When HIP is utilized to produce relatively large parts using materials that are difficult to form by other techniques, production runs as low as 20 parts could be economical.

7.8.1 Guidelines for Design

Unlike forging or casting processes, mechanical compaction of powders is restricted to two dimensions. It is impractical to apply pressure to the sides of mechanical dies, thus the flow of powders during compaction is almost entirely axial. It is also necessary to be able to eject the compact. These limitations give rise to certain design rules, which have been established by the Powder Metallurgy Parts Association and Metal Powder Industries Federation. These rules can be summarized as follows:

1. The shape of the part must permit ejection from the die (Figure 7.7a).
2. Parts with straight walls are preferred. No draft is required for ejection from lubricated dies.
3. Parts with undercuts or holes at right angles to the direction of pressing cannot be made (Figure 7.7b).
4. Straight serrations can be made easily but diamond knurls cannot (Figure 7.7c).
5. The shape of the part should be such that the powder is not required to flow into thin walls, narrow splines, or sharp corners. Sidewalls should be thicker than 0.75 mm (0.030 in.).
6. The shape of the part should permit the construction of strong tooling, and punches should have no sharp or feather edges (Figure 7.7d).
7. The part should be designed with as few changes in diameter and section thickness as possible.
8. Since pressure is not transmitted uniformly through a deep bed of powder, the length/diameter ratio of a mechanically pressed part should not exceed about 2.5:1.
9. Take advantage of the fact that certain materials, such as cermets and porous components, which are impossible, impractical, or uneconomical to obtain by any other method can be produced by P/M.
10. P/M parts may be bonded by assembling in the green condition and then sintering together to form a bonded assembly. Other joining methods are also possible and can be used to join P/M parts to castings or forgings.

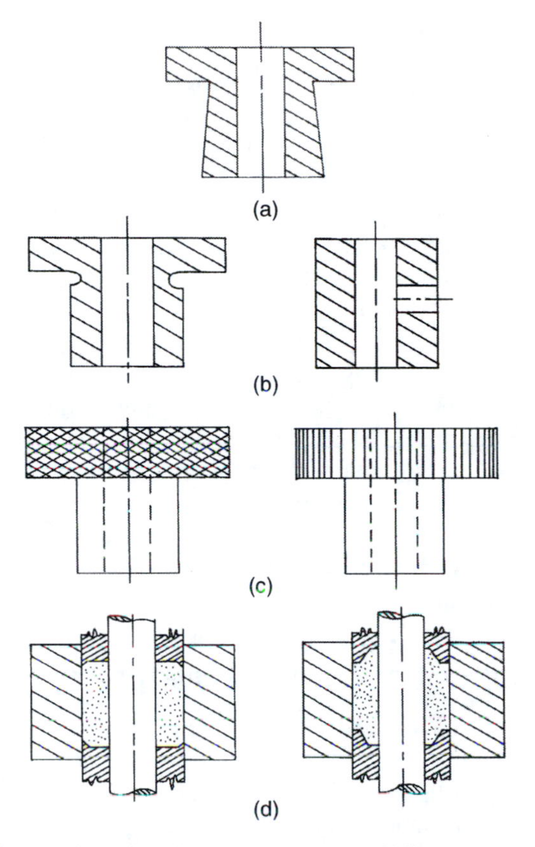

FIGURE 7.7 Design considerations for P/M components. (a) Reverse taper should be avoided, use parallel sides and machine the required taper after sintering; (b) undercuts and holes at right angles to pressing direction should be avoided; if necessary, such features are introduced by machining and sintering; (c) diamond knurls should be replaced by straight serrations; (d) part shapes that require featheredges on punches and similar weakening features should be avoided.

7.9 DESIGN OF SHEET METAL PARTS

Parts made from sheet metal cover a wide variety of shapes, sizes, and materials. Many examples are found in automotive, aircraft, and consumer industries. Generally, sheet metal parts are produced by shearing, bending, or drawing. The quality of the sheet material plays an important role in determining the quality of the finished product, as well as the life of tools and the economics of the process. Grain size of the sheet material is important and should be closely controlled. Steel of 0.035–0.040 mm (0.001–0.0016 in.) grain size is generally acceptable for deep drawing applications. When formability is the main requirement in a sheet material, drawing-quality low-carbon steels represent the most economic alternative. However, with the increasing demand for lighter energy-conserving products, other steel grades are increasingly used to make thinner and stronger components. Examples include control-rolled high-strength low-alloy (HSLA) steels and dual-phase steels.

7.9.1 Guidelines for Design

Metal sheets are usually anisotropic, which means that their strength and ductility vary when measured at different angles with respect to the rolling direction. This anisotropy is caused by the elongated inclusions, that is, stringers, and by preferred orientation in the grains, that is, texture.

The most important factor that should be considered when designing parts that are to be made by bending is bendability. This is related to the ductility of the material and is expressed in terms of the smallest bend radius that does not crack the material. Bendability of a sheet is usually expressed in multiples of sheet thickness, such as $2T$, $3T$, $4T$, etc. A $2T$ material has greater bendability than a $3T$ material. Because of anisotropy, bendability of a sheet is usually greater when tested such that the line of bend is at right angles to the rolling direction of the sheet.

Another factor that should be considered when designing for bending is springback, which is caused by the elastic recovery of the material when the bending forces are removed. One way of compensating for springback is to overbend the sheet. Another method is bottoming, which eliminates the elastic recovery by subjecting the bend area to high localized stresses. A tolerance of ±0.8 mm (±1/32 in.) or more should be allowed in bent parts.

7.10 DESIGNS INVOLVING JOINING PROCESSES

Joining can be considered as a method of assembly, where parts made by other processes are joined to make more complex shapes or larger structures. In this respect, joining extends the capabilities of processes like casting, forging, and sheet metal working; it allows the manufacture of products like machine frames, steel structures, motorcar bodies, beverage and food cans, storage tanks, and piping systems. Some joints are temporary and can be dismantled easily, as in the case of bolted joints, whereas others provide permanent assembly of joined parts, as in the case of rivets and weldments.

Normally, the major function of a joint is to transmit stresses from one part to another, and in such cases the strength of the joint should be sufficient to carry the expected service loads. In some applications, tightness of the joint is also necessary to prevent leakage. Because joints represent areas of discontinuities in the assembly, they should be located in low-stress regions—especially in dynamically loaded structures. Other design considerations that are applicable to the main joining processes are discussed in the following paragraphs.

7.10.1 Welding

Welding is defined by the American Welding Society (AWS) as a "localized coalescence of metals or nonmetals produced by either heating of the materials to a suitable temperature, with or without the application of pressure, or by the application of pressure alone, and with or without the use of filler metal." The various welding processes have been classified by the AWS as shown in Table 7.4. The different processes have been assigned letter symbols to facilitate their designation. The main factors that distinguish the different processes are (a) the source of the energy used

TABLE 7.4
Common Welding Processes and Their Recommended Use

Process	AWS Designation	Recommended Use				
		Carbon Steels	Low-Alloy Steels	Stainless Steels	Ni and Alloys	Al and Alloys
Fusion Welding						
Arc Welding	***W***					
Shielded metal arc welding	SMAW	a	a	a	a	—
Gas metal arc welding	GMAW	b	b	b	a	e
Pulsed arc	GMAW-P	a	a	a	a	d
Short-circuit arc	GMAW-S	d	d	d	d	—
Gas tungsten arc welding	GTAW	d	d	d	d	e
Flux-cored arc welding	FCAW	b	b	b	e	—
Submerged arc welding	SAW	a	a	a	c	—
Plasma arc welding	PAW	—	—	e	e	g
Stud welding	SW	a	a	a	—	g
Oxyacetylene welding	OAW	e	g	g	g	g
Plastic Welding						
Solid State Welding	***SSW***					
Forge welding	FOW	a	—	—	—	—
Cold welding	CW	c	—	—	—	—
Friction welding	FRW	b	b	b	b	b
Ultrasonic welding	USW	g	g	g	g	f
Explosion welding	EXW	c	e	e	e	e
Resistance Welding	***RW***					
Resistance spot welding	RSW	f	f	f	f	f
Resistance seam welding	RSW	f	f	f	f	f
Projection welding	RPW	f	f	f	f	f
Other Welding Processes						
Laser beam welding	LBW	e	e	e	e	f
Electroslag welding	ESW	i	i	i	i	—
Flash welding	FW	a	a	a	a	a
Induction welding	IW	g	—	—	—	—
Electron beam welding	EBW	a	a	a	a	a
Brazing and Soldering (B&S)						
Dip brazing	DB	f	g	g	g	e
Furnace brazing	FB	a	a	e	a	e
Torch brazing	TB	e	e	e	e	e
Induction brazing	IB	e	e	e	f	g
Dip soldering	DS	g	g	g	g	g
Furnace soldering	FS	g	g	g	g	g
Torch soldering	TS	g	g	g	g	g

Note: a = All thicknesses, b = 3 mm (1/8 in.) and up, c = 6 mm (¼ in.) and up, d = up to 6 mm (¼ in.), e = up to 18 mm (¾ in.), f = up to 6 mm (¼ in.), g = up to 3 mm (1/8 in.), and i = 18 mm (¾ in.) and up.

for welding and (b) the means of protection or cleaning of the welded metal. The processes given in Table 7.4 cover a wide range, which makes it possible to find an efficient and economic means of welding of almost all industrial metallic systems.

Welding has replaced riveting in many applications including steel structures, boilers, tanks, and motorcar chassis. This is because riveting is less versatile and always requires lap joints. Also, the holes and rivets subtract from strength, and a riveted joint can only be about 85% as strong, whereas a welded joint can be as strong as the parent metal. Welded joints are easier to inspect and can be made gas- and liquid-tight without the caulking, which has to be done in riveted joints. On the negative side, however, is that structures produced by welding are monolithic and behave as one piece. This could adversely affect the fracture behavior of the structure. For example, a crack in one piece of a multipiece riveted structure may not be serious, as it will seldom progress beyond that piece without detection. However, in the case of a welded structure, a crack that starts in a single plate or weld may progress for a large distance and cause complete failure. This is illustrated in Figure 4.6, which shows the catastrophic failure of the welded hull of a steel ship.

Another factor that should be considered when designing a welded structure is the effect of size on the energy-absorption ability of steels. A charpy impact specimen could show a much lower brittle–ductile transition temperature than a large welded structure made of the same material. The liberty ships that were made during the Second World War are examples of monolithic structures that were made out of unsuitable steel. Many of these ships failed, some while in harbor, as a result of one crack propagating across the whole structure. Thus the notch–ductility of the steel that is to be used in large welded structures should be carefully assessed.

7.10.1.1 Weldability of Materials

In fusion welding processes, the molten filler metal solidifies quite rapidly by heat conduction into the metal adjacent to the weld. Columnar grains are usually present in the weld bead while the base metal closest to it undergoes a considerable overheating and grain growth. This latter area, the heat affected zone (HAZ), is usually a source of failure in welded components. Thermal contraction of welded metals may cause residual stresses and distortions. Preheating of joints is an effective method of reducing the cooling rate of the weld, which reduces distortion and residual stresses. Postwelding heat treatment can be used to relieve internal stresses and to control the microstructure of the weld area. The need for preheating and postwelding heat treatment depends primarily on the weldability of the welded metal. Weldability can be considered to have two components:

1. Fabrication weldability, which is related to the ease with which a material can be welded
2. Service weldability, which is related to the ability of the process–material combination to form a weld that will perform the intended job successfully

In general, weldability of steel decreases as hardenability increases, because higher hardenability promotes the formation of microstructures that are more sensitive to cracking. Higher hardenability means more possibility of forming brittle martensite, which cannot withstand the shrinkage strains in the weld zone. Hydrogen-induced cracking is also more prevalent in welding of hardenable steels than in welding of low-carbon steels. Proper preheat, high-heat input, and maintenance of adequate interpass temperatures reduce the rate of cooling in the HAZ and this results in a softer, less-sensitive microstructure. The HAZ may also be softened by postweld heat treatment in the range of 480–670°C (895–1240°F). The carbon equivalent (CE) is often used to estimate the weldability of hardenable carbon and alloy steels. In this approach, the significant composition variables are reduced to a single number, CE, using one of several similar formulas as follows:

$$CE = \%C + \%Mn/6 + (\%Cr + \%Mo + \%V)/5 + (\%Si + \%Ni + \%Cu)/15 \quad (7.1)$$

Steels with CE less than 0.35% usually require no preheating or postheating. Steels with CE between 0.35 and 0.55% usually require preheating, and steels with CE greater than 0.55% may require both preheating and postheating. In addition to CE, other factors such as hydrogen level, restraint, and thickness must be considered simultaneously in relation to a specific application.

7.10.1.2 Tolerances in Welded Joints

Welding jigs and fixtures are frequently used in production to reduce distortion, warping, and buckling of the welded parts. The use of jigs also increases productivity, reduces costs, and results in higher accuracy. Typical dimensional tolerances that may be held on average weldments are

- 3 mm for small parts with little welding
- 6 mm for moderate-sized parts with a small amount of welding
- 9 mm for large parts with a moderate amount of welding
- 9–12 mm for large parts with a large amount of welding

7.10.1.3 Guidelines for Design of Weldments

In addition to cracking and residual stresses, defects like porosity, slag inclusions, incomplete fusion, and incorrect weld profile can also exist in the welded joint. Such defects can be eliminated by following the correct welding procedure and selecting the appropriate technique. Generally, strict quality control and nondestructive testing are essential if welding defects are to be eliminated and high reliability of welded joints is to be maintained. Other rules that should be considered when designing a welded structure include the following:

1. Welded structures and joints should be designed to have sufficient flexibility. Structures that are too rigid do not allow shrinkage of the weld metal, restrict the ability to redistribute stresses, and are subject to distortions and failure.
2. Accessibility of the joint for welding, welding position, and component matchup are the important elements of design.

3. Thin sections are easier to weld than thick ones.
4. Welded sections should be about the same thickness to avoid excessive heat distortion.
5. It is better to locate welded joints symmetrically around the axis of an assembly to reduce distortion.
6. If possible, welded joints should be placed away from the surfaces to be machined. Hard spots in the weld can damage the cutting tools.
7. An inaccessible enclosure in a weldment, or the mating surfaces of a lap joint, should be completely sealed to avoid corrosion.
8. Where strength requirements are not critical, short intermittent welds are preferable to long continuous ones as distortion is reduced.
9. Help shrinkage forces to work in the desired direction by presetting the welded parts out of position before welding so that shrinkage forces will bring them into alignment.
10. Use weld fixtures and clamps to reduce distortion.
11. Whenever possible, meeting of several welds should be avoided.
12. Balance shrinkage forces in a butt joint by welding alternately on each side.
13. Remove shrinkage forces by heat treatment or by shot peening.
14. Tolerances of ±1.5 mm (ca. ±1/16 in.) are possible in welded joints. Surfaces that need closer tolerances should be finished by machining after welding and postwelding heat treatment.
15. Parts that have been designed for casting or forging should be redesigned if they are to be made by welding. The new design should take advantage of the benefits of welding and avoid its limitations.

7.10.1.4 Types of Welded Joints

Metal plates can be joined by welding in five main types of joints, as shown in Figure 7.8. Lap, tee, and corner joints use fillet-type welds, as shown in Figure 7.9. Welding of thick plates requires edge preparation to ensure complete penetration. In such cases, one or both of the edges to be welded are chamfered in such a way as to minimize the amount of weld metal deposited. This is because the cost per unit weight of deposited weld metal is about 25–50 times as much as structural steel. In addition, the amount of shrinkage and distortion increases as the amount of deposited metal increases.

7.10.1.5 Strength of Welded Joints

Full-penetration butt welds are generally considered to have the same strength as the base metal. Hence there is no need to calculate the strength of the weld if the deposited metal is the same as the base metal. The strength of a fillet weld is inherently lower than a full-penetration butt weld. When the applied load is parallel to the weld line, the plane of rupture is at 45°, weld throat. The AWS code gives the allowable force per unit length of the weld as 30% of the tensile strength, S, of the

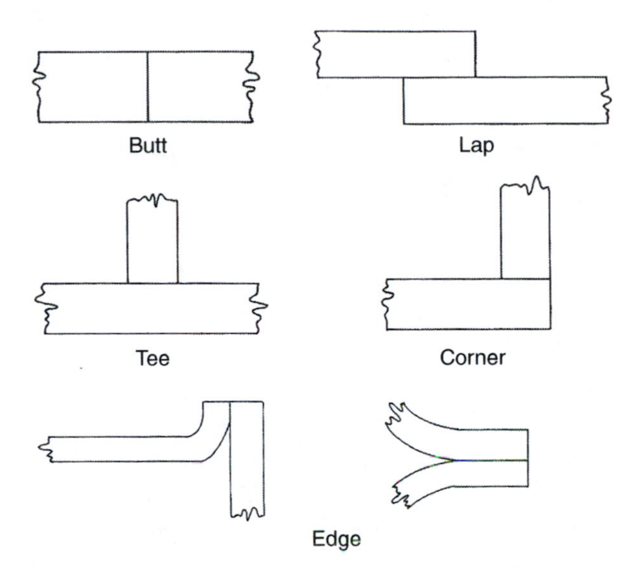

FIGURE 7.8 Types of welded joints.

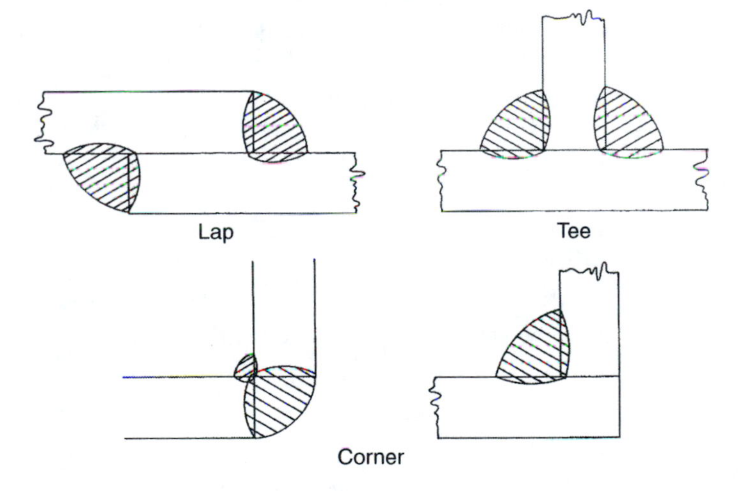

FIGURE 7.9 Use of fillet-type welds in lap, tee, and corner joints.

welding electrode. Thus, the load carrying capacity, P, of the two fillet welds shown in Figure 7.10 is

$$P = 2 \times 0.30S \times 0.707t \times L \tag{7.2}$$

where

L = Length of weld
t = Leg of weld—in this case, same as plate thickness

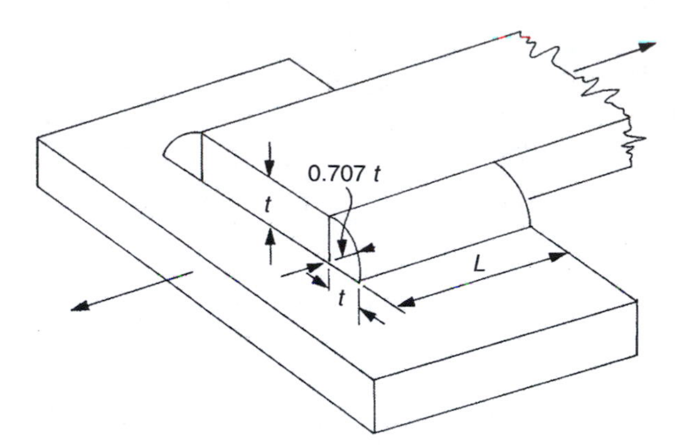

FIGURE 7.10 Parameters involved in calculating the load-carrying capacity of lap joints.

Design Example 7.2: Design of a Welded Joint

Problem

An AISI 1020 steel angle of dimensions 150 mm × 150 mm × 12 mm (6 in. × 4 in. × 1/2 in.) is to be welded to a steel plate by fillet welds along the edges of the 150 mm leg. The angle should support a load of 270 kN (60,000 lb) acting along its length. Determine the lengths of the welds to be specified. The welding electrode used is AWS–AISI E6012 with a tensile strength of 414 MPa (60 ksi).

Solution

From Equation 7.2,

$$270{,}000 = 2 \times 0.3 \times 414 \times 0.707 \times 12 \times L$$

where

L is 128 mm on either side parallel to the axis of the angle.

7.10.2 Adhesive Bonding

Adhesives represent an attractive method of joining and their use is increasing in many applications. Some of the main advantages in using adhesives are the following:

1. Thin sheets and parts of dissimilar thicknesses can be easily bonded.
2. Adhesive bonding is the most logical method of joining polymer–matrix composites.
3. Dissimilar or incompatible materials can be bonded.
4. Adhesives are electrical insulators and can prevent galvanic action in joints between dissimilar metals.
5. Flexible adhesives spread bonding stresses over wide areas and accommodate differential thermal expansion.

6. Flexible adhesives can absorb shocks and vibrations, which increases fatigue life.
7. Preparation of bonded joints requires no fastener holes, which gives better structural integrity and allows thinner-gage materials to be used.
8. Adhesives provide sealing action in addition to bonding.
9. The absence of screw heads, rivet heads, or weld beads in adhesive-bonded joints is advantageous in applications where interruption of fluid flow cannot be tolerated or where appearance is important.
10. Adhesive bonding can also be used in conjunction with other mechanical fastening methods to improve the strength of the joint.

The main limitations of adhesives are the following:

1. Bonded joints are weaker under cleavage and peel loading than under tension or shear.
2. Most adhesives cannot be used at service temperatures above 300°C (ca. 600°F).
3. Solvents can attack adhesive-bonded joints.
4. Some adhesives are attacked by UV light, water, and ozone.
5. The designer should also be aware of the adhesive's impact resistance and creep, or cold flow, strength.

7.10.2.1 Design of Adhesive Joints

The strength of the adhesive joint depends on the joint geometry, the direction of loading in relation to the joint, the adhesive material, surface preparation, and application and curing technique. As the strength of an adhesive joint is limited by the bonded area, lap and double strap joints are generally preferred to butt joints. If the geometry constraints do not allow for such joints, scarf or double-scarf joints should be made, as shown in Figure 7.11a.

When a lap joint is used to bond thin sections, tensile shear causes deflection, and this results in stress concentration at the end of the lap. Tapering the ends of the joint, as shown in Figure 7.11b, gives more uniform loading throughout the joint. Since adhesive joints are weaker under pealing forces, joint design should avoid this type of loading.

7.11 DESIGNS INVOLVING HEAT TREATMENT

Heat treatment represents an important step in the sequence of processes that are usually performed in the manufacture of metallic parts. Almost all ferrous and many nonferrous alloys can be heat treated to achieve certain desired properties. Heat treatment can be used to make the material hard and brittle, as in the case of quench-hardening of steels, or it can be used to make it soft and ductile, as in the case of annealing.

Generally, hardening of steels involves heating to the austenitic temperature range, usually 750–900°C (ca. 1400–1650°F), and then quenching to form the hard martensitic phase. The nonuniform temperature distribution that occurs during quenching and the volume change that accompanies the martensitic transformation

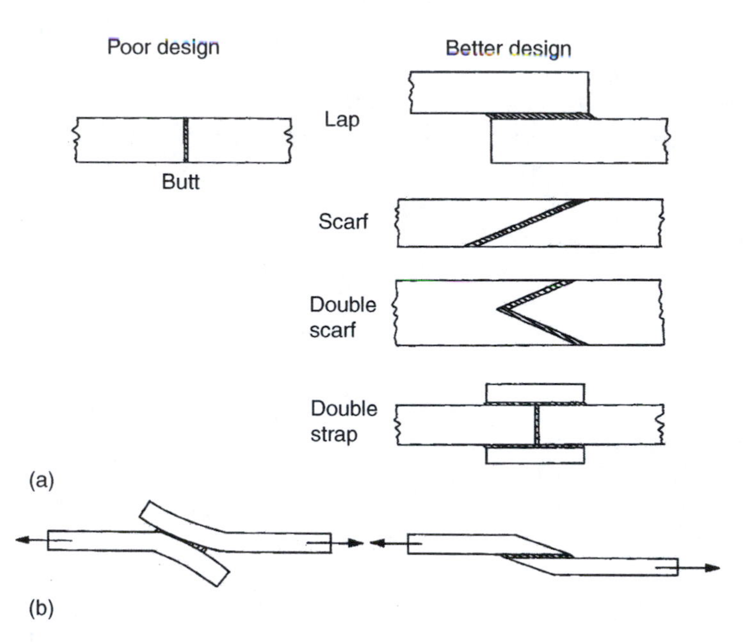

FIGURE 7.11 Adhesive joint design. Butt joint is weak and the bond area should be increased by using lap, scarf, double-scarf, or double-strap joints.

can combine to cause distortions, internal stresses, and even cracks in the heat-treated part. Internal stresses can cause warping or dimensional changes when the quenched part is subsequently machined or can combine with externally applied stresses to cause failure. Corrosion problems can also be aggravated due to the presence of internal stresses. These difficulties can be reduced or eliminated by selecting steels with high hardenability as they require a less severe cooling rate to achieve a given hardness value. Manganese, chromium, and molybdenum are commonly added to steels to increase their hardenability.

7.12 DESIGNS INVOLVING MACHINING PROCESSES

Machining operations are the most versatile and the most common manufacturing processes. Machining could be the only operation involved in the manufacture of a component, as in the case of shafts and bolts that are machined from bar stock, or it could be used as a finishing process, as in the case of cast and forged components. In all cases, it is important for the designer to ensure that the component will be machined conveniently and economically.

7.12.1 Machinability Index

As machining is relatively expensive, it should not be performed unless necessary, and tolerances that are closer than necessary should not be specified. The economics of metal cutting can be improved by using high cutting speeds and tools with long lives. If the material to be cut gives discontinuous chips and needs less power for cutting, the economics are further improved. The ease with which one or more of the discussed factors can be realized for a given material is taken as a measure of its machinability. Thus, a material with good machinability is one that requires less power consumption, causes less tool wear, and easily acquires a good surface finish. One of the methods for comparing machinability of materials is to determine the relative power required to cut them using single-point tools. Another method is to use the machinability index, which is defined as follows:

$$\text{Machinability index \%} = \frac{\text{Cutting speed of material for 20-min tool life} \times 100}{\text{Cutting speed of SALE 1112 steel for 20-min tool life}} \tag{7.3}$$

In this definition, the free machining steels SAE 1112 or AISI B 1112 is taken as the standard and its machinability index is arbitrarily fixed at 100%. The higher the machinability index, the easier and the more economical it is to finish the material by metal cutting. The machinability index of some common metallic materials is given in Table 7.5.

7.12.2 Guidelines for Design

The following discussion illustrates some component shapes and features that can cause difficulties in machining, take undue length of time to machine, call for precision and skill that may not be available, or that may even be impossible to machine by standard machining and cutting tools.

1. The workpiece must have a reference surface that is suitable for holding it on the machine tool or in a fixture. This could be a flat base or a cylindrical surface. If the final shape does not have such a surface, a supporting foot or tab could be added to the rough casting or forging for support purposes, and removed from the part after machining.
2. Whenever possible, the design should allow all the machining operations to be completed without resetting or reclamping.
3. Whenever possible, the radii between the different machined surfaces should be equal to the nose radius of the cutting tool.
4. If the part is to be machined by traditional cutting methods, deflection under cutting forces should be taken into account. For the same cutting force, the deflection is higher for thinner parts and for lower elastic moduli. Under these conditions, some means of support is necessary to ensure the accuracy of the machined part.
5. Twist drills should enter and exit at right angles to the drilled surface; Figure 7.12. Drilling at an angle to the surface causes deflection of the drill and could break it.
6. Features at an angle to the main machining direction should be avoided as they may require special attachments or tooling; Figure 7.12.
7. Flat-bottom drilled holes should be avoided as they involve additional operations and the use of bottoming tool.
8. To reduce the cost of machining, machined areas should be kept to a minimum. Two examples of methods of reducing the machined area are shown in Figure 7.13.

TABLE 7.5
Machinability Index of Some Common Metallic Materials

Material	Hardness (BHN)	Machinability Index
Steels		
AISI		
1015	121	50
1020	131	65
1030	149	65
1040	170	60
1050	217	50
1112	120	100
1118	143	80
1340	248	65
3140	262	55
4130	197	65
4340	363	45
18-8 stainless steel	150–160	25
Cast Irons		
Gray cast iron:		
soft	160–193	80
medium	193–220	65
hard	220–240	50
Malleable iron	110–145	120
Nonferrous Alloys		
Aluminum alloys	35–150	300–2000
Bronze	55–210	150–500
Magnesium alloys	50–75	500–2000
Zinc alloys	80–90	200

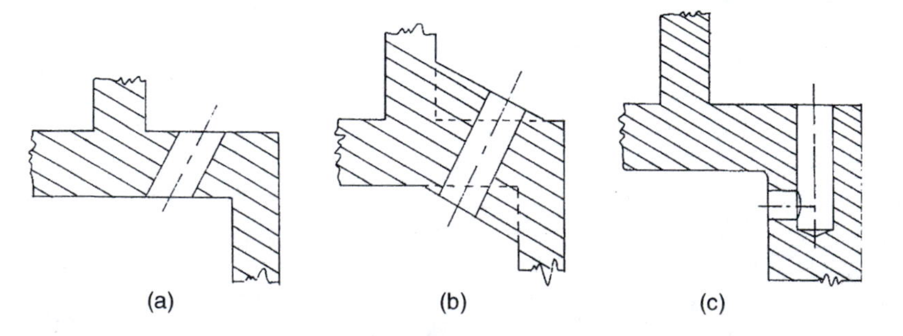

FIGURE 7.12 Design of drilled part: (a) Poor design as drill enters and exits at an angle to the surface; (b) better design, but drilling the hole needs a special attachment; (c) best design.

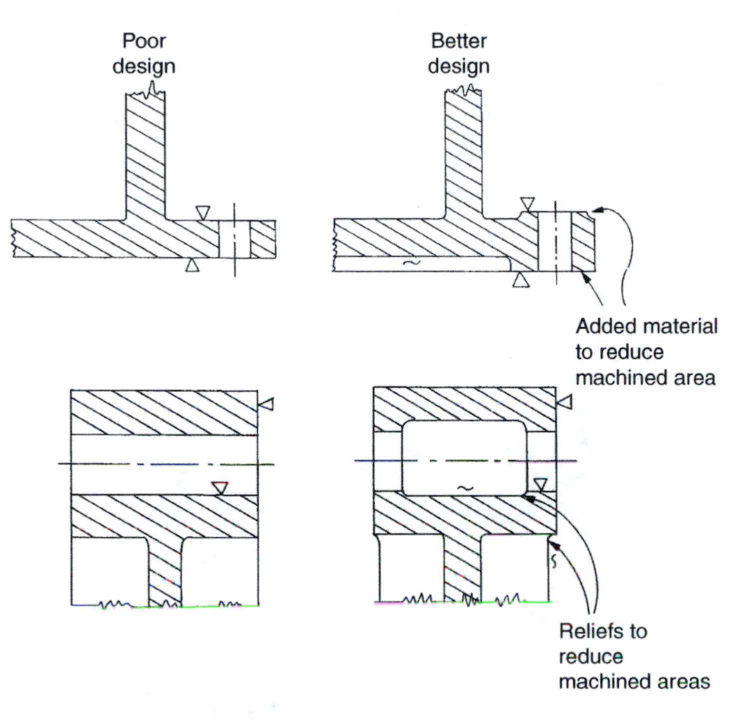

FIGURE 7.13 Some design details that can be introduced to reduce machining cost.

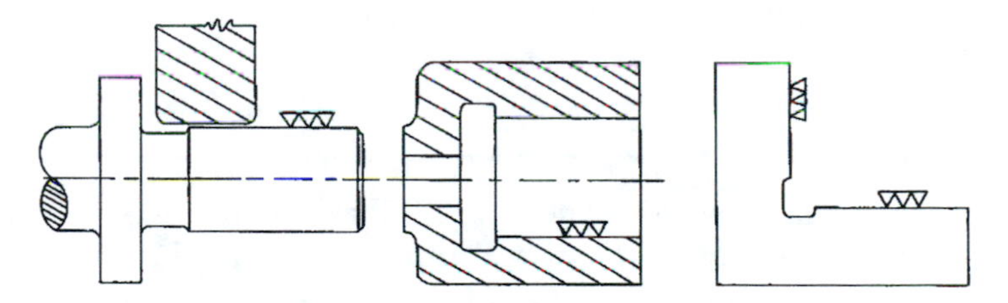

FIGURE 7.14 Some design details that can be introduced to give runout space for grinding wheels.

9. Cutting tools often require runout space as they cannot be retracted immediately. This is particularly important in the case of grinding, where the edges of the grinding wheel wear out faster than the center. Figure 7.14 gives some examples to illustrate this point.
10. Thread-cutting tools normally have a chamfer on their leading edge. This chamfer means that the first two pitches do not cut a full thread. If an external diameter ends at a shoulder, the mating screwed part cannot reach the shoulder unless an undercut or a countersink is provided, as shown in Figure 7.15. The length of the needed undercut or countersink is usually three pitches of the thread. Similar features are needed for internal screw threads, as shown in Figure 7.16.

Example 7.3 illustrates how a part can be redesigned to facilitate machining.

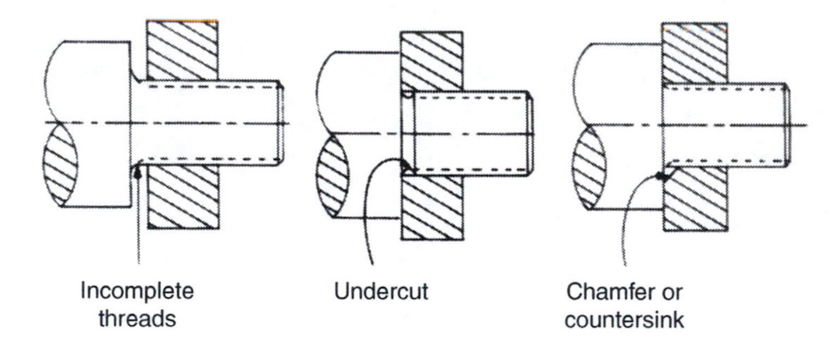

FIGURE 7.15 Some design details to account for the incomplete threads at the end of external screws.

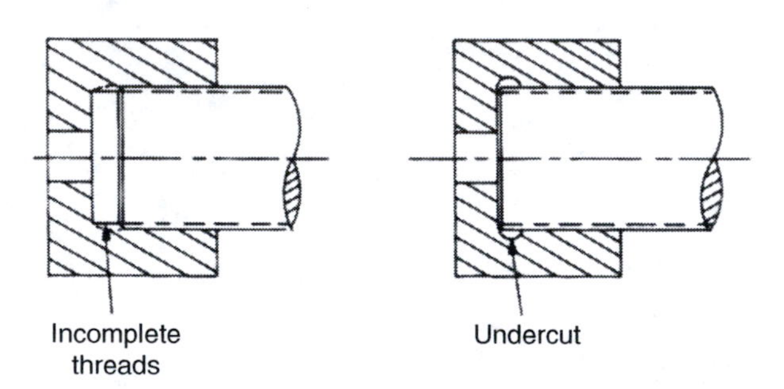

FIGURE 7.16 Some design details to account for the incomplete threads at the end of internal screws.

Design Example 7.3: Redesign of a Shaft Support Bracket (Based on Stephenson, Figures Reprinted with Permission of ASM International)

Problem

Figure 7.17 shows the initial design of the shaft support bracket that is intended to be bolted to a housing wall to provide support and lubrication to a rotating shaft. Accurate machining is needed for the bore, and high tolerance is expected in the location of bore relative to the dowel holes. The support bracket is made of nodular cast iron and large numbers will be machined on a horizontal spindle computer numerical control (CNC) machining center.

Analysis

The initial design had the following features that made it difficult to machine:

- Different diameters for the dowels and bolt holes, which require tool change and loss of time
- The bore and oil hole are long relative to their diameter, which require long processing steps

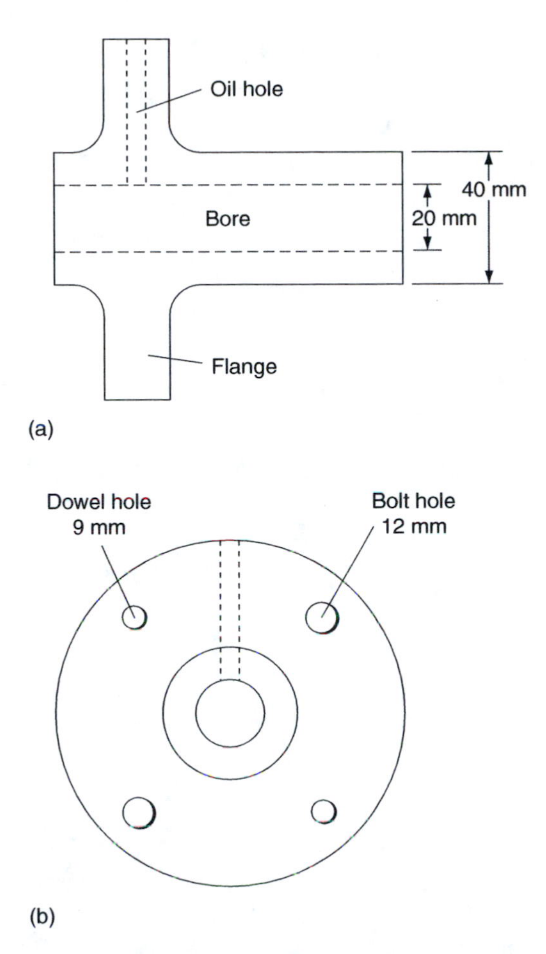

FIGURE 7.17 Side (a) and end (b) views of the initial design for a shaft support bracket. (Reprinted from Dieter, G.E., in *ASM Handbook Vol. 20 Materials Selection and Design, Materials Seletion and Design, Performance Indices pp. 281–290, Manufacture and Assembly Design pp. 676–686*, ASM International, Materials Park, OH, 1997. With permission.)

- There is no obvious features on the outer surface to fix the part and prevent rotation during machining

Solution

To avoid the drawbacks of the initial design, the part was redesigned as shown in Figure 7.18. The following features were changed:

- The dowels and bolt holes have the same diameter.
- The center of the bore has a larger diameter than the ends to reduce the length to be machined. This will also eliminate the necessity to the exit burr at the end of the oil hole.

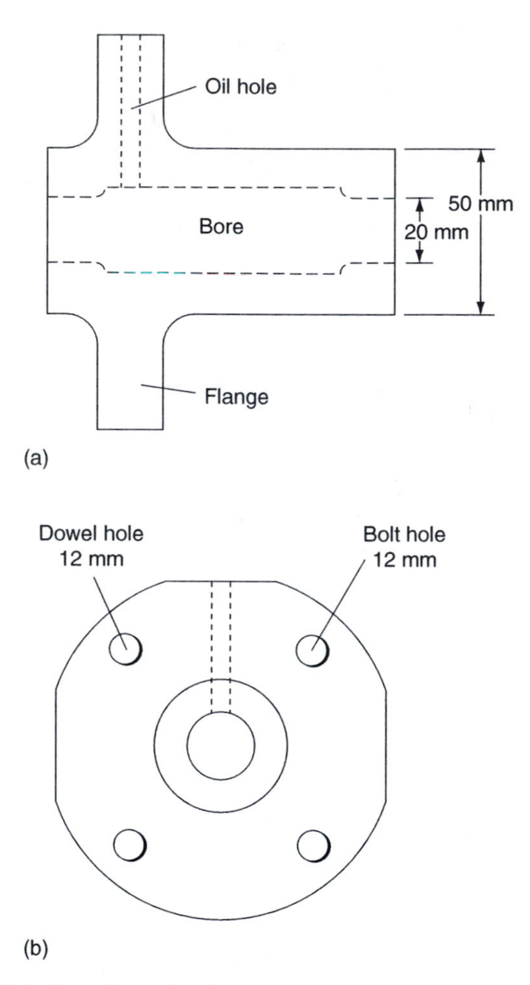

FIGURE 7.18 Side (a) and end (b) views of a shaft support bracket redesigned to simplify machining. (Reprinted from Dieter, G.E., in *ASM Handbook Vol. 20 Materials Selection and Design, Materials Seletion and Design, Performance Indices pp. 281–290, Manufacture and Assembly Design pp. 676–686*, ASM International, Materials Park, OH, 1997. With permission.)

- The length of the oil hole is reduced.
- Flat surfaces were cast on outer surfaces, making it possible to locate the part and to use less fixing force while machining.

Conclusion

According to Stephenson, these changes made it possible to reduce the machining time from 173 to 119 s, which is about 33%. The changes are also expected to improve the quality by making it possible to achieve the required tolerances.

7.13 SUMMARY

1. As the design progresses from concept to configuration and the material choices get narrower, manufacturing processes, which have initially been broadly defined, also need to be better identified. The compatibility between the candidate materials and the manufacturing processes is used to narrow down the available alternatives.
2. DFMA seeks to minimize the cost of a product through DFM and DFA.
3. Casting is particularly suited for parts that contain internal cavities that are inaccessible, too complex, or too large to be easily produced by machining. Cast parts are isotropic but could contain shrinkage cavities if not designed well.
4. Compression, transfer, and injection molding processes are commonly used for molding plastic parts. Incorrect design could lead to internal stresses and warpage in the finished product.
5. Forged parts have wrought structures that are usually stronger and more ductile than cast products. However, cast products are more isotropic. Rapid changes in thickness of forged components could result in cracks and surface laps.
6. P/M techniques can be used to produce a large number of small parts to the final shape in few steps, with little or no machining, and at high rates. Many metallic alloys, ceramic materials, and particulate-reinforced composites can be processed by P/M techniques. Porosity is normally undesirable and can be reduced by hot isostatic pressing.
7. Welding has replaced riveting in many applications including steel structures, boilers, and motorcar chassis. However, welded structures are monolithic and can suffer catastrophic failure. Because they represent areas of discontinuities, welded joints should be located away from highly stressed regions, especially in dynamically loaded structures.
8. Adhesives represent an attractive method of joining and are increasingly used for thin sheets, polymer–matrix composites, and dissimilar or incompatible materials. In addition, adhesives are electrically insulating, which can prevent galvanic corrosion in joints between dissimilar metals. However, they are relatively weaker and can be attacked by organic solvents.
9. Almost all ferrous and many nonferrous alloys can be heat treated to achieve certain desired properties. Heat treatment can be used to make the material hard and brittle, as in the case of quench-hardening of steels, or it can be used to make it soft and ductile, as in the case of annealing.
10. Machining operations are the most versatile and the most common manufacturing processes. Machining could be the only operation involved in the manufacture of a component, as in the case of shafts and bolts that are machined from bar stock, or it could be used as a finishing process, as in the case of cast and forged components.

7.14 REVIEW QUESTIONS

7.1 Recommend materials and manufacturing processes, illustrated by sketches, for the following products: (a) plastic bottle for mineral water, (b) fiberglass bath tub, (c) inner lining of a refrigerator door, and (d) steel bars for reinforcing concrete structures.

7.2 Indicate how the treatment or the composition of the following pairs of materials is expected to influence their mechanical behavior. Use (H) for higher and (L) for lower. For example, YS of AA 2014 O is lower than AA 2014 T6.

Material	YS	UTS	Elastic Modulus	Ductility	Hardness	Toughness
AA 2014 O	L					
AA 2014 T6	H					
AISI 1015						
AISI 1040						
AISI 1060 as quenched						
AISI 1060 quenched and tempered						
Low-density polyethylene						
High-density polyethylene						

7.3 Suggest the sequence of primary, secondary, and finishing manufacturing processes for the crankshaft. Draw details of the crankshaft end that will ensure prolonged fatigue life.

7.4 What are the main material requirements and manufacturing processes for the following products: (a) railway line, (b) electrical resistance heater, and (c) small passenger airplane wing structure.

7.5 Suggest possible manufacturing processes for the following items in an internal combustion engine: (a) piston, (b) connecting rod, (c) cylinder head, and (d) camshaft.

7.6 From the manufacturing point of view, what are the main attractive features of plastics in comparison with metals?

7.7 Recommend suitable plastics and manufacturing processes for the following products: (a) telephone, (b) 2-L (0.5-gal) lubricating oil container, (c) safety shield for a mechanical press, and (d) hard hat for construction workers.

7.8 In making a milling machine frame, gray cast iron and steel AISI 1015 were considered as candidate materials. Compare the use of the two materials indicating the advantages and disadvantages of using each of them in this application. What are the most suitable manufacturing processes in each case?

7.9 It is required to select a material and manufacturing processes for tie-rods of a suspension bridge. The rods are 10 m long and should carry a tensile load of 50 kN without yielding. The maximum extension should not exceed

18 mm. Which one of the materials listed in the following table will give (a) the lightest rod and (b) the least-expensive rod? Recommend the manufacturing processes for the materials selected in (a) and (b).

Material	Yield Strength (MPa)	Elastic Modulus (GPa)	Specific Gravity	Relative Cost/kg
ASTM A675 grade 60	205	212	7.8	1.00
High-strength steel	485	212	7.8	1.50
Aluminum 5052 H38	259	70.8	2.7	5.00
Polyester—65% glass fibers	340	19.6	1.8	10.00

7.10 What are the problems that are likely to arise when heat treating a steel part of nonuniform sections? How are these problems overcome?

7.11 What are the advantages of casting in comparison with welding in terms of flexibility of shape design?

7.12 What are the advantages of P/M in comparison with casting when manufacturing small gears?

7.13 An AISI 1020 steel angle of dimensions 150 mm × 100 mm × 12 mm (6 in. × 4 in. × 1/2 in.) is to be welded to a steel plate by fillet welds along the edges of the 150 mm leg. The angle should support a load of 270 kN (60,000 lb) acting along its length. Determine the lengths of the welds to be specified. The welding electrode used is AWS-AISI E6012 with a tensile strength of 414 MPa (60 ksi). (Answer: 128 mm on each side parallel to the axis of the angle)

7.14 Compare welding and casting as methods of fabrication of 500 mm (20 in.) diameter gears. The total number required is 10 units.

7.15 Compare the use of spot welding and adhesive bonding in the assembly of the steel metal components of motorcar bodies.

7.16 Brazing alloys and adhesives are known to be relatively weak. Suggest methods of strengthening joints made by these techniques.

7.17 It is required to produce a folding chair for use at the seashore. Draw neat sketches of the folding chair showing the different elements. Suggest candidate materials and manufacturing processes for each of the elements, as well as method of assembly. It is estimated that 10,000 chairs will be produced per year.

7.18 It is required to produce water storage tanks that are to be placed on top of buildings to supply water for household use. The capacity of the tank is 1 m^3. It is estimated that 1000 tanks will be produced per year. (a) What are the main material requirements for the tank? (b) Suggest some candidate materials for the water storage tank. (c) Draw neat sketches showing the main dimensions to illustrate how the material and manufacturing processes affect the design. (d) Suggest suitable sequence of manufacturing processes for each of the candidate materials.

7.19 Pipes carrying steam at high pressure in a power station are being designed and joined by welding. What are the most important material properties that need to be considered and the quality control measure that need to be taken to ensure that the pipes will not fail in service?

7.20 Suggest suitable manufacturing processes and indicate the main distinguishing characteristics of the materials used in manufacturing the following components: (a) nodular cast iron crankshaft for an internal combustion engine, (b) gas turbine blade made of superalloys, (c) 2-L water bottle made of polyethylene, and (d) cutting tool made of cemented carbides.

BIBLIOGRAPHY AND FURTHER READING

Biles, W.E., Plastic part processing, in *Handbook of Materials Selection*, Kutz, M., Editor. Wiley, New York, 2002, pp. 969–1036.

Boothroyd, G., Design for manufacture and assembly, in *Materials Selection and Design, ASM Handbook*, Vol. 20, Dieter, G.E., Editor. ASM International, Materials Park, OH, 1997, pp. 676–686.

Boothroyd, G., Dewhurst, P., and Knight, W.A., *Product Design for Manufacture and Assembly*, Marcel Dekker, New York, 1994.

DeGarmo, E.P., Black, J.T., and Kohser, R.A., *Manufacturing Processes in Manufacture*, 7th Ed., Macmillan, New York, 1988.

Dieter, G.E., *Engineering Design: A Materials and Processing Approach*, 2nd Ed., McGraw-Hill, New York, 1991.

Dowling, W.E., Design for heat treatment, in *Materials Selection and Design, ASM Handbook*, Vol. 20, Dieter, G.E., Editor. ASM International, Materials Park, OH, 1997, pp. 774–780.

Doyle, L.E., Keyser, C.A., Leach, J.L., Schrader, G.F., and Singer, M.B., *Manufacturing Processes and Materials for Engineers*, 3rd Ed., Prentice-Hall, Englewood Cliffs, NJ, 1985.

Farag, M.M., *Selection of Materials and Manufacturing Processes for Engineering Design*, Prentice Hall, London, 1989.

Farag, M.M., *Materials Selection for Engineering Design*, Prentice Hall, London, 1997.

Ferguson, B.L., Design for deformation processes, in *Materials Selection and Design, ASM Handbook*, Vol. 20, Dieter, G.E., Editor. ASM International, Materials Park, OH, 1997, pp. 730–744.

Kalpakjian, S., *Manufacturing Engineering and Technology*, 3rd Ed., Addison-Wesley, Reading, MA, 1995.

Lindberg, R.A., *Processes and Materials of Manufacture*, 3rd Ed., Allyn and Bacon, Boston, MA, 1983.

Muccio, E.A., Design for plastics processing, in *Materials Selection and Design, ASM Handbook*, Vol. 20, Dieter, G.E., Editor. ASM International, Materials Park, OH, 1997, pp. 793–803.

Piwonka, T.S., Design for casting, in *Materials Selection and Design, ASM Handbook*, Vol. 20, Dieter, G.E., Editor. ASM International, Materials Park, OH, 1997, pp. 723–729.

Sampath, K., Design for joining, in *Materials Selection and Design, ASM Handbook*, Vol. 20, Dieter, G.E., Editor. ASM International, Materials Park, OH, 1997, pp. 762–773.

Sanderow, H.I., Design for powder metallurgy, in *Materials Selection and Design, ASM Handbook*, Vol. 20, Dieter, G.E., Editor. ASM International, Materials Park, OH, 1997, pp. 745–753.

Schey, J.A., Manufacturing processes and their selection, in *Materials Selection and Design, ASM Handbook*, Vol. 20, Dieter, G.E., Editor. ASM International, Materials Park, OH, 1997, pp. 687–704.

Stephenson, D.A., Design for machining, in *Materials Selection and Design, ASM Handbook*, Vol. 20, Dieter, G.E., Editor. ASM International, Materials Park, OH, 1997, pp. 754–761.

Stoll, H.W., Introduction to manufacturing and design, in *Materials Selection and Design, ASM Handbook*, Vol. 20, Dieter, G.E., Editor. ASM International, Materials Park, OH, 1997, pp. 696–675.

Zohdi, M.E. and Biles, W.E., Metal forming, shaping and casting, in *Handbook of Materials Selection*, Kutz, M., Editor. Wiley, New York, 2002, pp. 925–967.

Zohdi, M.E., Biles, W.E., and Webster, D.B., Production processes and equipment for metals, in *Handbook of Materials Selection*, Kutz, M., Editor. Wiley, New York, 2002, pp. 847–923.

Part III

Selection and Substitution of Materials in Industry

Discussion in earlier parts of this book illustrates the interdependence of the various activities involved in developing a concept into a finished product. It is shown in Part II that the materials and processes used in making a component have great influence on its design and performance in service.

In addition to designing and manufacturing products at the required level of quality, organizations must also be able to sell their products at competitive prices and to make profits. To achieve these objectives, it is important that the different materials and processes involved in the design and manufacture should be evaluated in terms of their economic as well as their technical merits. Understanding the elements that make up the cost of a finished product is vital in making this evaluation.

Unlike the exact sciences, where there is normally only one correct solution to a problem, materials selection and substitution decisions require the consideration of conflicting advantages and limitations, necessitating compromises and trade-offs; as a consequence, different satisfactory solutions are possible. This is illustrated by the fact that similar components performing similar functions, but produced by different manufacturers, are often made from different materials and even by different manufacturing processes.

This part of the book integrates both the technical and economic aspects in materials selection and substitution. Chapter 8 gives an introduction to the economics and environmental aspects of materials and processes, and Chapters 9 and 10 explore a variety of procedures that can be followed when selecting or substituting materials for a given product. Chapter 11 gives several detailed case studies drawn from widely different areas to illustrate the use of design calculations together with materials selection and substitution procedures in arriving at the optimum choice of materials and processes.

8 Economics and Environmental Impact of Materials and Processes

8.1 INTRODUCTION

The success of the product in terms of its marketability and competitiveness depends to a large extent on its selling price and acceptance by the society. In a wide range of engineering industries, the cost of materials and manufacturing represents about 30–70% of the product cost and needs to be minimized if the product is to be economically viable.

The cost of materials and processing consists of (a) cost of standard components that are purchased from a supplier, (b) cost of components that are custom designed and made for the product, (c) cost of assembly, and (d) overheads. The cost of standard components can be estimated from previous company experience or from price quotes from suppliers. The cost of custom-made components is determined by the cost of design, materials, processing, and overheads. According to Kalpakjian and Schmid (2001), approximate breakdown of costs in manufacturing of a wide variety of components is 5% for design, 50% for materials, 15% for labor, and 30% for overheads. The factors that affect the cost of materials and processing, which account for 65% of the total cost, are discussed in the following sections of this chapter. The cost of assembly is discussed in Chapter 7. The overheads are usually assigned to a product on the basis of cost elements in (a), (b), and (c).

In addition to selling the product at a competitive price, manufacturers are now attempting to gain social acceptance by reducing the environmental impact of their products and complying with the increasingly tighter environmental legislation. These issues are also discussed in this chapter.

The goal of this chapter is to review ways of optimizing the economic and environmental parameters associated with the production of custom-made components in industry. The main objectives are to get a better understanding about

1. elements of the cost of materials and factors affecting their prices,
2. taking cost into account when comparing materials,
3. economics of manufacturing processes,
4. environmental impact of materials and processes, and
5. life cycle cost and recycling economics.

8.2 ELEMENTS OF THE COST OF MATERIALS

From the economic point of view, engineering materials can generally be classified into two main categories, depending on their cost. The first category contains the commonly used materials, like plain carbon steels and polyethylene, which are manufactured by cost-effective large-scale processes. The second category contains the special or high-performance materials, like the superalloys and silicones, which are manufactured to meet special needs. The materials in the second category are more expensive than the materials in the first, and are only used to meet special requirements that cannot be met by the less-expensive, commonly used materials. This division must not be too rigid because a material developed for a particular application may prove to have properties that eventually lead to its widespread use. For example, aluminum and titanium moved in a few decades from being special and expensive materials to being moderately priced items of everyday industrial use.

Regardless of whether the material used in making a product is common and inexpensive or special and expensive, it is expected to have a considerable influence on the final cost of the product. This is because the cost of materials usually represents a high proportion of the total product cost, and also because material processability affects manufacturing costs. It is, therefore, necessary to analyze the various elements of the cost of materials to find possible means of minimizing it. This can be done by considering the sequence of operations in which raw materials are progressively converted into final products. As an example, Figure 8.1 shows the buildup of the cost of steel and aluminum with the progress of processing operations from ore to finished product. The main elements of the cost of materials can be grouped as follows.

8.2.1 Cost of Ore Preparation

The main elements of the cost of ore preparation are the cost of ore at the mine and the cost of beneficiation. The cost of ore depends on its location and the method of mining it, whereas the cost of beneficiation depends on the concentration of the required material in the ore, the degree of complexity in mineralogical association, and the type of gangue materials. For example, the cost of iron ore preparation is relatively low because commercial ores are usually mined in open-pit mines and contain 50–65% Fe. However, preparation of copper and gold ores is much more expensive as the concentration of the metal is about 1–1.5% and 0.0001–0.001%, respectively.

The cost of transporting the ore from the mine to the extraction site can be considerable and can be reduced by performing the beneficiation process at the mine. Regardless, ores are known to be transported across countries and continents and the cost can be considerable.

8.2.2 Cost of Extraction from the Ore

The main elements of the cost of extraction include the cost of power and the cost of auxiliary materials. In metals, the more stable the compound in which the element is found, the greater is the amount of energy and the cost needed for reduction. For example, the large amount of electric power and the high cost of auxiliary materials needed for extraction of aluminum are the main reasons for its high cost, as can be seen in Figure 8.1.

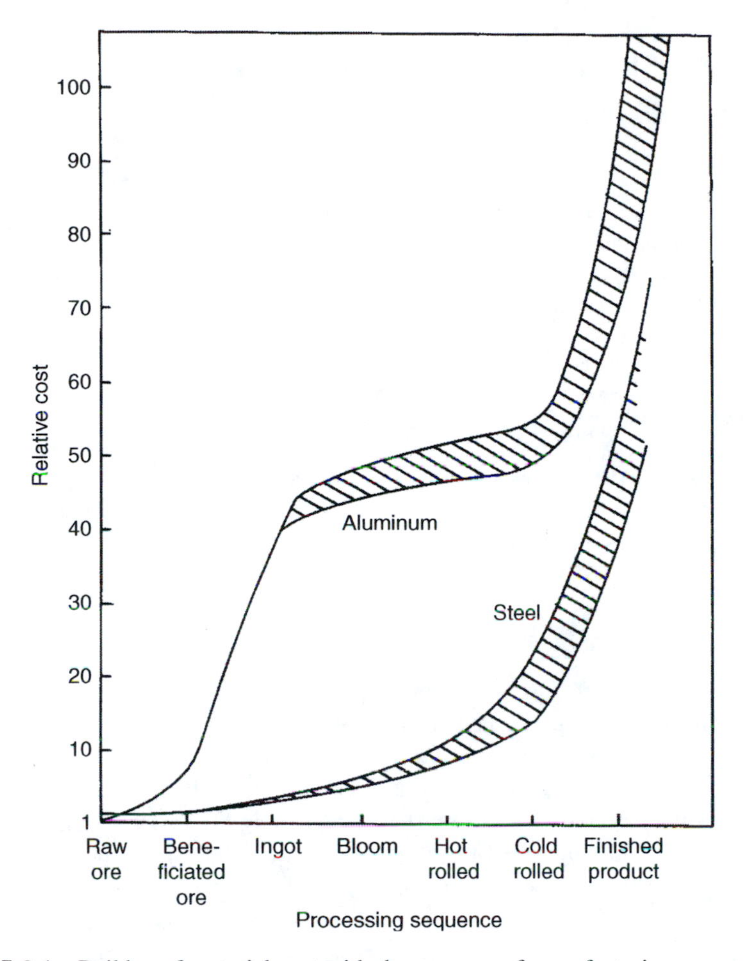

FIGURE 8.1 Buildup of material cost with the progress of manufacturing processes from ore to finished product.

8.2.3 Cost of Purity and Alloying

Impurity level is known to have an important influence on the cost of materials and, in general, the higher the permissible impurity level the lower is the cost. For example, the relative cost of aluminum ingots nearly doubles as the purity increases from 99.5% to 99.99%. The cost of an alloy is not only affected by the purity of the elements used, but also by their nature and by the degree of complexity of alloy structure. The cost of an alloy is not simply the cost of its constituents; because in the majority of cases, more sophisticated techniques of production have to be employed to make full use of the alloying elements. For example, the less-demanding specifications of SAE 326 (LM4) aluminum alloy permit the use of lower purity aluminum as the base metal, which makes its cost about 50% less than the cost of SAE 324 (LM10) alloy that specifies at least 99.7% purity for its base aluminum.

TABLE 8.1
Cost Analysis in a Cast Iron Foundry

Item	Cost ($/casting)	Percentage of Total Cost
Metal (weight of casting 15 kg [33 lb], 55% yield, 12% scrap, sprue)	4.40	35.9
Mold and core	1.80	14.7
Clean and sort	0.53	4.3
Heat treat	0.56	4.5
Hand tool finish and grind	1.30	10.6
Inspect	0.12	1.0
Ship	0.05	0.4
Total direct cost	8.76	71.4
Overheads and administration	3.50	28.6
Total cost	12.26	100.0

8.2.4 Cost of Conversion to Semifinished Products

The cost of converting ingots to semifinished products ready for delivery to the manufacturer of end products includes the costs of casting, forging, and rolling. The main cost elements in this case are labor, energy, overheads, and the cost of material losses. This last element can be considerable in metal industries where material losses can range from 25% to more than 50%. Table 8.1 gives an example of the costs in a cast iron foundry. Although this is a hypothetical case, the figures can be considered as representatives of those items found in industry.

8.2.5 Cost of Conversion to Finished Product

The final stage in production is to convert the semifinished material into a finished product ready for delivery to the end user. Manufacturing processes that are usually involved in this stage are pressing, machining, surface finishing, assembly, and packaging. Many of these processes are expensive and wasteful of materials. Material losses of 200–300% are not uncommon at this stage. With mass production, the costs of direct labor and overheads are usually small in comparison with the material cost. In such cases, it becomes even more important to reduce the material costs and to optimize material utilization.

8.3 FACTORS AFFECTING MATERIAL PRICES

As with most other commodities, the price of engineering materials is affected by a variety of factors. Inflation, economic recessions, supply and demand, amount of material purchased, inventory costs, and quality of material are among the major factors that can influence the price, and they are discussed in this section.

8.3.1 General Inflation and Price Fluctuations

Generally, the prices of most established engineering materials show a steady increase when considered over a relatively long period of time. The main reasons for such price increases are rising costs of raw materials, energy, and labor. Governmental antipollution policies and similar legislation have also contributed to the increase in prices in recent years. Over the last two decades, most metallic and polymeric materials showed an average price inflation rate of about 5–15% per year. In addition, material prices are also known to suffer short-term price fluctuations. Political factors, local wars, industrial strikes, and world recessions are mostly responsible for such price fluctuations that occur as a result of changes in supply and demand.

8.3.2 Supply and Demand

In a free market economy, the price of a commodity is fixed by the equilibrium between supply and demand. When the supply increases, prices should decrease as competing producers pare their profit margins to maintain their market share.

8.3.3 Order Size

The cost of materials is usually affected by the size of order. The larger the size of an order for a given material, the smaller will be the unit cost. This is because administrative expenses and delivery charges remain almost unchanged and tend to represent a higher proportion of the total cost of a smaller order. For example, the unit prices of most popular plain carbon steels can nearly double as the amount of material purchased decreases from 5 t lots to about 1/2 t. The unit price can be even higher for smaller order sizes.

8.3.4 Standardization of Grades and Sizes

If selection of the material were based entirely on using the most suitable grade and size for each part, almost as many materials, grades, and sizes as parts would be chosen. If all the parts were produced in equal and large quantities, it might be practical to make each part from a different material. However, production is seldom in equal and large quantities, and selecting a different material for each part can lead to serious problems in material cost, storage, inventory, and equipment required by each material. The other extreme is to make all parts from material with same grade and size. This is obviously not practical, except in rare instances. A compromise is to standardize on the smallest number of material grades and sizes that will satisfy plant-wide needs at the lowest cost. Fabrication requirements must be checked to ensure that cost savings in eliminating grades and sizes are not more than offset by added manufacturing costs. Minor design changes can often be made to permit standardization.

8.3.5 Inventory Costs

It has been shown that significant savings may be realized by ordering in large quantities. Inventory costs, however, can quickly offset the initial saving in large-quantity purchases. For example, the cost of storing steel for 1 year can range from 10% to

25% of the initial cost. Inventory costs include interest on investment, storage, taxes, insurance, obsolescence, deterioration, inventory taking, record keeping, rehandling, and reinspection.

8.3.6 Cost Extras for Special Quality

In the case of most steels, a base price that represents the lowest cost is usually given to the most often used quality, for example, merchant quality for hot-rolled bars and commercial quality for sheets. Cost extras are then added to account for the customer's special requirements. Some of the special requirements that usually increase the base price, cost extras, are the following:

1. *Grade.* Standard American Iron and Steel Institute (AISI) quality is supplied as the base material. Semikilled or killed grades are extra.
2. *Restrictions.* Standard quality usually allows for wider chemistry limits. Extra cost is usually charged for specifying a narrower range of carbon, a minimum manganese content, etc. Resulfurized or free machining steel is supplied at an extra cost. Specifying the grain size or hardenability is also an extra.
3. *Size and form.* Special sections or closer dimensional tolerances than specified in standards are cost extras.
4. *Treatment.* Specifying treatments like annealing, normalizing, quenching and tempering, stress relieving, or pickling is usually considered as a cost extra.
5. *Length.* Specifying a certain length of the stock is usually considered as a cost extra.
6. *Cutting.* Hot cut and cold shear are standard. Machine cutting is an extra.
7. *Packaging.* Wrapping, burlap, or boxing are extras.

The price after adding the extras can be more than twice the base price in some cases.

8.3.7 Geographic Location

Material prices are usually given free on board (FOB) of the supplier. The customer pays the cost of transportation. This cost item can be considerable for longer distances and smaller quantities.

8.4 COMPARISON OF MATERIALS ON COST BASIS

Most engineering materials are sold on the basis of cost per unit weight, although some semifinished and finished products are sold on other bases. For example, the prices of pipe and tubing are usually given per unit length basis, whereas the prices of paint are given on the basis of cost of unit liquid volume. Figure 8.2 compares some metallic and plastic materials on the basis of relative cost per unit weight. The

Polymers	Relative cost	Metals
	80	
ETFE, ECTFE	70	
	60	
FEP	50	Zirconium, tungsten, bismuth
	40	
Silicone (general purpose) PTFE		Cobalt Vanadium
	30	Tin Titanium
Polyphenylene sulfide Polysulfene Nylon 6/12	20	Chromium 99.8% Nickel, Inconel 600 Tool steel
	15	
Polycarbonate Nylons (6, 6/6, glass reinforced) Polyesters	10	Copper–nickel
	9	Brass
	8	Magnesium ingot 316 stainless steel sheet
Acetals, cellulose acetate		
	7	
Chlorinated vinyls Olefins		Duralumin
	6	
		Electrolytic copper
ABS, acrylics, melamine	5	
		304 stainless steel sheet
Alkyds, phenolic, urea Styrene butadiene	4	
		Aluminum ingot 99.5% Zinc
Polyethylenes, polystyrenes	3	
Polypropylenes, rigid vinyls	2	Lead HSLA steel cold-rolled sheet Galvanized sheet steel
	1.5	Hot-rolled carbon steel bar Nodular cast iron Cold-rolled carbon steel bar Structural shapes, gray cast iron
	1	Hot-rolled carbon steel sheet

FIGURE 8.2 Comparison of some engineering materials on the basis of cost per unit mass. Comparison is made relative to the cost of hot-rolled low-carbon steel sheet.

cost of hot-rolled plain carbon steel is taken as unity and all other costs are given relative to it. All forms of plain carbon steels, cast irons, and low-alloy steels are less expensive than other materials, which explains their widespread use. The figure shows the wide difference between the prices of the different materials. For example, nickel and Inconel 600 are about 20 times as expensive as plain carbon steel, tin and titanium 30 times as expensive, whereas zirconium and tungsten are 50 times as expensive.

In many applications, engineering materials are not highly stressed, mainly because the amount of material used is determined by the size and shape, method of production, or rigidity of the part. Examples include machine frames, motor bodies, household appliances, and fittings. In such cases, it may be more appropriate to compare materials on the basis of their cost per unit volume, as shown in Figure 8.3. As in Figure 8.2, the different materials are related to hot-rolled plain carbon steel whose cost per unit volume is taken as unity. Some plastics now appear to be cheaper than steel because of their low density. The cost of aluminum also gets close to steel.

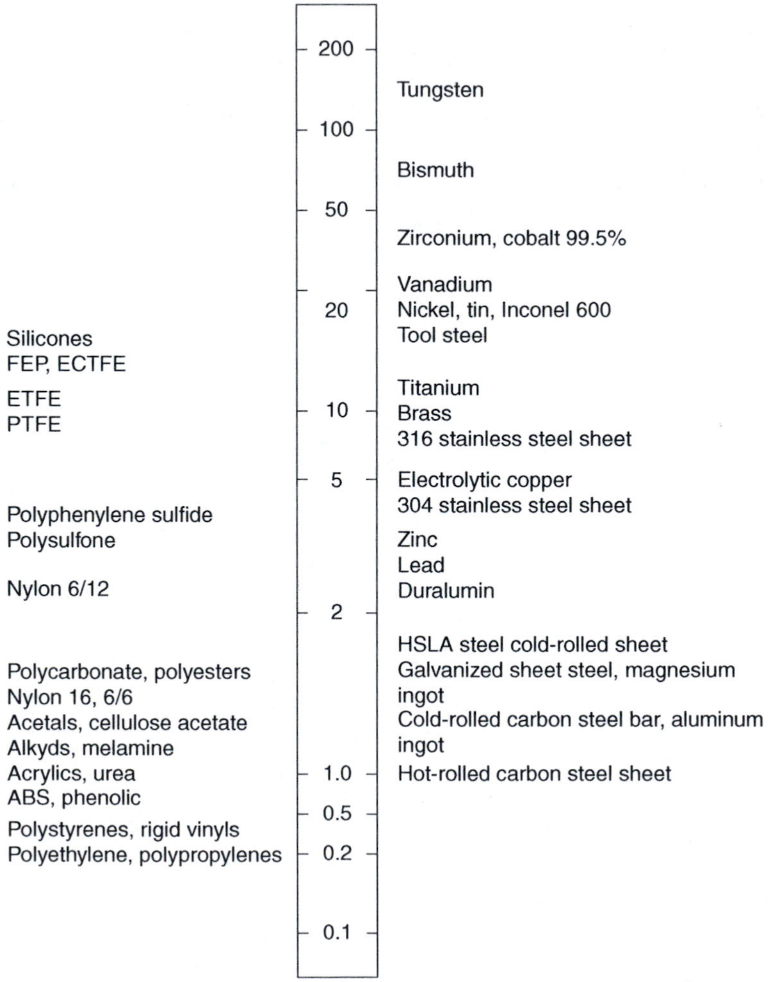

200
Tungsten
100
Bismuth
50
Zirconium, cobalt 99.5%
Vanadium
20
Nickel, tin, Inconel 600
Tool steel
Silicones
FEP, ECTFE
Titanium
ETFE
10
Brass
PTFE
316 stainless steel sheet
5
Electrolytic copper
304 stainless steel sheet
Polyphenylene sulfide
Polysulfone
Zinc
Lead
Nylon 6/12
Duralumin
2
HSLA steel cold-rolled sheet
Polycarbonate, polyesters
Galvanized sheet steel, magnesium ingot
Nylon 16, 6/6
Acetals, cellulose acetate
Cold-rolled carbon steel bar, aluminum ingot
Alkyds, melamine
Acrylics, urea
1.0
Hot-rolled carbon steel sheet
ABS, phenolic
0.5
Polystyrenes, rigid vinyls
Polyethylene, polypropylenes
0.2
0.1

FIGURE 8.3 Comparison of some engineering materials on the basis of cost per unit volume. Comparison is made relative to the cost of hot-rolled low-carbon steel sheet.

8.5 VALUE ANALYSIS OF MATERIAL PROPERTIES

One of the important applications of value analysis is to assess the value of any product by reference to the cheapest available or conceivable product that will perform the same function. This technique can be adapted to material selection and substitution. In the case of steel, for example, plain carbon steels should be considered as a reference point for estimating the value. Additional steel prices above those of plain carbon steel should be critically analyzed. The value of each item of cost extras needs to be examined in relation to the function that the part has to perform in service.

When comparing materials to select the one that will perform the required function at the least cost, the engineer has two basic alternatives:

- To select the least expensive material
- To select a more expensive material that will simplify processing or eliminate steps in manufacturing

An example of cheaper materials that perform the job of more expensive alternatives are the manganese-containing grades of stainless steels like 201, 202, 203, and 216, which are less expensive than their type 300 counterparts because they contain less nickel. Cladding, galvanizing, and tinning offer cheaper alternatives to using stainless steels if the corrosion conditions are not severe.

An example of a higher cost material resulting in a lower-cost component, because of savings in processing, is precoated sheet steel that enable fabricators to omit the finishing step. The coating can be alkyd, polyester, acrylic, vinyl, epoxy, or phenolic paint, or it can be a plastic, such as vinyl, fluorocarbon, or polyethylene. The coatings can be applied to most metals, for example, steel, tin, zinc, and aluminum, and each type of coating provides a different set of properties. Aluminized and chromized coatings on carbon steel resist moderate heat and do not need expensive finishing by enameling.

As the cost of materials is so important, efforts should be made to optimize their use to achieve an overall reduction of the product cost. However, selecting a cheaper material may not always be the answer to a less-expensive product. For example, as materials are usually priced on the basis of cost per unit weight, it may be more economical to pay the extra cost of higher strength since less material will be needed to carry the load. This is illustrated in Example 8.1.

Design Example 8.1: Selecting the Least Expensive Alternative Material for a Cable Car Suspension Member

Problem

Select the least expensive alternative out of the candidate materials in Table 8.2 for making a suspension member in the cable car system. The member is 1 m long and is expected to carry a tensile load of 50 kN without yielding.

TABLE 8.2
Candidate Materials for Suspension Member

Material	YS (MPa)	Specific Gravity	Relative Cost	Cross-Sectional Area (mm^2)	Relative Cost of Member
AISI 1015	329	7.8	1	152	1
AISI 1040	380	7.8	1.1	132	0.95
AISI 4820	492	7.8	1.8	102	1.2
Brass	532	8.5	7	94	4.7

Solution

The cross-sectional area is obtained by dividing the load by the strength of the material. The volume and then the weight are calculated. The cost of the member using different materials is calculated relative to AISI 1015.

Conclusion

Table 8.2 shows that although the cost of AISI 1040 is slightly higher than that of AISI 1015, its higher strength and the resulting smaller cross-sectional area result in a lower total cost. The higher prices of AISI 4820 and brass are not compensated by their higher strengths.

Another example is the case where manufacturing of a product involves a large amount of machining. In such cases, it may be more economical to select a more expensive material with better machinability than to select a cheaper material that is difficult to machine. This is illustrated in Example 8.2.

Design Example 8.2: Selection of the Most Economical Material for a Bolt

Problem

A large number of bolts is to be machined on a high-speed turret lathe. Compare the economics of manufacturing the bolts from AISI 1112 steel, cartridge brass, and 2014 aluminum. The dimensions of the bolt are such that it will need 10 cc of the stock material for its manufacture, the differences in material utilization due to differences in stock sizes are being ignored. Table 8.3 gives the characteristics of the different materials and the estimated cost of manufacture.

Solution

The calculations show that the superior machinability of aluminum more than offsets the savings in material cost of steel, which makes the aluminum bolt less expensive than the steel bolt.

TABLE 8.3
Characteristics and Estimated Cost of Manufacturing of Some Materials

	AISI 1112 Steel	Cartridge Brass	2014 Aluminum
Cost of stock material ($/kg)	0.6	4.2	3.6
Machinability index	100	200	400
Density of material (g/cc)	7.8	8.5	2.8
Weight of material required (g)	78	85	28
Cost of material required ($)	0.047	0.36	0.10
Number of machined bolts per hour	52	85	90
Labor and overhead rate ($/h)	15	15	15
Labor cost per bolt ($)	0.29	0.18	0.17
Cost of material and labor ($)	0.337	0.54	0.27

8.6 ECONOMICS OF MATERIAL UTILIZATION

Manufacturing a part or a product at a competitive cost can only be accomplished when materials and processes are used as effectively as possible. Ideally, the manufacturer should use the cheapest material and not pay for properties that are not needed for successful performance of the part. As discussed earlier, this could lead to cost and inventory problems as a result of stocking a very large number of materials, grades, and sizes. Standardization of material grades and sizes as well as judicious design are answers to this problem. Where production involves more than one size, it may become necessary to buy base quantities of large-size stock, and use it for a variety of smaller size parts. However, this can increase the amount of resulting scrap and, consequently, the final cost of the products since the selling price of scrap is usually only 10–40% of the stock material price. Table 8.4 gives approximate amounts of scrap generated in some manufacturing processes.

Another form of inefficiency in material utilization is encountered in metal casting. The yield in casting is expressed as the percentage of the weight of good castings obtained from the charged metal weight. With high yield there is less return metal to remelt, which in turn reduces the net cost of conversion, melting loss, and molten metal treatments. The metal cost of castings with a high yield is therefore less than that of low-yield castings. Typical yields in cast iron foundries range from 40% to 70%.

The efficiency of utilizing materials can be measured by a material utilization factor, m, which can be defined as

$$m = \frac{\text{weight of the finished part}}{\text{weight of material used to make the part}} \tag{8.1}$$

The nearer the value of m to unity, the less waste will be incurred and hence the lower will be the direct material cost. The cost of scrap material, C_s, that results from

TABLE 8.4
Approximate Amount of Scrap Generated in Some Manufacturing Processes

Process	Scrap (%)
Powder metallurgy	5
Permanent mold casting	10
Closed-die forging	20–25
Extrusion (hot or cold)	15
Sheet metal forming	10–25
Machining	10–60

Note: Based on data reported by Kalpakjian, S. and Schmid, R., *Manufacturing Engineering and Technology*, Prentice-Hall, Upper Saddle River, NJ, 2001.

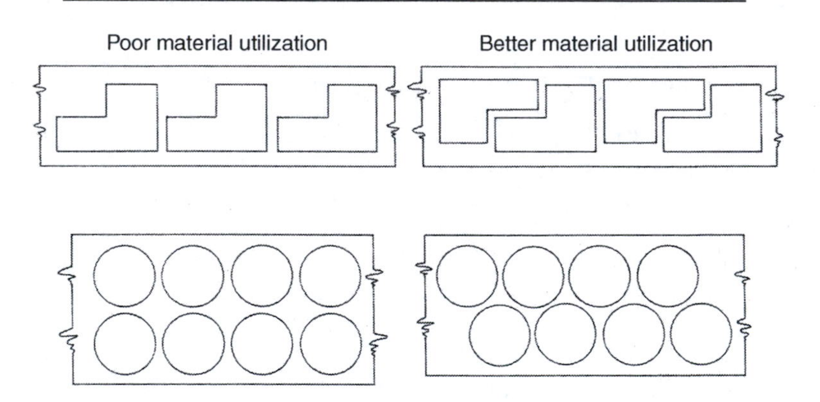

FIGURE 8.4 Effect of blanking die design on material utilization.

manufacturing N components is given by

$$C_s = (1 - m)WC_wN \tag{8.2}$$

where

W = Weight of the component
C_w = Cost of stock material per unit weight

Figure 8.4 shows an example of how the value of m can be increased by a change in the blanking die design and Example 8.3 explains the economic factors involved.

Design Example 8.3: Minimizing the Cost of a Punched Component

Problem

It is required to produce 150,000 L-shaped brass parts similar to those shown in Figure 8.4 by punching from a strip of 100 mm width and 1 mm thickness. The dimensions of the outer sides of the L shape are 60 mm each, and the dimensions of the inner sides of the L shape are 30 mm each, which means that the width of the L shape is 30 mm. In placing the L shapes within the strip, a minimum distance of 10 mm is needed all round the shape. Figure 8.4 shows two possible layouts for the L shape within the strip; choose the more economical one. The cost of the brass strip is \$4.2/kg and the scrap material can be sold at \$0.8/kg. The density of brass can be taken as 8.4 g/cc.

Analysis

With the layout on the L shape on left-hand side of Figure 8.4, a relatively simple punch and die costing \$12,000 can be used to cut a single L shape in each stroke. However, it leads to more waste of material than the layout on the right-hand side of the figure, which uses a more complicated punch and die costing \$18,000 to cut two shapes at a time.

$$\text{Weight of one L shape} = [(6 \times 6) - (3 \times 3)]0.1 \times 8.4 = 22.68 \text{ g}$$

$$\begin{aligned}&\text{Weight of the strip for one L shape for the simple layout}\\&= 10 \times (6 + 0.5 + 0.5) \times 0.1 \times 8.4 = 58.8 \text{ g}\end{aligned}$$

$$\begin{aligned}&\text{Weight of the strip for two L shapes for the complex layout}\\&= 10 \times (6 + 1 + 3 + 0.5 + 0.5) \times 0.1 \times 8.4 = 92.4 \text{ g}\end{aligned}$$

$$m_{\text{simple}} = \frac{22.68}{58.8} = 0.386$$

$$m_{\text{complex}} = \frac{22.68 \times 2}{92.4} = 0.49$$

$$\begin{aligned}&\text{Cost of material for simple layout}\\&= 58.8 \times 4.2 \times 10^{-3} - 58.8(1 - 0.386)0.8 \times 10^{-3} = \$0.218\end{aligned}$$

$$\begin{aligned}&\text{Cost of material for complex layout}\\&= 46.2 \times 4.2 \times 10^{-3} - 46.2(1 - 0.49)0.8 \times 10^{-3} = \$0.175\end{aligned}$$

$$\text{Cost saving in the complex layout} = (0.218 - 0.175)150{,}000 = \$6450$$

$$\text{Extra cost for complex punch and die} = 18{,}000 - 12{,}000 = \$6000$$

Conclusion

Cost saving due to better utilization of material is higher than the additional cost of making a complex die and punch.

$$\text{Break-even point} = \frac{6000}{(0.218 - 0.175)} = 139{,}535 \text{ L shapes}$$

Note:

If the material of the L shapes was cheaper than brass, plain carbon steel, for example, the cost saving due to better material utilization would have been less and the break-even point would have occurred at a higher number of L shapes.

8.7 ECONOMIC COMPETITION IN THE MATERIALS FIELD

Although the average world consumption of engineering materials is increasing with time, the consumption of some materials is increasing at a much faster rate than others. Generally, the consumption of the older materials like steel, copper, concrete, and timber is growing at a slower rate than aluminum and plastics. The slow growth of the older materials reflects the increasing efficiency in their utilization, as well as competitive inroads made by aluminum and plastics. The more efficient utilization of materials is illustrated by the fact that the use of copper in electrical generators has been reduced from about 100 kg (220 lb)/MW to 25 kg (55 lb)/MW during the last decade. The more efficient design and better alloys used in present-day aircraft have reduced the use of metals from 3.5 kg (7.7 lb) per passenger mile in the Boeing stratocruiser to about 1.4 kg (3 lb) in the Boeing 707.

The substitution of a new material for an established one usually involves overcoming the inertia that tends to preserve the existing structure of the industry, that is, the investment in capital plant and labor skills. A good example of this is the difficulty some car companies might find in introducing all plastics injection or blow-molded car seats in place of the current labor-intensive tubular metal frame construction. The main forces that can overcome the inertia against change are

- legislation,
- cost saving, and
- superior performance.

8.7.1 Legislation

Legislation can be a major driving force for change, as in the case of motorcar industry. For example, legislation on crash padding in car interiors led to a sudden increase in the amount of plastics employed in European cars. Similarly, the introduction of legislation in the United States requiring that new cars should average 29 mi to the gallon, resulted in the initiation of development programs to reduce the weight of the car. This caused the average amount of plastics in the motorcar to increase from about 55 kg (120 lb) in 1980 to more than 100 kg (220 lb) in 1990. Legislation, however, can also oppose change, as in the case of side-impact-resistance legislation that restricted the design and the introduction of all-plastics car doors and other structural members.

8.7.2 Cost Saving

Cost saving is a major driving force for materials substitution. For example, aluminum has taken a sizable fraction of the beverage cans and oil containers market from steel, but the steel industry has now reduced the thickness of steel in a can to two-thirds in the hope of holding the market. Plastics have started competing with metals in this application and the body of some beverage cans is now made of either plastic, aluminum, or steel, as shown in Figure 8.5.

FIGURE 8.5 An example of the competition between materials. The beverage can body can be made of aluminum (*right*), steel (*middle*), and plastic (*left*).

8.7.3 Superior Performance

The superior performance of new materials is also a major cause of change. The sports-equipment industry offers many examples, like tennis rackets, rowing oars, and vaulting poles, where fiber-reinforced composites have replaced traditionally used materials. The subject of materials substitution is discussed in more detail in Chapter 10 and some detailed case studies are presented in Chapter 11.

8.8 PROCESSING TIME

The processing time represents an important parameter in most of the major elements of manufacturing costs. For example, the direct labor cost for a given process is usually calculated by multiplying the time required for the process by a labor rate. Overhead costs are also commonly calculated by multiplying operation time by an overhead rate.

8.8.1 Elements of Processing Time

The total time required to perform an operation may be divided into four parts as follows:

1. *Setup time*. This is the time required to prepare for operation and may include the time to get tools from the crib and arrange them on the machine. Setup time is performed once for each lot of parts and should, therefore, be listed separately from the other elements of the operation time. If 45 min is required for a setup and only 10 parts are made, an average of 4.5 min must

be charged against each part. However, if 90 parts are made from the same setup, only 0.5 min is charged per part. Setup time is usually estimated from previous performance of similar operations.

2. *Man or handling time*. This is the time the operator spends loading and unloading the part, manipulating the machine and tools, and making measurements during each cycle of operation. Personal and fatigue allowances as well as time to change tools are also included in this part.
3. *Machine time*. This is the time during each cycle of the operation that the machine is working or the tools are cutting. Many organizations have developed standard data for various machine classes based on accumulated time studies. In some cases, the machine time can be calculated from the process parameters.
4. *Down or lost time*. This is the unavoidable time lost by the operator because of breakdowns, waiting for tools and materials, etc.

Floor-to-floor time (FFT) is the time that elapses between picking up a part to load on the machine and depositing it after unloading from the machine. FFT includes the time for loading, manipulation, machining or processing, and unloading the part. Allowances for tool setup and changes, fatigue, and delays are added to FFT to give the basic production time for the operation.

8.9 PROCESSING COST

There are several accounting methods for estimating the cost of manufacturing, and selecting the most appropriate method depends on the organization of the company and type of products they manufacture. Some examples of cost estimation methods are discussed in the following sections.

8.9.1 Rules of Thumb

Rules of thumb employ simple conversions of a material parameter into total cost of component. For example, the cost of a sand-cast part can be given as a function of its weight, and the cost of an injection-molded component can be related to the cost of plastic used in making it. This method is based on past experience and is useful for routinely produced components. However, it has limited predictive capabilities in the case of new products.

8.9.2 Standard Costs

Standard-cost systems are based on the assumption that there is a certain amount of material and a given amount of labor that go into the manufacture of a component. Implementation of this system involves advance preparation of standard rates for materials, labor, and expenses. Standard time is defined as the total time in which a job should be completed at standard performance of a qualified worker. Information on standard times and costs can be collected by the company for their particular processes or found in the literature where a large amount of data for special or general work has been published.

Comparing the actual materials, labor, and times against the corresponding standard values makes it possible to isolate the reason for any deviation from the overall standard cost. The variance is expressed as the difference between the actual performance against the standard base. The variance can be related to materials, labor, or cost. For example, performance variance of labor may be calculated by comparing actual time worked with standard time. The ratio of actual performance to standard is sometimes called labor-effectiveness ratio or labor-efficiency ratio. Case study 8.4 illustrates the use of variance in analyzing performance.

Case Study 8.4: Using Standard Costs Method to Evaluate Manufacturing Effectiveness

Problem

A foundry has the following standard costs for a cast product for a rate of production of 2000 castings per month:

Labor = \$1.00
Materials = \$0.50
Overheads = \$2.00
Total cost = \$3.50

A scheme for increasing labor efficiency caused the production to increase to 2200 castings per month. The actual costs during this month are shown in Table 8.5.

Result

The results show that the increased labor effectiveness caused a favorable variance of \$200. However, the decreased efficiency of using materials caused an unfavorable variance of \$700, which is more than the gain due to increased labor effectiveness. The net result is unfavorable variance of \$500.

TABLE 8.5
Comparison of Standard and Actual Costs for the Cast Product

Cost Element	Actual Costs (\$)	Calculated According to Standard Costs (\$)	Variance (\$)
Labor	2000	2200	−200
Materials	1800	1100	+700
Overheads	4400	4400	—
Total cost	8200	7700	+500

8.9.3 Technical Cost Modeling

In the technical cost modeling method, elements of cost involved in the manufacturing of the component are calculated separately. For example, processing cost of a component can be divided into the following components:

a. Direct labor cost, which is usually calculated by multiplying the labor hourly rate times the FFT.
b. Direct cost of using the equipment, which can be calculated by multiplying the hourly rate of using the processing machines times the FFT. This cost accounts for the depreciation, maintenance, and utilities cost of equipment.
c. Cost of tooling and production aids, which includes the share of the component in the total cost of designing and fabricating dies, molds, cutters, jigs, and fixtures that are needed to process it. For example, if the tooling cost is $20,000 and the number of components that will be processed is 50,000, then the tooling cost per component is $0.4.
d. Overheads is usually charged as a fraction of items (a) and (b).

In some cases, and for ease of accounting, a company may develop composite hourly rate for a process, which includes the cost of labor, cost of using equipment, and overheads. According to Ulrich and Eppinger (1995), the composite hourly rates can be about $25 for a simple stamping press, $30 for a small injection-molding machine, $44 for sand-casting, $50 for investment casting, and $75 for medium-sized computer-controlled milling machine. Case study 8.5 illustrates the use of the technical cost modeling in selecting the optimum processing route.

Case Study 8.5: Selection of the Least Expensive Route for Manufacturing a Shaft

Problem

A small shaft can be machined on either a turret lathe or a single-spindle automatic lathe. Select the least expensive route for a 100-piece order and a 1000-piece order.

Analysis

Table 8.6 compares the different times and costs for machining a small shaft on a turret lathe and a single-spindle automatic lathe. In addition to illustrating how the different costs are calculated, the table shows that the cost of processing decreases as the number of pieces per order increases. The table also shows that the turret lathe is more economical for smaller orders, whereas the automatic lathe is more economical for larger orders.

TABLE 8.6
Comparison of Times and Costs for a Small Shaft

Parameter	Turret Lathe	Single-Spindle Automatic Lathe
Cost of machine and standard tools ($)	19,318	45,000
Annual depreciation on 15-year basis of machine and standard tools ($)	1,287.93	3,000
Over 2000 h/year, cost of depreciation/h ($) (a)	0.644	1.5
Overhead rate/h not including depreciation ($) (b)	9.00	9.00
Setup time (h)	2.5	3.5
Labor rate/h for setup time ($/h)	7.50	7.50
Setup cost per order ($)	18.75	26.25
Operation time per piece (min)	4.0	3.0
Production per 50-min h (pieces/h) (c)	12.5	16.67
Labor rate/h for operating lathe ($/h) (d)	7.50	3.75 (attends to two lathes simultaneously)
Composite rate/h ($) (a + b + d)	17.14	14.25
Cost per piece without setup and special tools ($) (a + b + d)/c = (e)	1.37	0.86
Setup cost for 100-piece order ($) (f)	1.88	2.63
Cost per piece for 100-piece order ($) (e + f)	3.25	3.49
Setup cost for 1000-piece order ($) (g)	0.19	0.26
Cost per piece for 1000-piece order ($) (e + g)	1.56	1.12

Note: Based on data in Doyle, L.E., Keyser, C.A., Leach, J.L., Schrader, G.F., and Singer, M.B., *Manufacturing Processes and Materials for Engineers*, Prentice-Hall, Englewood Cliffs, NJ, 1985.

8.10 ECONOMICS OF TIME-SAVING DEVICES

Jigs and fixtures are special production tools that are specially designed for quick and accurate location of the workpiece during manufacture. A fixture is a special work-holding device that holds the workpiece during machining, welding, assembly, etc. It is usually designed to facilitate setup or holding of a particular part or shape. A jig, however, not only holds the workpiece but also guides the tools, as in drill jigs, or accurately locates the parts of the work relative to each other, as in welding jigs. Such production tools are expensive and their cost adds to the total production cost. It is, therefore, important to make sure that they can be justified economically by the saving in production time that will result from their use. The following factors must be considered when considering the economics of special tooling:

1. The cost of the special tooling
2. Interest rate on the cost of the tooling
3. Savings in labor cost as a result of using the tooling

4. Savings in machine cost as a result of increased productivity
5. The number of units that will be produced using the tooling

From the economic point of view, the use of special tooling can only be justified if the savings in production costs per piece is greater than, or at least equal to, the tooling cost per piece. The savings per piece, S_p, can be calculated from the following relationship:

$$S_p = (Rt + R_o t) - (R't' + R_o t') \tag{8.3}$$

where

R = Labor rate/h without tooling ($/h)
R' = Labor rate/h using tooling ($/h)
t = Production time per piece without tooling (h)
t' = Production time per piece with tooling (h)
R_o = Machine cost/h, including overheads ($/h)

The total tooling cost, C_T, is the sum of the initial tooling cost, C_t, plus the interest on the tooling cost. Taking the number of years over which the tooling will be used as (n), the rate of interest as (i), and assuming straight-line depreciation

$$C_T = C_t + \frac{(C_t n i)}{2} \tag{8.4}$$

This relationship is only approximate, but it is sufficiently accurate for our purpose because the tooling life is usually relatively short. When the time over which the tooling will be used is less than 1 year, the interest on the tooling may be ignored.

The tooling cost per piece, C_p, is calculated as

$$C_p = \frac{C_T}{N} \tag{8.5}$$

where N is the number of pieces that will be produced with the tooling.

For the tooling to be economically justified, $S_p > C_p$.

Example 8.6 illustrates the use of the mentioned procedure.

Design Example 8.6: Economic Justification for Using a Drill Jig

Problem

Using a drill jig is expected to reduce the drilling time from 30 to 12 min. If a jig is not used, a skilled worker will be needed with an hourly rate of $12. Using the jig makes it possible for a less-skilled worker to do the job at an hourly rate of $10. The hourly rate for using the drilling machine is $8. The cost of designing and manufacturing the jig is estimated as $1200. The interest rate is 12%, and the expected life of the jig is 2 years. It is estimated that the jig will be used for the production of 400 parts during its useful life. Is the use of the jig economically justifiable? How many parts need to be produced for the jig to break even?

Solution

The saving per piece, S_p, as a result of using the jig is calculated from Equation 8.3 as

$$S_p = \left[12 \times \left(\frac{30}{60}\right) + 8 \times \left(\frac{30}{60}\right)\right] - \left[10 \times \left(\frac{12}{60}\right) + 8 \times \left(\frac{12}{60}\right)\right] = 10 - 3.60 = \$6.40$$

The total jig cost, C_T, is calculated from Equation 8.4 as

$$C_T = 1200 + \frac{(1200 \times 2 \times 0.12)}{2} = \$1344$$

The jig cost per part is calculated from Equation 8.5 as

$$C_p = \frac{1344}{400} = \$3.36$$

As the cost of jig per part is less then the savings per part, the jig is economically justifiable.

The break-even number of parts, N', is calculated as

$$6.40 = \frac{1344}{N'}$$

Thus,

$N' = 210$ parts

This means that at least 211 parts need to be manufactured to justify the use of the jig.

8.11 ENVIRONMENTAL IMPACT ASSESSMENT OF MATERIALS AND PROCESSES

8.11.1 ENVIRONMENTAL CONSIDERATIONS

With the increasing awareness that many of the attempts to develop society can have a negative impact on environment, there is an increasing pressure on manufacturers to reduce the environmental burden associated with their products. The production of engineering materials and their processing into products can have a considerable impact on this burden. However, assessing this impact is not always easy as there is a large number of emissions and waste products associated with these activities. For example, emissions to air can include CO_2, CO, SO_2, and NO_2, and emissions to water can include organics, metals, nitrates, and phosphates. Several aggregation systems have been proposed to make it easy for designers to incorporate the environmental impact in their design and the following two methods are examples.

8.11.2 ENERGY CONTENT OF MATERIALS

Ashby uses energy associated with the production of 1 kg of a material, H_p, as an indication of its environmental impact. Table 8.7 gives the value of H_p for some materials and Example 8.7 illustrates its use in design.

TABLE 8.7
Energy Content of Some Engineering Materials

Material Group	Material	Energy Content H_p (MJ/kg)
Ferrous metals	Cast iron	16.4–18.2
	Carbon steels	23.4–25.8
	Stainless steels	77.2–85.3
Nonferrous alloys	Aluminum alloys	184–203
	Titanium alloys	885–945
Ceramics and glasses	Soda-lime glass	13.0–14.4
	Alumina	49.5–54.7
	Silicon carbide	70.2–77.6
Polymers	Polypropylene and polyethylene	76.2–84.2
	PVC	63.5–70.2
	Epoxy	90–100
	Polyester	84–90
Composites	CFRP	258–286
	GFRP	107–118

Note: Based on data reported by Ashby, M.F., *Materials Selection in Mechanical Design*, Elsevier, Amsterdam, 2005.

Design Example 8.7: Accounting for Weight and Environmental Impact in Selecting a Material for a Tie Bolt

Problem

Aluminum alloy 7075T6 (YS = 511 MPa, ρ = 2.7 g/cc) and titanium 6Al4V (YS = 939 MPa, ρ = 4.5 g/cc) are being considered for making a tensile member (tie bolt) of length 200 mm that will carry a load of 50 kN. Which of the two materials will give a lighter member and which will have less impact on environment?

Analysis

Taking a factor of safety of 1.5,

$$\text{Weight of the aluminium member} = \frac{50000 \times 1.5 \times 200 \times 2.7}{511 \times 1000} = 79.3\ \text{g}$$

$$\text{Weight of the titanium member} = \frac{50000 \times 1.5 \times 200 \times 4.5}{939 \times 1000} = 71.9\ \text{g}$$

From Table 8.7, an average energy content for aluminum and titanium alloys can be taken as 193.5 and 915 MJ/kg, respectively.

The energy content of the aluminum member = 0.0793 × 193.5 = 15.4 MJ.
The energy content of the titanium member = 0.0719 × 915 = 65.7 MJ.

Conclusion

The titanium alloy member is lighter but has higher energy content.From the foregoing analysis, it is seen that the weight of a tensile member is proportional to ρ/YS, where ρ is the density and YS the yield strength. The environmental impact of the material in the tensile member is proportional to the parameter ($H_p\rho$/YS), which needs to be minimized for an environmental conscious design.

8.11.3 Life Cycle Assessment

Another method of aggregating the environmental impact is the use of the Eco-Indicator 99 (EI 99). According to ISO 14001, life cycle assessment (LCA), the environmental impact of a given product over its entire life cycle can be divided into three main phases:

a. Production phase, including energy requirements for primary and secondary materials used and all the processes involved in manufacturing them into a finished product ($EI_{prod} = EI_{mat} + EI_{mfct}$)
b. Use or operation phase, including the energy, fuel, and emissions over the entire lifetime of the product (EI_{use})
c. End-of-life phase, including the energy used in disposal of the discarded product and whatever energy is gained from its recycling (EI_{eol})

In this case, the total environmental impact of the product over its entire life cycle (EI_{LC}) is given by

$$EI_{LC} = EI_{mat} + EI_{mfct} + EI_{use} + EI_{eol} \tag{8.6}$$

According to Giudice et al. (2005) the environmental impact of the production phase can be expressed as follows:

$$EI_{prod} = EI_{mat} + EI_{mfct} = EI_{mat}W + EI_{prss}\mu \tag{8.7}$$

where

EI_{mat} = Eco-indicator per unit weight of material as estimated by the EI 99 method
W = Weight of the material
EI_{prss} = Eco-indicator of process as estimated by the EI 99 method
μ = Characteristic parameter of the process or quantity of the material processed

$$EI_{eol} = EI_{dsp}(1 - \xi)W + EI_{rcl}\xi W \tag{8.8}$$

where

EI_{dsp} and EI_{rcl} = Environmental impacts of disposal and recycling processes per unit weight of material, respectively

ξ = Recyclable fraction

Case study 8.8 illustrates the use of the above mentioned parameters in LCA. A more detailed case study in materials substitution is discussed in Section 11.5.

Case Study 8.8: LCA for Motorcar Brake Disk

Problem

Currently, a disk brake is made of gray cast iron (GCI), and aluminum matrix composite (AlMC) is being considered as a substitute material. Use LCA to analyze this decision.

Analysis

The following analysis is based on the case study by Giudice et al. (2005). The main performance requirements of a disk brake include resistance to the thermal and mechanical loading resulting from the braking action, light weight, and compliance to geometric and volume constraints. The disk of the brake is currently made from GCI processed by sand-casting. The substitute material under consideration is AlMC processed by squeeze casting. According to Giudice et al. (2005), for equivalent performance, the geometry and weight of disks made of the two materials are given in Table 8.8. The table also gives the different environmental impact components according to Equations 8.6 through 8.8.

The data in Table 8.8 show that the production of AlMC has greater environmental impact than GCI and recovers less points at the end of its life because of its poor recyclability. Giudice et al. (2005) calculated the EI_{use} component assuming the weight of a motorcar with GCI disks to be 1000 kg, mean fuel consumption to be 0.085 L/km, reduction in fuel consumption to be 4.5% for a 10% reduction in weight, and expected traveling distance to be 150,000 km.

TABLE 8.8

Comparison of GCI and AlMC as Motorcar Brake Disk Materials

	Volume (dm^3)	Weight (kg)	EI_{prod} (mPt)	EI_{eol} (mPt)	EI_{use} (mPt)	EI_{LC} (mPt)
GCI	0.83	6.00	208.9	−165.4	2,729,884	2,729,927
AlMC	1.36	3.83	2293.3	−21.1	2,719,201	2,721,479

Note: Data based on case study by Giudice, F., La Rosa, G., and Risitano, A., *Mater. Design*, 26, 9–20, 2005.

The results in Table 8.8 show the predominant influence of EI_{use} compared to the other environmental impact components. Because of its lightweight and subsequent savings in fuel consumption, AlMC has lower EI_{LC} in spite of its higher EI_{prod} and lower EI_{eol} recovery.

Conclusion

This case study illustrates the importance of including all the components of EI_{LC} when comparing the environmental impact of materials in a product.

8.12 RECYCLING ECONOMICS

The public concern for the environment and the increasing cost of landfill fees provide major incentives for reuse of components and recycling of materials. The process normally starts with the last owner delivering the retired product to a dismantler who then separates reusable components, and shreds the rest. Metals and other useful materials are then separated from the shredded material and the remainder is sent to the landfill. Economic incentives for those taking part in this process would ensure that it would work. For example, the dismantler has to sell the materials salvaged at a sufficient price to make a net profit after compensating the last owner for bringing in the product, paying the expenses of shredding, and paying the landfill fees. This net financial gain can also be used as a factor in selecting the most economic material for a given component. Case study 8.9 is used to illustrate the important role played by recycling in determining the total cost of a product and is based on studies by Sanders et al. (1990) and Berry (1992).

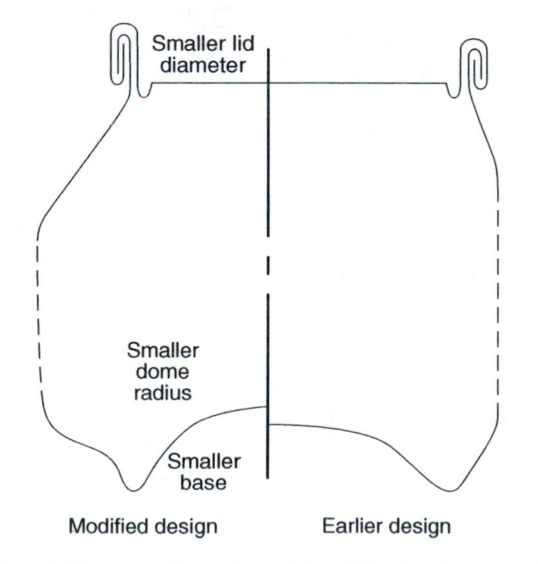

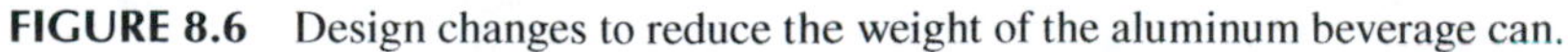

FIGURE 8.6 Design changes to reduce the weight of the aluminum beverage can.

Case Study 8.9: Recycling Economics of Aluminum Beverage Containers

Metal Content

The aluminum industry has been trying to reduce the metal content of beverage containers through innovative designs and material selection. This is because the metal value constitutes about 70% of the container cost. The early two-piece seamless aluminum 12-ounce (355 mL) beverage container was 15–20 g. Today, the can weighs only about 10–11 g. This reduction is achieved by reducing the starting thickness of the can stock sheet from 0.38 mm (0.015 in.) to about 0.3 mm (0.012 in.). The can is required to withstand an internal pressure of 90 psi. This reduction in thickness was made possible by reducing the can neck diameter, that is, the can lid. The neck diameter of early lids was 68.3 mm (2 11/16 in.) but today it is 60.3 mm (2 6/16 in.). Cans with smaller lid diameter withstand more internal pressure without buckling, therefore, lid thickness can be reduced. As the lid diameter was reduced, the base diameter was also reduced to maintain stackability of the cans. This has also allowed the thickness of the base metal to be reduced. Figure 8.6 shows the changes in lid and base design of the aluminum beverage can. No change in the can alloy composition was necessary. The can body is made of 3004 alloy (0.18 Si, 0.45 Fe, 0.13 Cu, 1.1 Mn, and 1.1 Mg), and the lid is made of 5182 alloy (0.10 Si, 0.24 Fe, 0.03 Cu, 0.35 Mn, and 4.5 Mg). These alloys are compatible in composition, which is an important factor in recycling. With small adjustments in composition, the recycled material can be reused in making either the body or lid stock.

Cost Analysis

Decreasing the metal weight in the can through the design changes and higher recovery rates of used cans for recycling have allowed the aluminum can to remain competitive in relation to other packaging materials as shown in Table 8.9.

TABLE 8.9
Comparison of Costs Involved in Manufacturing of 355 mL Containers (in cents) in 1987

Cost Element	Aluminum	Glass	Plastic	Steel
Raw material	4.6	1.7	1.71	3.18
Manufacturing cost	3.19	7.8	6.27	3.81
Distribution cost	0.05	0.50	0.05	0.08
Recycling credit	(1.29)	(2.25)	(0.00)	(0.33)
Total net cost	6.55	7.75	8.03	6.74

Note: Based on Berry, D., *JOM*, 44, 21–25, 1992.

The table shows that the aluminum raw material used in a beverage can costs more than any of the alternative materials. This explains the motive for design changes to reduce the weight of material in the can. An aluminum can has an advantage in manufacturing cost over other materials and its distribution cost is as low as that of plastic in view of its lightweight. The other major cost advantage of aluminum is derived from its recyclability. The recycling credit is the result of recycling manufacturing scrap, skeletons and stampings, and postconsumer scrap, that is, used cans. At present, 63% of used aluminum cans are collected and recycled. This value contrasts to 33% for glass beverage bottles and 24.7% for steel beverage cans. Without this higher recycling credit, the aluminum can would be more expensive than the steel can.

8.13 LIFE CYCLE COST

Life cycle costing (LCC) gives the total cost of a product throughout its life and includes

- the selling price of the product,
- running cost, and
- postuse cost.

The elements of cost that determine the selling price of a product were discussed in Section 1.7. Running costs include all the costs paid by the owner throughout the life of the product. These include cost of energy used, maintenance cost, and cost of spare parts. Postuse costs include disposal cost minus resale and returns from recycling. Other life cycle costs include support, training, site preparation, and operating costs generated by the acquisition.

Life cycle costs can be estimated for new projects at either total plant level or at equipment item level. The main aims of LCC are the following:

1. To provide a comprehensive understanding of the total commitment of asset ownership
2. To identify areas of the life cycle where improvement can be achieved through redesigning or reallocation of resources
3. To improve profitability and industrial efficiency

The applications of LCC in industry include comparing mutually exclusive projects, checking future performance of an asset, and trade-offs between design and operation parameters. Case study 8.10 illustrates how LCC is used to decide whether or not aluminum alloys should replace steel in making the body of motorcars.

Case Study 8.10: LCC Cost Comparison of the Use of Aluminum and Steel in Making Motorcar Body

Problem

Motorcar manufacturers are facing the challenge of increasingly stringent regulations on levels of emission and fuel consumption. The challenge can be met by increasing the power train efficiency, or reducing the weight by either downsizing or using lighter materials. Of these alternatives, weight reduction by using lighter materials is the most promising. This case study explores the possibility of substituting the lighter aluminum alloys for steel in the motorcar body, made of a structure covered by panels.

Analysis

Table 8.10 gives the parameters involved in the LCC analysis as presented for a rate of production of 200,000 units per year by Dieffenbach and Mascarin (1993).

The steel unibody represents today's motorcar. The aluminum unibody is similarly manufactured with the main difference being the use of aluminum sheets instead of steel. Because the same processes are used in fabrication and assembly of the two materials, the costs are roughly the same. The cost differences are, therefore, due to difference in material prices only. Figure 8.7 shows that the cost of the aluminum unibody is about 1.4 times that of steel.

However, the operating cost in the case of aluminum is about 0.6 times that of steel. The main operating cost difference between the two bodies is due to fuel consumption. Repair costs do not vary significantly from one material to another since labor and insurance costs are predominant.

The recycling value of steel is very small compared to manufacturing costs and is ignored here. The recycling value of aluminum is about 10 times higher than steel and is shown in Figure 8.7 as credit, below the zero line.

TABLE 8.10
Comparison of LCC Parameters for Steel and Aluminum Unibodies at a Rate of 200,000 Units Per Year

Parameter	Steel Unibody	Aluminum Unibody
Weight of structure (kg)	225	135
Weight of panels (kg)	79	39
Steel content of structure (%)	100	60
Steel content of panels (%)	100	50
Fabrication	Stamping	Stamping
Assembly	Spot welding	Spot welding
Relative cost of extrusions and castings	1	2
Relative cost of sheets	1	4
Relative selling cost of scrap	1	10

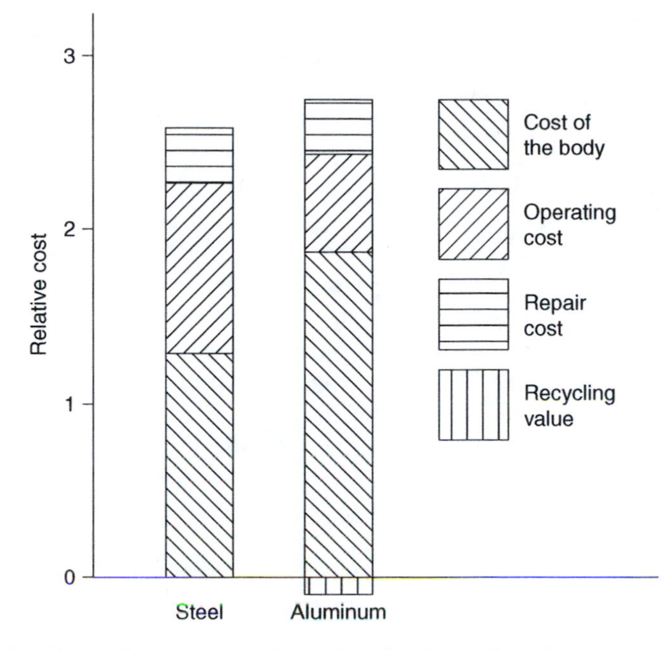

FIGURE 8.7 Life cycle cost comparison of steel and aluminum in motorcar body. (From Dieffenbach, J.R. and Mascarin, A.E., *JOM*, 45, 16–19, 1993.)

The total life cycle cost is arrived at by subtracting the recycling value from the costs above the zero line. The LCC ratio of steel to aluminum is about 1:1.02, that is, 2% in favor of steel. This result indicates that manufacturers will continue to use steel for the motorcar body. Aluminum unibody, however, holds promise if additional importance is placed on weight and fuel economy.

8.14 SUMMARY

1. In addition to designing and manufacturing products at the required level of quality, manufacturers must also be able to sell their products at competitive prices and make profits. It is, therefore, important that the different materials and processes involved in the design and manufacture should be evaluated in terms of their economic as well as their technical merits.
2. The cost of materials and manufacturing usually represents about 30–70% of the total product cost in a wide range of engineering industries. As the cost of materials is so important, efforts should be made to optimize their use to achieve an overall reduction of the product cost. However, selecting a cheaper material may not always be the answer to a less-expensive product. In some cases it may be more economical to pay the extra cost of higher strength since less material will be needed to carry the load. It may also be more economical to select a more expensive material with better machinability than to select a cheaper material, that is difficult to machine.

3. The main elements that make up the cost of materials include cost of ore preparation, cost of extraction from the ore, cost of alloying, cost of conversion to semifinished products, and cost of conversion to finished product.
4. Plain carbon steels and cast irons are less expensive than other materials when compared on the basis of cost per unit weight. However, plastics become less expensive if compared on the basis of cost per unit volume.
5. FFT is the time between picking up a part to load on the machine and depositing it after unloading from the machine. FFT includes loading time, manipulation time, machining or processing time, and unloading time. Allowance for tool setup and changes, fatigue allowance, and delays are added to the FFT to give the basic production time for the operation.
6. Processing cost consists of direct labor cost, cost of using equipment, cost of tooling and production aids, and overheads.
7. Time-saving devices can be expensive and should be justified economically by the saving in production time that will result from their use.
8. The energy content of materials (H_p) can be used to assess the environmental impact of a product. The parameter $H_p\rho$/YS, where ρ is the density and YS the yield strength, can be taken as a measure of the environmental impact of the material in a tensile member.
9. LCA of the impact of a product over its entire life consists of the energy used in its production phase, use or operation phase, and end-of-life phase.
10. Recycling economics play an important role in determining the total cost of a product, especially if high-cost materials are involved in its manufacture.
11. Life cycle cost of a product includes the selling price of the product, running cost, and postuse cost.

8.15 REVIEW QUESTIONS

8.1 Explain why stainless steels are more expensive than plain carbon steels.

8.2 Why has the price of titanium decreased in the last 20 years?

8.3 What are the economic reasons for the increased use of carbon fiber-reinforced materials?

8.4 Although plastics are more expensive than many metals, on the basis of cost per unit weight, plastic products are generally less expensive. Discuss this statement.

8.5 Although powdered alloys are more expensive than similar solid alloys, powder metallurgy products are competitive on a cost basis.

8.6 Why is aluminum more expensive than carbon steel, although its raw material is more abundant in nature?

8.7 A company which manufactures shelving units for libraries is considering the possibility of replacing steel shelves with a composite material (epoxy—75% glass fabric). Shelves are 150 cm long and 40 cm wide with the thickness depending on the type of material used in making it. The company is expecting to make 200,000 shelves per year. Given the following information, determine whether the substitution is economically justified.

	Steel	Epoxy–75% Glass Fabric
Shelf thickness for equal stiffness (mm)	0.7	1.8
Density (g/cc)	7.8	1.9
Cost of material ($/kg)	0.7	1.5
Cost of new equipment ($; will last 5 years)	—	500,000
Cost of dies and tools ($; will last 1 year)	30,000	15,000
Cycle time per shelf (min)	1	3
Labor rate ($/h)	20	20
Cost saving in packaging and handling due to lighter weight ($)	—	1.2 per shelf

8.8 Using a drill jig is expected to reduce the drilling time from 25 to 15 min. If a jig is not used, a skilled worker will be needed with an hourly rate of $15. Using the jig makes it possible to use a less-skilled worker with an hourly rate of $11. The hourly rate for using the drilling machine is $8. The cost of designing and making the jig is $1500. The expected number of parts to be produced during the lifetime of the jig is 500. The interest on the capital can be ignored, as the duration of the job is short. Is the use of the jig economically justifiable?

8.9 Design a simple jig for drilling three holes in a rectangular workpiece that is 250 mm long, 150 mm wide, and 50 mm thick.

8.10 Calculate the number of parts that will justify introducing the jig in question 8.9 given the following information:

Labor rate without jig	=	$12/h
Labor rate with jig	=	$10/h
Production time per part without jig	=	15 min
Production time per part with jig	=	10 min
Machine cost/h	=	$20
Cost of designing and making the jig	=	$500
The expected length of production run	=	7 months
Rate of interest	=	12%

(Answer: number of parts = 173)

8.11 The manufacture of a given component requires eight operations in the machine shop. The average setup time per operation is 2.5 h. Average operation time per machine is 7 min. The average nonoperation time needed for handling of workpieces and tools, inspection, delays, and temporary storage is 50 min per operation. Estimate the total time required to get a batch of 100 components through the workshop. (Answer: 120 h)

8.12 The component in question 8.9 is required in batches of 500, and a turret lathe is being considered for manufacturing the component. The machining time will be unchanged, but the total handling time will be reduced to 1.5 min per operation. If the setup time of the turret lathe is estimated as 6 h, calculate

a. total time to process the batch on the engine lathe,
b. total time to process the batch on the turret lathe, and
c. average production time in both cases.
(Answers: (a) 493.3 h; (b) 478.87 h; and (c) 59.2 min for engine lathe, 57.5 for turret lathe)

BIBLIOGRAPHY AND FURTHER READING

Ashby, M.F., *Materials Selection in Mechanical Design*, 3rd Ed., Elsevier, Amsterdam, 2005.

Berry, D., Recyclability and selection of packaging materials, *JOM*, December, **44**, 21–25, 1992.

DeGarmo, E.P., Black, J.T., and Kohser, R.A., *Materials and Processes in Manufacturing*, 7th Ed., Collier Macmillan, London, 1988.

Dieffenbach, J.R. and Mascarin, A.E., *JOM*, June, **45**, 16–19, 1993.

Doyle, L.E., Keyser, C.A., Leach, J.L., Schrader, G.F., and Singer, M.B., *Manufacturing Processes and Materials for Engineers*, 3rd Ed., Prentice-Hall, Englewood Cliffs, NJ, 1985.

Farag, M.M., *Materials Selection for Engineering Design*, Prentice-Hall, London, 1997.

Giudice, F., La Rosa, G., and Risitano, A., Materials selection for life-cycle design process: a method to integrate mechanical and environmental performances in optimal choice, *Mater. Design*, **26**, 9–20, 2005.

Haslehurst, M., *Manufacturing Technology*, 3rd Ed., Hodder and Stoughton, London, 1981.

Humphreys, K.K. and Katell, S., *Basic Cost Engineering*, Marcel Dekker, New York, 1981.

Kalpakjian, S. and Schmid, R., *Manufacturing Engineering and Technology*, 4th Ed., Prentice-Hall, Upper Saddle River, NJ, 2001.

Ludema, K.C., Caddell, R.M., and Atkins, A.G., *Manufacturing Engineering: Economics and Processes*, Prentice-Hall, London, 1987.

Sanders, R.E., Jr, Tragester, A.B., and Rollings, C.S., Recycling of lightweight aluminum containers, paper presented at the 2nd Int. Symp. On Recycling of Metals and Engineered Materials, Van Linden, J.H.C., Editor. The Minerals, Metals & Materials Society, Warrendale, PA, pp. 187–201, 1990.

Ulrich, K.T. and Eppinger, S.D., *Product Design and Development*, McGraw-Hill, New York, 1995.

White, J.A., Agee, M.H., and Case, K.E., *Principles of Engineering Economic Analysis*, 2nd Ed., Wiley, New York, 1984.

9 The Materials Selection Process

9.1 INTRODUCTION

Selecting the appropriate material and manufacturing process are important requisites for the development of a product that will perform successfully in service. It is estimated that there are more than 40,000 currently useful metallic alloys and probably close to that number of nonmetallic engineering materials such as plastics, ceramics and glasses, composite materials, and semiconductors. This large number of materials and the many manufacturing processes available to the engineer often make the selection process a difficult task. If the selection process is carried out haphazardly, there will be the risk of overlooking a possible attractive alternative solution. This risk can be reduced by adopting a systematic selection procedure. Rigorous and thorough approach to materials selection is, however, often not followed in industry and much selection is based on past experience—"when in doubt make it stout out of the stuff you know about." Although it is unwise to totally ignore past experience, since what worked before is obviously a solution, such solution may not be the optimum solution. The increasing pressure to produce more economic and competitive products, in addition to the frequent introduction of new materials and manufacturing processes, makes it necessary for the engineer to be always on the lookout for possible improvement.

This chapter analyzes the nature of the selection process and proposes procedures for gradually narrowing down the available choices until an optimum combination of material and manufacturing process is identified. Because of the large amount of data involved in the selection process, a variety of quantitative selection procedures have been developed to analyze such data so that a systematic evaluation can be easily made. Several of the quantitative procedures can be adopted to computer-aided selection from a data bank of material properties and process characteristics.

Although the materials and process selection is most often thought of in terms of new product development, there are many other reasons for reviewing the type of materials and processes used in manufacturing an existing product. These reasons include taking advantage of new materials or processes; improving service performance, including longer life and higher reliability; meeting new legal requirements, accounting for changed operating conditions; reducing cost; and making the product more competitive. The issues involved in materials substitution are discussed in Chapter 10.

The overall goal of this chapter is to illustrate how systematic selection procedures can be used to select optimum materials and processes for a given component.

The main objectives are to illustrate how to

1. analyze material performance requirements for a given application;
2. create alternative solutions, screen them, and then rank the viable candidates;
3. use quantitative methods in materials selection;
4. incorporate computer methods in the selection process; and
5. find reliable sources of material properties.

9.2 THE NATURE OF THE SELECTION PROCESS

Unlike the exact sciences, where a problem normally has one single correct solution, material selection processes are open-ended and normally lead to several possible solutions to the same problem. This is illustrated by the fact that similar components performing similar functions, but produced by different manufacturers, are often made from different materials and even by different manufacturing processes. However, selecting the optimum combination of material and process is not a simple task that can be performed at one certain stage in the history of a project, it should gradually evolve during the different stages of product development. Figure 9.1 shows that ideas developed on the basis of marketing surveys are first translated into industrial design, leading to broad description of the product in terms of the following parameters:

- What is it?
- What does it do?
- How does it do it?
- How much should it be?

As discussed in Chapters 1 and 5, answering these questions allows the product development team to formulate the product specifications, develop various concepts, and then select the optimum solution. After that, the product is decomposed into subassemblies and the different components of each subassembly are then identified. Experience in most industries has shown that it is desirable to adopt the holistic decision-making approach of concurrent engineering in product development. With concurrent engineering, materials and manufacturing processes are considered in the early stages of design and are more precisely defined as the design progresses from the concept to the embodiment and finally the detail stages, as shown in Figure 9.1.

Although each material and process selection decision has its own individual character and its own sequence of events, there is a general pattern common to the selection process. In the first stages of development of a new component, questions such as the following are posed:

- What are the important, or primary, design and material requirements?
- What are the secondary requirements and are they necessary?

Answering these questions makes it necessary to specify the performance requirements of the component and to broadly outline the main materials characteristics and

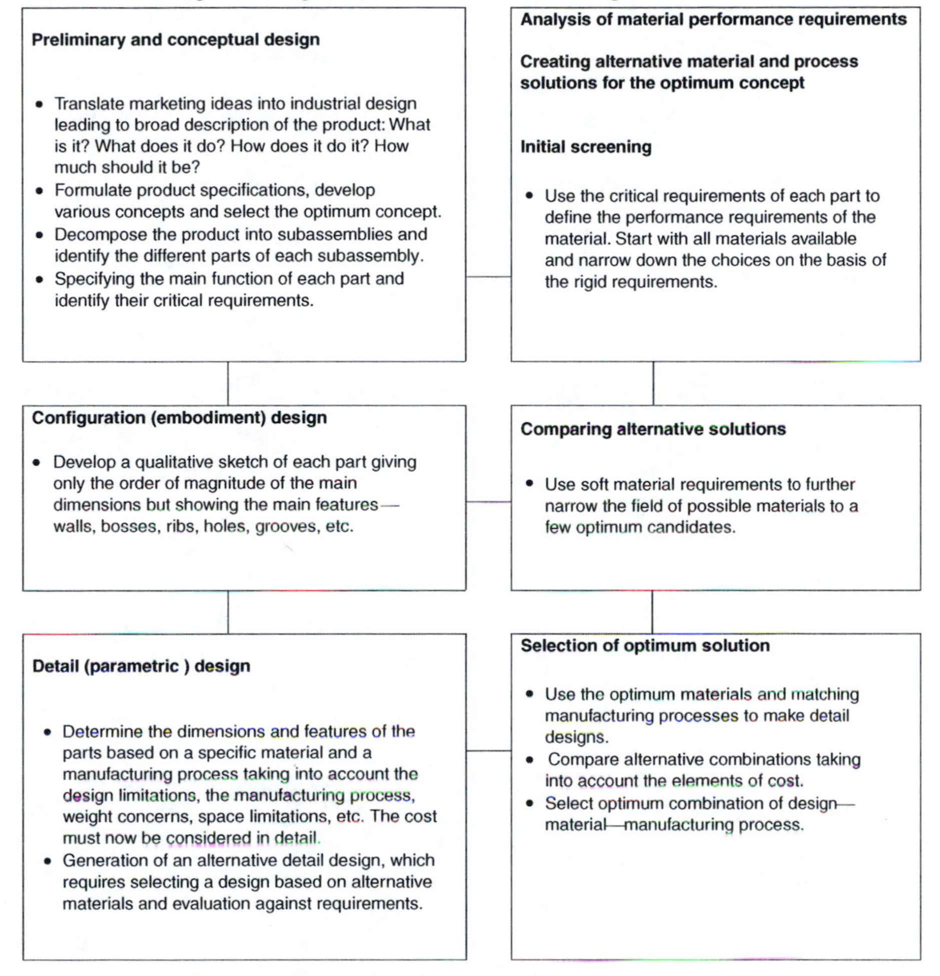

FIGURE 9.1 Major stages of design and the related stages of materials selection.

processing requirements. On this basis, certain classes of materials and manufacturing processes may be eliminated and others chosen as likely candidates for making the component. The relevant material properties are then identified and ranked in the order of importance. Candidate materials that possess these properties are then graded according to their expected performance and cost. Processing details are also examined at this stage. Optimization techniques may then be used to select the optimum material and processing route. This may result in design modifications to achieve production economy or to suit the available production facilities and equipment. Even in the stage of product manufacture, some changes in materials may be necessary. Processing problems may arise causing the replacement of an otherwise satisfactory material. For example, heat treating, joining, or finishing difficulties may require material substitution which, in turn, may result in different service performance characteristics requiring some redesign.

The processes discussed can generally be grouped into the following four steps:

1. Analysis of the performance requirements and creating alternative solutions
2. Initial screening of solutions
3. Comparing and ranking alternative solutions
4. Selecting the optimum solution

The four steps mentioned are discussed in Sections 9.3 through 9.6.

9.3 ANALYSIS OF THE MATERIAL PERFORMANCE REQUIREMENTS AND CREATING ALTERNATIVE SOLUTIONS

The material performance requirements can be divided into five broad categories, namely, functional requirements, processability requirements, cost, reliability, and resistance to service conditions.

9.3.1 Functional Requirements

Functional requirements are directly related to the required characteristics of the component, subassembly, or the product. For example, if the component carries a uniaxial tensile load, the yield strength of a candidate material can be directly related to the load-carrying capacity of the product. However, some characteristics of the component or product may not have simple correspondence with measurable material properties, as in the case of thermal shock resistance, reliability, and esthetic qualities. Under these conditions, the evaluation process can be quite complex and may depend upon predictions based on simulated service tests, upon the most closely related material properties, or even upon consumer impressions. For example, thermal shock resistance can be related to thermal expansion coefficient, thermal conductivity, modulus of elasticity, ductility, and tensile strength. However, esthetic characteristics can be related to warmth and tactile feel in some cases, which largely depend on consumer impressions about certain materials in a given application.

9.3.2 Processability Requirements

The processability of a material is a measure of its ability to be worked and shaped into a finished component. With reference to a specific manufacturing method, processability can be defined as castability, weldability, machinability, etc. Ductility and hardenability can be relevant to processability if the material is to be deformed or hardened by heat treatment, respectively. The closeness of the stock form to the required product form can be taken as a measure of processability in some cases.

It is important to remember that processing operations will almost always affect the material properties so that processability considerations are closely related to functional requirements.

9.3.3 Cost

Cost is usually a controlling factor in evaluating materials, because in many applications there is a cost limit for a given component. When the cost limit is exceeded, the design may have to be changed to allow for the use of a less expensive material or manufacturing process sequence. The cost of processing often exceeds the cost of the stock material. In some cases, a relatively more expensive material may eventually yield a less expensive product than a low-priced material, which is more expensive to process, as discussed in Section 8.5.

9.3.4 Reliability Requirements

Reliability of a material can be defined as the probability that it will perform the intended function for the expected life without failure. Material reliability is difficult to measure, because it is not only dependent upon the material's inherent properties, but is also greatly affected by its production and processing history. Generally, new and nonstandard materials will tend to have lower reliability than established, standard materials.

Despite difficulties of evaluating reliability, it is often an important selection factor that must be taken into account. Failure analysis techniques are usually used to predict the different ways in which a product can fail, and can be considered as a systematic approach to reliability evaluation. The causes of failure of a component in service can usually be traced back to defects in materials and processing, faulty design, unexpected service conditions, or misuse of the product. Failure of materials in service is discussed in Chapters 2 and 3. Reliability of components is discussed in Section 5.7.

9.3.5 Resistance to Service Conditions

The environment in which the product or component will operate plays an important role in determining the material performance requirements. High or low temperatures, as well as corrosive environments, can adversely affect the performance of most materials in service, as discussed in Chapters 2 and 3, respectively. Whenever more than one material is involved in an application, compatibility becomes a selection consideration. In a thermal environment, for example, the coefficients of thermal expansion of all the materials involved may have to be similar to avoid thermal stresses. In wet environments, materials that will be in electrical contact should be chosen carefully to avoid galvanic corrosion. In applications where relative movement exists between different parts, wear resistance of the materials involved should be considered. The design should provide access for lubrication, otherwise, self-lubricating materials have to be used. The selection of materials to resist failure under different service conditions is discussed in Chapter 4.

9.3.6 Creating Alternative Solutions

Having specified the material requirements, the rest of the selection process involves the search for the material that would best meet those requirements. The starting point is the entire range of engineering materials. At this stage, creativity is essential to open up channels in different directions and not let traditional thinking interfere with the exploration of ideas. Steel may be the best material for one design concept, whereas plastic is best for a different concept, although the two designs provide similar functions. The importance of this phase is that it creates alternatives without much regard to their feasibility.

9.4 INITIAL SCREENING OF SOLUTIONS

After all the alternatives have been suggested, the ideas that are obviously unsuitable are eliminated and attention is concentrated on those that look practical. Quantitative methods can be used for initial screening to narrow down the choices to a manageable number for subsequent detailed evaluation. Following are some of the quantitative methods for initial screening of materials and manufacturing processes.

9.4.1 Rigid Materials and Process Requirements

Initial screening of materials can be achieved by first classifying their performance requirements into two main categories:

- Rigid, or go-no-go, requirements
- Soft, or relative, requirements

Rigid requirements are those that must be met by the material if it is to be considered at all. Such requirements can be used for the initial screening of materials to eliminate the unsuitable groups. For example, metallic materials are eliminated when selecting materials for an electrical insulator. If the insulator is to be flexible, the field is narrowed further as all ceramic materials are eliminated. Other examples of the material rigid requirements include behavior under operating temperature, resistance to corrosive environment, ductility, electrical and thermal conductivity or insulation, and transparency to light or other waves.

Examples of manufacturing process rigid requirements include batch size, production rate, product size and shape, tolerances, and surface finish. Whether or not the equipment or experience for a given manufacturing process exists in a plant can also be considered as a hard requirement in many cases. Compatibility between the manufacturing process and the material is also an important screening parameter. For example, cast irons are not compatible with sheet metal–forming processes and steels are not easy to process by die-casting. In some cases, eliminating a group of materials results in automatic elimination of some manufacturing processes. For example, if plastics are eliminated because service temperature is too high, injection and transfer molding should be eliminated as they are unsuitable for other materials. Compatibility between materials and manufacturing processes is discussed in Section 7.3.

Soft, or relative, requirements are those that are subject to compromise and trade-offs. Examples of soft requirements include mechanical properties, specific gravity, and cost. Soft requirements can be compared in terms of their relative importance, which depends on the application under study.

9.4.2 Cost per Unit Property Method

The cost per unit property method is suitable for initial screening in applications where one property stands out as the most critical service requirement. As an example, consider the case of a bar of a given length (L) to support a tensile force (F). The cross-sectional area (A) of the bar is given by

$$A = \frac{F}{S} \tag{9.1}$$

where S is the working stress of the material, which is related to its yield strength divided by an appropriate factor of safety.

The cost of the bar (C') is given by

$$C' = C\rho AL = \frac{C\rho FL}{S} \tag{9.2}$$

where

C = Cost of the material per unit mass
ρ = Density of the material

Since F and L are constant for all materials, comparison can be based on the cost of unit strength, which is the quantity $(C\rho)/S$.

Materials with lower cost per unit strength are preferable. If an upper limit is set for the quantity $(C\rho)/S$, then materials satisfying this condition can be identified and used as possible candidates for more detailed analysis in the next stage of selection.

The working stress of the material in Equations 9.1 and 9.2 is related to the static yield strength of the material since the applied load is static. If the applied load is alternating, it is more appropriate to use the fatigue strength of the material. Similarly, the creep strength should be used under loading conditions that cause creep.

Equations similar to Equation 9.2 can be used to compare materials on the basis of cost per unit stiffness when the important design criterion is the deflection in the bar. In such cases, S is replaced by the elastic modulus of the material, E. Equations 9.1 and 9.2 can also be modified to allow comparison of different materials under loading systems other than uniaxial tension. Table 9.1 gives some formulas for

TABLE 9.1
Formulas for Estimating Cost per Unit Property

Cross-Section and Loading Condition	Cost per Unit Strength	Cost per Unit Stiffness
Solid cylinder in tension or compression	$C(\rho/S)$	$C(\rho/E)$
Solid cylinder in bending	$C(\rho/S)^{2/3}$	$C(\rho/E)^{1/2}$
Solid cylinder in torsion	$C(\rho/S)^{2/3}$	$C(\rho/G)^{1/2}$
Solid cylindrical bar as slender column	—	$C(\rho/E)^{1/2}$
Solid rectangle in bending	$C(\rho/S)^{1/2}$	$C(\rho/E)^{1/3}$
Thin-walled cylindrical pressure vessel	$C(\rho/S)$	—

the cost per unit property under different loading conditions based on either yield strength or stiffness. Case study 9.1 illustrates the use of the cost per unit property method.

Case Study 9.1: Selecting a Beam Material for Minimum Cost

Problem

Consider a structural member in the form of a simply supported beam of rectangular cross section. The length of the beam is 1 m (39.37 in.), the width is 100 mm (3.94 in.), and there is no restriction on the depth of the beam. The beam is subjected to a concentrated load of 20 kN (4409 lb), which acts on its middle. The main design requirement is that the beam should not suffer plastic deformation as a result of load application. Use the information given in Table 9.2 to select the least expensive material for the beam.

Solution

Based on Table 9.2 and the appropriate formula from Table 9.1, the cost of unit strength for the different materials is calculated and the results are given in the last column of Table 9.2. The results show that steels AISI 1020 and 4140 are equally suitable, whereas aluminum 6061 and epoxy glass are more expensive. This answer is reasonable since steels are usually used in structural applications unless special features, such as lightweightedness or corrosion resistance, are required. If the weight of the beam that can carry the given load is calculated for the different materials, it can be shown that the 1020 steel beam is the heaviest, followed by 4140 steel, then aluminum, with epoxy glass being the lightest. In some cases, as in aerospace applications, it may be worth paying the extra cost of the lighter structural member. This issue is discussed in more detail in the case study of materials substitution in aerospace industry (Chapter 11).

TABLE 9.2
Characteristics of Candidate Materials for the Beam

Material	Working Stress[a] MPa	Working Stress[a] ksi	Specific Gravity	Relative Cost[b]	Cost of Unit Strength
Steel AISI 1020, normalized	117	17	7.86	1	0.73
Steel AISI 4140, normalized	222	32	7.86	1.38	0.73
Aluminum 6061, T6 temper	93	13.5	2.7	6	1.69
Epoxy + 70% glass fibers	70	10.2	2.11	9	2.26

[a] The working stress is computed from yield strength using a factor of safety of 3.

[b] The relative cost per unit weight is based on AISI 1020 steel as unity. Material and processing costs are included in the relative cost.

9.4.3 Ashby's Method

Ashby's material selection charts are useful for the initial screening of materials. Figure 9.2 plots the elastic modulus and strength against density for a variety of materials. For simple axial loading, the relationships are E/ρ or S/ρ. For other types of loading and component geometry, different $E-\rho$ and $S-\rho$ relationships apply, as shown in Table 9.1. Lines with the appropriate slopes are shown in Figure 9.2 to

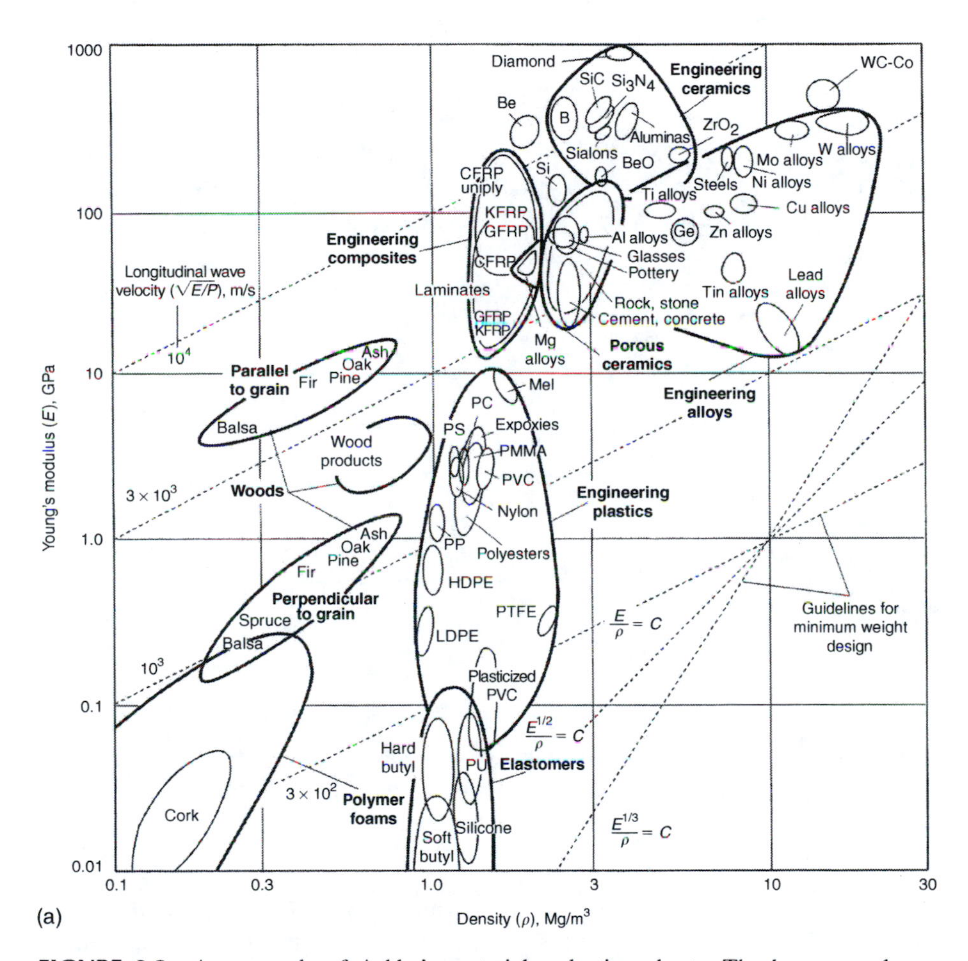

FIGURE 9.2 An example of Ashby's materials selection charts. The heavy envelopes enclose data for a given class of material. (a) Young's modulus, E, plotted against density, ρ, for various engineered materials. The guidelines of constant E/ρ, $E^{1/2}/\rho$, and $E^{1/3}/\rho$ allow selection of materials for minimum weight, deflection-limited design. (b) Strength, σ_f, plotted against density, ρ, for various engineered materials. Strength is yield strength for metals and polymers, compressive strength for ceramics, tear strength for elastomers, and tensile strength for composites. The guidelines of constant σ_f/ρ, $\sigma_f^{2/3}/\rho$, and $\sigma_f^{1/2}/\rho$ are used in minimum weight, yield limited, design. (Reprinted from Ashby, M.F., Materials selection charts, in *ASM Metals Handbook, Materials Selection and Design*, Vol. 20, Dieter, G., Volume Chair. ASM International, Materials Park, OH, 1997, pp. 266–280. With permission.)

(*Continued*)

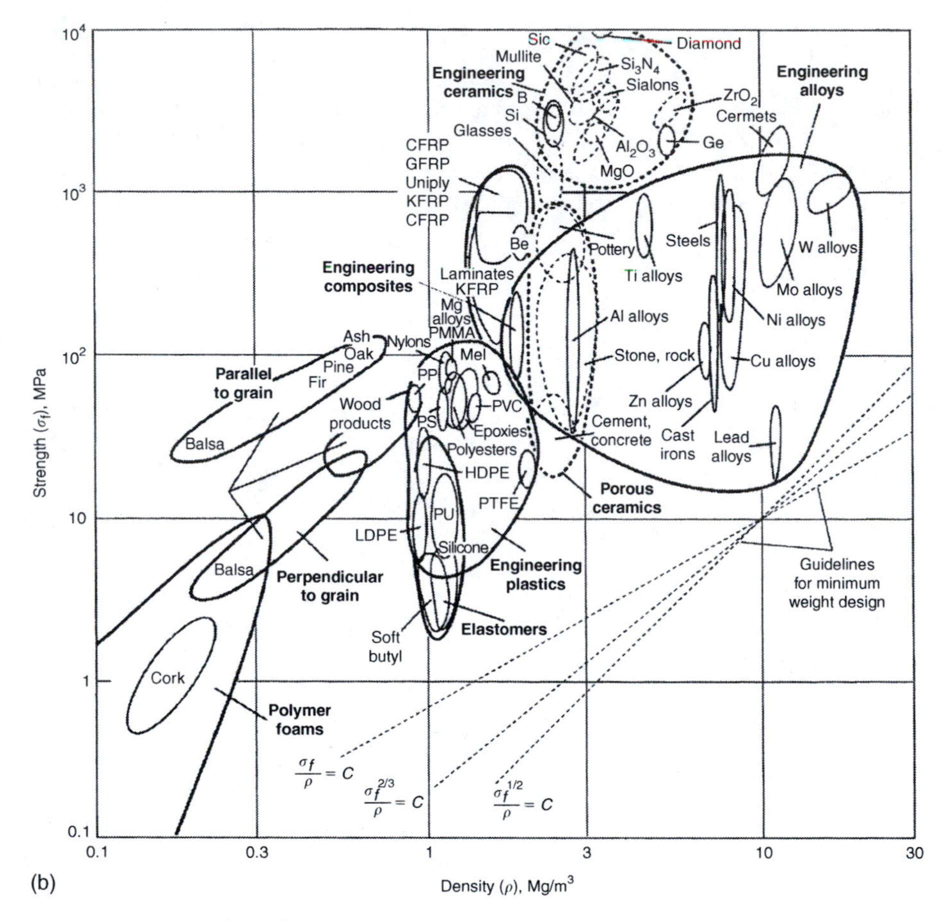

FIGURE 9.2 (Continued)

represent the different conditions. All the materials that lie on a given line will perform equally well, those above the line are better, and those below it are worse.

9.4.4 Dargie's Method

The initial screening of materials and processes can be a tedious task if performed manually from handbooks and supplier catalogs. This difficulty has prompted the introduction of several computer-based systems for materials and process selection (Dargie, Esawi and Ashby, and Weiss are examples). The system proposed by Dargie et al. (1982) (MAPS 1) will be briefly described here and the method proposed by Esawi and Ashby in the next section. Dargie's system proposes a part classification code similar to that used in group technology. The first five digits of the MAPS 1 code are related to the elimination of unsuitable manufacturing processes. The first digit is related to the batch size. The second digit characterizes the bulk and depends on the major dimension and whether the part is long, flat, or compact. The third digit characterizes the shape, which is classified on the basis of being prismatic, axisymmetric, cup shaped, nonaxisymmetric, and nonprismatic. The fourth digit is related to tolerance and the fifth digit is related to surface roughness. The next three digits of the MAPS 1 code are related to the elimination of unsuitable materials. The sixth digit is related to service temperature. The seventh digit is related to the acceptable corrosion rate. The eighth digit characterizes the type of environment to which the part is exposed. The system uses two types of databases for preliminary selection:

- The suitability matrices
- The compatibility matrix

The suitability matrices deal with the suitability of processes and materials for the part under consideration. Each of the code digits has a matrix. The columns of the matrix correspond to the value of the digit and the rows correspond to the processes and materials in the database. The elements of the matrix are either 0, indicating unsuitability, or 2 indicating suitability. The compatibility matrix expresses the compatibility of the different combinations of processes and materials. The columns of the matrix correspond to the materials, whereas the rows correspond to the processes. The elements of the matrix are either 0 for incompatible combinations, 1 for difficult or unusual combinations, or 2 for combinations used in usual practice.

Based on the part code, the program generates a list of candidate combinations of materials and processes to produce it. This list helps the designer to identify possible alternatives early in the design process and to design for ease of manufacture.

9.4.5 Esawi and Ashby's Method

Another quantitative method of initial screening is proposed by Esawi and Ashby. The method compares the approximate cost of resources of materials, energy, capital, time, and information needed to produce the component using different combinations of materials and manufacturing processes. The method can be used early in the design process and is capable of comparing combinations of materials and processes such as the cost of a polymer component made by injection molding with that of a competing design in aluminum made by die-casting.

According to this method, the total cost of a component has three main elements—material cost, tooling cost, and overhead cost. The material cost is a function of the cost per unit weight of the material and the amount of the material needed. Since the cost of tooling (dies, molds, jigs, fixtures, etc.) is normally assigned to a given production run, the tooling cost per component varies as the reciprocal of the number of components produced in that run. The overhead per component varies as the reciprocal of the production rate. Application of this method in initial screening requires a database, such as CES 4, which lists material prices, attributes of different processes, production rates, tool life, and approximate cost of equipment and tooling. CES 4 software contains records for 112 shaping processes such as vapor deposition, casting, molding, metal forming, machining, and composite forming. The software output can be in the form of graphs giving the variation of cost with the batch size for competing process/material combinations. Another type of output is relative cost per unit when a given component is made by different processing routes.

9.5 COMPARING AND RANKING ALTERNATIVE SOLUTIONS

After narrowing down the field of possible materials using one or more of the quantitative initial screening methods described in Section 9.4, quantitative methods can then also be used to further narrow the field of possible materials and matching manufacturing processes to a few promising candidates, which have good combinations of soft requirements. Following is a description of some such methods. It should be emphasized at this stage that none of the proposed quantitative methods is meant to replace the judgment and experience of the engineer. The methods are only meant to help the engineer in ensuring that none of the viable solutions are neglected and in making sounder choices and trade-offs for a given application.

9.5.1 Weighted Properties Method

In the weighted properties method each material requirement, or property, is assigned a certain weight, depending on its importance to the performance of the part in service. A weighted property value is obtained by multiplying the numerical value of the property by the weighting factor (α). The individual weighted property values of each material are then summed to give a comparative materials performance index (γ). Materials with the higher performance index (γ) are considered more suitable for the application.

9.5.2 Digital Logic Method

In the cases where numerous material properties are specified and the relative importance of each property is not clear, determinations of the weighting factors, α, can be largely intuitive, which reduces the reliability of selection. The digital logic approach can be used as a systematic tool to determine α. In this procedure, evaluations are arranged such that only two properties are considered at a time. Every possible combination of properties or goals is compared and no shades of choice are required, only a yes or no decision for each evaluation. To determine the relative importance of each property or goal, a table is constructed, the properties or goals are listed in the left-hand column and comparisons are made in the columns to the right, as shown in Table 9.3.

In comparing two properties or goals, the more important goal is given numerical one (1) and the less important is given zero (0). The total number of possible decisions $N = n(n - 1)/2$, where n is the number of properties or goals under consideration. A relative emphasis coefficient or weighting factor, α, for each goal is obtained by dividing the number of positive decisions for each goal (m) into the total number of possible decisions (N). In this case $\Sigma\alpha = 1$.

To increase the accuracy of decisions based on the digital logic approach, the yes–no evaluations can be modified by allocating gradation marks ranging from 1 (no difference in importance) to 3 (large difference in importance). In this case, the total gradation marks for each selection criterion are reached by adding up the individual gradation marks. The weighting factors are then found by dividing these total gradation marks by their grand total. A simple interactive computer program can be written to help in determining the weighting factors. A computer program will also make it

TABLE 9.3
Determination of the Relative Importance of Goals Using the Digital Logic Method

Goals	Number of Positive Decisions, $N = n(n-1)/2$										Positive Decisions	Relative Emphasis Coefficient, α
		2	3	4	5	6	7	8	9	10		
1	1	1	0	1							3	0.3
2	0				1	0	1				2	0.2
3		0			0			1	0		1	0.1
4			1			1		0		0	2	0.2
5				0			0		1	1	2	0.2
	Total number of positive decisions										10	$\Sigma\alpha = 1.0$

easier to perform a sensitivity analysis, whereby several runs of the process are performed to test the sensitivity of the final ranking to changes in some of the decisions.

9.5.3 Performance Index

In its simple form, the weighted properties method has the drawback of having to combine unlike units, which could yield irrational results. This is particularly true when different mechanical, physical, and chemical properties with widely different numerical values are combined. The property with higher numerical value will have more influence than is warranted by its weighting factor. This drawback is overcome by introducing scaling factors. Each property is so scaled that its highest numerical value does not exceed 100. When evaluating a list of candidate materials, one property is considered at a time. The best value in the list is rated as 100 and the others are scaled proportionally. Introducing a scaling factor facilitates the conversion of normal material property values to scaled dimensionless values. For a given property, the scaled value, B, for a given candidate material is equal to

$$B = \text{Scaled property} = \frac{\text{Numerical value of property} \times 100}{\text{Maximum value in the list}} \tag{9.3}$$

For properties such as cost, corrosion or wear loss, and weight gain in oxidation, a lower value is more desirable. In such cases, the lowest value is rated as 100 and B is calculated as

$$B = \frac{\text{Minimum value in the list} \times 100}{\text{Numerical value of property}} \tag{9.4}$$

For material properties that can be represented by numerical values, application of the procedure discussed is simple. However, with properties such as corrosion and wear resistance, machinability and weldability, and aesthetic quality, numerical values are rarely given and materials are usually rated as very good, good, fair, poor, etc. In such cases, the rating can be converted to numerical values using a relative scale. For example, corrosion resistance rating: excellent, very good, good, fair, and poor can be given numerical values of 5, 4, 3, 2, and 1, respectively. After scaling the different properties, the material performance index (γ) can be calculated as

$$\text{Material performance index, } \gamma = \sum_{i=1}^{n} B_i \alpha_i \tag{9.5}$$

where i is summed over all the n relevant properties.

Cost (stock material, processing, finishing, etc.) can be considered as one of the properties and given the appropriate weighting factor. However, if there are a large number of properties to consider, the importance of cost may be emphasized by considering it separately as a modifier to the material performance index (γ). In the cases where the material is used for space filling, cost can be introduced on per unit volume basis. A figure of merit (M) for the material can then be defined as

$$M = \frac{\gamma}{C\rho} \tag{9.6}$$

where

C = Total cost of the material per unit weight (stock, processing, finishing, etc.)
ρ = Density of the material

When an important function of the material is to bear stresses, it may be more appropriate to use the cost of unit strength instead of the cost per unit volume. This is because higher strength will allow less material to be used to bear the load and the cost of unit strength may be a better representative of the amount of material actually used in making the part. In this case, Equation 9.6 is rewritten as

$$M = \frac{\gamma}{C'} \tag{9.7}$$

where C' is determined from Table 9.1 depending on the type of loading.

This argument may also hold in other cases where the material performs an important function such as electrical conductivity or thermal insulation. In these cases, the amount of the material, and consequently the cost, are directly affected by the value of the property.

When a large number of materials with a large number of specified properties are being evaluated for selection, the weighted properties method can involve large number of tedious and time-consuming calculations. In such cases, the use of computer would facilitate the selection process. The steps involved in the weighted properties method can be written in the form of a simple computer program to select materials from a data bank. The type of material information needed for computer-assisted ranking of alternative solution is normally structured in the form of databases of properties such as those published by ASM, as will be described in Section 9.8. An interactive program can also include the digital logic method to help in determining the weighting factors.

Case study 9.2 illustrates the use of the weighted property method.

Case Study 9.2: Selecting the Optimum Material for a Cryogenic Storage Tank

Problem

It is required to select the optimum material for a large cryogenic storage tank to be used in transporting liquid nitrogen gas.

Analysis

An important rigid requirement for materials used in cryogenic applications is that the material must not suffer ductile–brittle transition at the operating temperature, which is about −196°C (−320.8°F) in this case. This rules out all carbon and low alloy steels and other bcc materials, which suffer ductile–brittle transformation at low temperatures, as discussed in Section 4.5. Fcc materials do not usually become unduly brittle at low temperatures. Many plastics are also excluded on this basis.

Processability is another rigid requirement. As welding is normally used in manufacturing metal tanks, good weldability becomes a rigid requirement. Availability of materials in the required plate thickness and size is also another screening factor.

As a first step, the performance requirements of the storage tank should be translated into material requirements. In addition to having adequate toughness at the operating temperature, the material should be sufficiently strong and stiff. With a stronger material, thinner walls can be used, which means a lighter tank and lower cooldown losses. Thinner walls are also easier to weld. Lower specific gravity is also important as the tank is used in transportation. Lower specific heat reduces cooldown losses, lower thermal expansion coefficient reduces thermal stresses, and lower thermal conductivity reduces heat losses. The cost of material and processing will be used as a modifier to the material performance index, as given in Equation 9.7.

The digital logic method is used to determine the weighting factors. With seven properties to evaluate, the total number of decisions $= N(N - 1)/2 = 7(6)/2 = 21$. The different decisions are given in Table 9.4.

The weighting factor can be calculated by dividing the number of positive decisions given to each property by the total number of decisions. The resulting weighting factors are given in Table 9.5.

Toughness is given the highest weight followed by density. The least important properties are Young's modulus, thermal conductivity, and specific heat; other properties are in between.

The properties of a sample of the candidate materials are listed in Table 9.6. The yield strength and Young's modulus correspond to room temperature that is conservative as they generally increase with decreasing temperature.

The next step in the weighted properties method is to scale the properties given in Table 9.6. For the present application, materials with higher mechanical properties are more desirable and highest values in toughness, yield strength, and Young's modulus are considered as 100. Other values in Table 9.6 are rated in proportion. However, lower values of specific gravity, thermal expansion

TABLE 9.4
Application of Digital Logic Method to Cryogenic Tank Problem

	Decision Numbers																				
Property	**1**	**2**	**3**	**4**	**5**	**6**	**7**	**8**	**9**	**10**	**11**	**12**	**13**	**14**	**15**	**16**	**17**	**18**	**19**	**20**	**21**
Toughness	1	1	1	1	1	1															
Yield strength	0						1	0	0	1	1										
Young's modulus		0					0					0	0	0	1						
Density			0					1				1				1	1	1			
Expansion				0					1				1			0			1	1	
Conductivity					0					0				1			0		0		0
Specific heat						0					0				0			0		0	1

TABLE 9.5
Weighting Factors for Cryogenic Tank

Property	Positive Decisions	Weighting Factor
Toughness	6	0.28
Yield strength	3	0.14
Young's modulus	1	0.05
Density	5	0.24
Expansion	4	0.19
Conductivity	1	0.05
Specific heat	1	0.05
Total	21	1.00

TABLE 9.6
Properties of Candidate Materials for Cryogenic Tank

	1	2	3	4	5	6	7
Material	Toughness Index[a]	Yield Strength (MPa)	Young's Modulus (GPa)	Specific Gravity	Thermal Expansion[b]	Thermal Conductivity[c]	Specific Heat[d]
Al 2014-T6	75.5	420	74.2	2.8	21.4	0.37	0.16
Al 5052-O	95	91	70	2.68	22.1	0.33	0.16
SS 301-FH	770	1365	189	7.9	16.9	0.04	0.08
SS 310-3/4H	187	1120	210	7.9	14.4	0.03	0.08
Ti-6Al-4V	179	875	112	4.43	9.4	0.016	0.09
Inconel 718	239	1190	217	8.51	11.5	0.31	0.07
70Cu-30Zn	273	200	112	8.53	19.9	0.29	0.06

[a] Toughness index (TI), is based on UTS, yield strength (YS), and ductility e, at −196°C (−321.8°F) TI = (UTS + YS)e/2.

[b] Thermal expansion coefficient is given in 10^{-6}/°C. The values are averaged between RT and −196°C.

[c] Thermal conductivity is given in cal/cm^2/cm/°C/s.

[d] Specific heat is given in cal/g/°C. The values are averaged between RT and −196°C.

coefficient, thermal conductivity, and specific heat are more desirable for this application. Accordingly, the lowest values in the table were considered as 100 and other values rated in proportion according to Equation 9.4. The scaled values are given in Table 9.7. The table also gives the performance index that is calculated according to Equation 9.5.

The performance index shows the technical capability of the material without regard to the cost. In this case, stainless steels are the optimum materials. It now remains to consider the cost aspects by calculating the figure of merit (M). In the present case, it is more appropriate to use Equation 9.7 as the primary

TABLE 9.7
Scaled Values of Properties and Performance Index

Material	Scaled Properties							Performance Index (γ)
	1	2	3	4	5	6	7	
Al 2014–T6	10	30	34	96	44	4.3	38	42.2
Al 5052-O	12	6	32	100	43	4.8	38	40.1
SS 301-FH	100	100	87	34	56	40	75	70.9
SS 310-3/4H	24	82	97	34	65	53	75	50.0
Ti–6Al–4V	23	64	52	60	100	100	67	59.8
Inconel 718	31	87	100	30	82	5.2	86	53.3
70Cu–30Zn	35	15	52	30	47	5.5	100	35.9

TABLE 9.8
Cost, Figure of Merit, and Ranking of Candidate Materials

Material	Relative Cost[a]	Cost of Unit Strength × 100	Performance Index	Figure of Merit	Rank
Al 2014-T6	1	0.67	42.2	62.99	2
Al 5052-O	1.05	3.09	40.1	12.98	6
SS 301-FH	1.4	0.81	70.9	87.53	1
SS 310-3/4H	1.5	1.06	50.0	47.17	3
Ti–6Al–4V	6.3	3.20	59.8	18.69	4
Inconel 718	5.0	3.58	53.3	14.89	5
70Cu–30Zn	2.1	8.96	35.9	4.01	7

[a] The cost includes stock materials and processing cost. The relative cost is obtained by considering the cost of Al 2014 as unity and relating the cost of other materials to it.

function of the tank material is to bear stresses. The formula for a thin-wall pressure vessel is given in Table 9.1 as

$$\text{Cost of unit strength} = \frac{C\rho}{S}$$

where S is the yield strength.

The values of the relative cost, cost of unit strength, performance index, figure of merit, M, and the ranking of the different materials is shown in Table 9.8. The results show that full hard stainless steel grade 301 is the optimum material followed by Al 2014-T6.

In the procedure discussed, the strength and density were considered twice, once in calculating the performance index (γ) and another time in calculating the cost of unit strength. This procedure may have overemphasized their effect on the final selection. This could be justifiable in this case as higher strength and lower density are advantageous from the technical and economic points of view.

9.5.4 Limits on Properties Method

In the limits on properties method, the performance requirements are divided into three categories:

- Lower limit properties
- Upper limit properties
- Target value properties

For example, if it is desired to have a strong light material, a lower limit on the strength and an upper limit on the density are specified. When compatibility between materials is important, a target value for the thermal expansion coefficient or for the position in the galvanic series may be specified to control thermal stresses or galvanic corrosion, respectively. Whether a given property is specified as an upper or lower limit may depend upon the application. For example, when selecting materials for an electrical cable, the electrical conductivity will be specified as a lower limit property for the conducting core and as an upper limit property for the insulation outer layer.

The limits on properties method are usually suitable for optimizing material and process selection when the number of possible alternatives is relatively large. This is because the limits that are specified for the different properties can be used for eliminating unsuitable materials from a data bank. The remaining materials are those whose properties are above the lower limits, below the upper limits, and within the limits of target values of the respective specified requirements. After the screening stage, the limits on properties method can then be used to optimize the selection from among the remaining materials.

As in the case of weighted properties method, each of the requirements or properties is assigned a weighting factor, α, which can be determined using the digital logic method, as discussed earlier. A merit parameter, m, is then calculated for each material according to the relationship:

$$m = \left(\sum_{i=1}^{n_l} \alpha_i \frac{Y_i}{X_i}\right)_l + \left(\sum_{j=1}^{n_u} \alpha_j \frac{X_i}{Y_i}\right)_u + \left(\sum_{k=1}^{n_t} \alpha_k \left|\frac{X_k}{Y_k} - 1\right|\right)_t \tag{9.8}$$

where

l, u, and t = Lower limit, upper limit, and target value properties, respectively

n_l, n_u, and n_t = Numbers of lower limit, upper limit, and target value properties, respectively

α_i, α_j, and α_k = Weighting factors for the lower limit, upper limit, and target value properties, respectively

X_i, X_j, and X_k = Candidate material lower limit, upper limit, and target value properties, respectively

Y_i, Y_j, and Y_k = Specified lower limits, upper limits, and target values, respectively

According to Equation 9.8, the lower the value of the merit parameter, m, the better the material.

As in the weighted properties method, the cost can be considered in two ways:

1. Cost is treated as an upper limit property and is given the appropriate weight. When the number of properties under consideration is large, this procedure may obscure its importance.
2. Cost is included as a modifier to the merit parameter as follows:

$$m' = \frac{\text{CX}}{\text{CY}} m \tag{9.9}$$

where

CY and CX = Specified cost upper limit and the candidate material cost, respectively

m = Merit parameter calculated without taking the cost into account

In this case, the material with the lowest cost-modified merit parameter, m', is the optimum.

Case study 9.3 illustrates the use of the limits on properties method in selection.

Case Study 9.3: Selecting an Insulating Material for a Flexible Electrical Conductor

Problem

Consider the case of selecting an insulating material for a flexible electrical conductor for a computer system. Space saving and adaptability to special configurations are important. Service temperature will not exceed 75°C (167°F). Cost is an important consideration because large quantities of these cables will be used in installing the system.

Analysis

Rigid requirements in this case are flexibility, or ductility, of the insulating material and operating temperature. The requirement for ductility eliminates all ceramic insulating materials, and the operating temperature eliminates some plastics such as low-density polyethylene.

The next step is to analyze the electrical and physical design requirements. They are as follows:

1. Dielectric strength, which is related to the breakdown voltage; it is a lower-limit property in this case. Owing to space limitations, the dielectric strength should be more than 10,000 V/mm.
2. Insulating resistance depends on both the resistivity of the material and the geometry of the insulator; it is the lower limit property. The minimum acceptable value is 10^{14} Ω/cm.

3. Dissipation factor affects the power loss in the material due to the alternating current and is an upper limit property. The maximum acceptable value is 0.0015 at 60 Hz.
4. Dielectric constant is a measure of the electrostatic energy stored in the material and affects the power loss. This property is an upper limit requirement in this case, although it is taken as a lower limit property in applications such as capacitors. The maximum allowable value is 3.5 at 60 Hz.
5. Thermal expansion coefficient is a target value to ensure compatibility between the conductor and insulator at different temperatures. As the conductor is made of aluminum, the target value for the expansion coefficient is 2.3×10^{-5}/°C.
6. Specific gravity is an upper limit property to ensure lightweightedness. This property will not be considered here because weight is not critical.

Table 9.9 gives the properties of some candidate materials, which do not violate the rigid requirements of ductility and operating temperature and also satisfy the upper or lower limits of design.

The first step is to determine the weighting factors for the different properties. Because the number of properties is relatively small, the cost will be included as one of the properties. The digital logic method is used as shown in Table 9.10. The number of properties under consideration is 6 and the total number of decisions $= 6(6 - 1)/2 = 15$.

The next step in the selection process is to calculate the merit parameter, m, using Equation 9.8 for the different materials. In the present case, the dielectric strength and volume resistance are lower limit properties; dissipation factor, dielectric constant, and cost are upper limit properties; and thermal expansion coefficient is a target value. In calculating the relative merit of the different materials, the log value of the volume resistivity was used. As no upper limit is given to the cost, the cost of the most expensive material in Table 9.9 is taken as the upper limit. The relative merit parameter, m, and the rank of the different materials are given in Table 9.11.

TABLE 9.9
Properties of Some Candidate-Insulating Materials

Material	Dielectric Strength (V/mm)	Volume Resistance (Ω/cm)	Dissipation Factor (60 Hz)	Dielectric Constant (60 Hz)	Thermal Expansion (10^{-5}/°C)	Relative Cost[a]
PTFE	14,820	10^{18}	0.0002	2.1	9.5	4.5
CTFE	21,450	10^{18}	0.0012	2.7	14.4	9.0
ETFE	78,000	10^{16}	0.0006	2.6	9.0	8.5
Polyphenylene oxide	20,475	10^{17}	0.0006	2.6	6.5	2.6
Polysulfone	16,575	10^{14}	0.0010	3.1	5.6	3.5
Polypropylene	21,450	10^{16}	0.0005	2.2	8.6	1.0

[a] Cost includes material and processing cost. Relative cost is based on the cost of material and processing of polypropylene.

TABLE 9.10
Weighting Factors for an Electrical Insulator

Property	Decision Numbers															Total	Weighting Factor
	1	2	3	4	5	6	7	8	9	10	11	12	13	14	15		
Dielectric strength	0	1	1	0	1											3	0.20
Volume resistance	1					1	1	1	1							5	0.33
Dissipation factor		0				0				1	1	0				2	0.13
Dielectric constant			0				0			0			1	0		1	0.07
Thermal expansion				1				0			0		0		0	1	0.07
Cost					0				0			1		1	1	3	0.20
Total																15	1.00

TABLE 9.11
Evaluation of Insulating Materials

Material	Merit Parameter (m)	Rank
PTFE	0.78	3
CTFE	1.07	6
ETFE	0.81	5
Polyphenylene oxide	0.66	1
Polysulfone	0.78	3
Polypropylene	0.66	1

Conclusion

The results show that polypropylene and polyphenylene oxide have equal merit parameters. The final selection between the two materials may depend on availability and possibility of coloring.

9.5.5 Analytic Hierarchy Process

Because arriving at an optimum material involves decision making in the presence of multiple and often conflicting criteria, the models of multiple criteria decision making can be adopted for solving materials selection problems. Such models require complex hierarchical comparisons among candidates, which make it necessary for the decision maker to adopt one of several decision support tools and methodologies such as statistical techniques, outranking techniques, and analytic hierarchy process (AHP). The AHP was developed by Thomas Saaty in 1980 and provides an effective tool to solving multicriteria decision-making problems where there are a limited number of alternatives but each has a number of attributes, which make it suitable for adoption to materials selection problems. AHP is particularly suited for the cases where some or all of the selection criteria (decision elements) are difficult to quantify.

AHP starts by building a hierarchy consisting of a goal and subordinate elements of the problem. The alternatives among which the choice is to be made are placed at the lowest level in the hierarchy. Starting from the top of the hierarchy, pairwise comparisons are then made among all the elements at a particular level in the hierarchy. This is achieved by comparing each possible pair of the performance requirements on a scale of 9 to 1 as follows:

- 9 is given to one requirement when it is judged to be extremely more important to the performance than the other
- 7 when it is very strongly more important
- 5 when it is strongly more important
- 3 when it is moderately more important
- 1 when it is equally important or when comparing a requirement to itself

Intermediate values can be used for intermediate preferences.

Case study 9.4 gives a simple example to illustrate how the processes involved in AHP can be used in materials selection. A more realistic example of using AHP in selection is given in a case study in Chapter 11—selection of materials for tennis rackets.

Case Study 9.4: Using AHP to Select the Optimum Material for a Roof Truss

Problem

It is required to select the optimum material for a roof truss of a small warehouse. The candidate materials are AISI 1020, AISI 4130, AA 6061, and a composite material (epoxy–70% glass fabric). The performance requirements for the truss material are assumed to be high strength (σ), high elastic modulus (E), low density (ρ), and low cost (C). The properties of the candidate materials are given in Table 9.12.

The hierarchy diagram for this example is shown in Figure 9.3.

Table 9.13 illustrates the pairwise comparison for the material requirements of truss. The table is constructed by comparing each of the requirements to all others noting that when the importance of E relative to σ is 5, then the importance of σ relative to E is 1/5.

The next step is to construct the table of weights (Table 9.14) by normalizing the pairwise comparisons in the table of preferences (Table 9.13) by dividing the number in a given cell by the sum of the numbers in its column, for example, the number in the cell $\sigma - \sigma$ is $[1/(1 + 5 + 3 + 2) = 0.091]$. With no inconsistency in decision making, the numbers in a given row should be equal and represent the weight of the property. The sum of each column should also be 1.00. Averaging

TABLE 9.12
Properties of the Candidate Materials for the Truss

	Yield Strength (σ), MPa	Elastic Modulus (E), GPa	Density (ρ), g/cc	Cost Category (C)[a]
AISI 1020	280	210	7.8	5
AISI 4130	1520	212	7.8	3
AA 6061	275	70	2.7	4
Epoxy–70% glass fabric	1270	28	2.1	2

[a] 5, Very inexpensive; 4, inexpensive; 3, moderate price; 2, expensive; 1, very expensive.

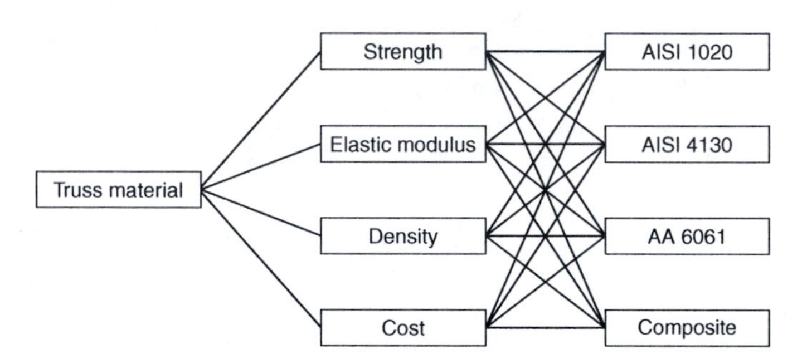

FIGURE 9.3 A simple AHP diagram for selecting a material for a truss.

TABLE 9.13
Pairwise Comparison of Material Requirements

	σ	E	ρ	C
σ	1	1/5	1/3	1/2
E	5	1	2	4
ρ	3	1/2	1	3
C	2	1/4	1/3	1

TABLE 9.14
Calculation of Weights

	σ	E	ρ	C	Average/ Weight	Consistency Measure
σ	0.091	0.102	0.091	0.059	0.086	4.02
E	0.455	0.513	0.545	0.471	0.496	4.07
ρ	0.273	0.256	0.273	0.353	0.289	4.09
C	0.182	0.128	0.091	0.118	0.129	4.04
Total/average	1.001	0.999	1.000	1.001	1.000	4.055

the values of each row should result in a better estimate of the weight of the property in this row and the sum of all the weights should be 1.00.

One of the advantages of AHP is that it allows subjective decisions to be formalized and gives the degree of consistency of such decisions. In the present case, the consistency of the pairwise comparison can be assured before adopting the estimated weights. Consistency measures for the weights allocated to the different properties is calculated as

$$\text{Consistency, measure for } \sigma = \frac{(0.086 \times 1 + 0.496 \times 1/5 + 0.289 \times 1/3 + 0.129 \times 1/2)}{0.086} = 4.02$$

$$\text{Consistency measure for } E = \frac{(0.086 \times 5 + 0.496 \times 1 + 0.289 \times 2 + 0.129 \times 4)}{0.496} = 4.07$$

$$\text{Consistency measure for } \rho = \frac{(0.086 \times 3 + 0.496 \times 1/2 + 0.289 \times 1 + 0.129 \times 3)}{0.289} = 4.09$$

$$\text{Consistency measure for } C = \frac{(0.086 \times 2 + 0.496 \times 1/4 + 0.289 \times 1/3 + 0.129 \times 1)}{0.129} = 4.04$$

If the decision-making process is perfectly consistent, each consistency measure will equal the number of properties, which is 4 in this case. The foregoing results show that there is some inconsistency. Whether this inconsistency is acceptable depends on the value of the consistency index (CI) and consistency ratio (CR), which are calculated as follows:

$$\text{CI} = \frac{\lambda - n}{n - 1} \quad \text{and} \quad \text{CR} = \frac{\text{CI}}{\text{RI}} \tag{9.10}$$

where

λ = Average consistency measure for all properties, equal to 4.055 in this case
n = Number of properties, equal to 4 in this case
RI = Random index, which depends on n as shown in Table 9.15.

RI values in Table 9.15 give the average values of CI if the preference entries in the table of preferences were chosen at random. If the decision-making process is perfectly consistent, $\lambda = n$ and CR = 0. Normally, CR < 0.10 are considered acceptable. With CR > 0.10, the AHP may not give meaningful results.

For the present case

$$\text{CI} = \frac{4.055 - 4}{4 - 1} = 0.018 \quad \text{and} \quad \text{CR} = \frac{0.018}{0.9} = 0.02$$

TABLE 9.15
Random Index as a Function of the Number of Properties (n)

n	2	3	4	5	6	7	8	9	10
RI	0.00	0.58	0.90	1.12	1.24	1.32	1.41	1.45	1.49

TABLE 9.16
Results of AHP for the Truss Materials

			Major Contributions to the Score			
Material	**Rank**	**Score**	σ (%)	E (%)	ρ (%)	C (%)
AISI 1020	2	0.286		77		23
AISI 4130	1	0.293	16	76		8
AA 6061	3	0.231		22	59	19
Composite	4	0.191	20		80	

These results show that the consistency of the pairwise comparison in Table 9.13 is acceptable and that the weights allocated to the different properties are, therefore, consistent.

Comparing Alternatives

The final step in the AHP is to compare each of the alternative materials with respect to each of the properties. In this case, a matrix similar to Table 9.13 is constructed in which the materials are compared pairwise relative to each of the properties to determine their relative priorities. The total score for a given material is obtained by summing up its priorities with respect to the properties, taking the weight of each of the properties into account. The material with highest total score is selected.

Conclusion

As this simple example shows, the decision-making process using AHP can involve a large number of tedious calculations, especially in the cases where the number of properties/alternatives is large. Fortunately, however, such calculations can be performed on a spreadsheet (e.g., as described by Ragsdale), or using one of several of the free web-based (e.g., easymind) or commercial software products (e.g., Decision Plus by Info Harvest or Hiview by Catalyze).

A student version of Decision Plus software, which is available at no cost from Info Harvest, is used to solve this case study. The properties given in Table 9.12 and the weights that were calculated in Table 9.14 were used. The results are given in Table 9.16 and show that AISI 4130 achieved the highest score; its main assets are high strength and elastic modulus. A close second is AISI 1020, its main assets are high elastic modulus and low cost.

9.6 SELECTING THE OPTIMUM SOLUTION

Having ranked alternative materials using one of the methods in Section 9.5, the final step is selection of the optimum material. Candidates that have the most promising performance indices can now be used to develop a detail design each. Each detail design will exploit the points of strength of the material, avoid the weak points, and reflect the requirements of the manufacturing processes needed for the material.

The type of material information needed for detail design is different from that needed for initial screening and ranking. What is needed at this stage is detailed high-quality information about the highest-ranking candidates. As shown in Section 9.8, such information is usually unstructured and can be obtained from handbooks, publications of trade organizations, and technical reports in the form of text, pdf files, tables, graphs, photographs, etc. There are instances where some of the desired data may not be available or may be available for slightly different test conditions. In such cases educated judgment is required.

After completing the different designs, solutions are then compared, taking the cost elements into consideration to arrive at the optimum design–material–process combination, as illustrated in Case study 9.5.

Case Study 9.5: Reaching a Final Decision on the Optimum Material for a Sailing-Boat Mast Component

Problem

Select the least expensive component that satisfies the requirements for a simple structural component for a sailing boat mast in the form of a hollow cylinder of length 1000 mm, which is subjected to compressive axial forces of 153 kN. Because of space and weight limitations, the outer diameter of the component should not exceed 100 mm, the inner diameter should not be less than 84 mm, and the mass should not exceed 3 kg. The component will be subjected to mechanical impact and spray of water. Assembly to other components requires the presence of relatively small holes.

Material Performance Requirements

Analysis in this case study is based on the paper by Farag and El-Magd (1992). Possible modes of failure and the corresponding material properties that are needed to resist failure for the present component include the following:

- Catastrophic fracture due to impact loading, especially near assembly holes, is resisted by high fracture toughness of the material. This is a rigid material requirement and will be used for initial screening of materials.
- Plastic yielding is resisted by high yield strength. This is a soft material requirement but a lower limit will be determined by the limitation on the outer diameter.
- Local and global buckling are resisted by high elastic modulus. This is a soft material requirement but a lower limit will be determined by the limitation on the outer diameter.
- Internal fiber buckling for fiber-reinforced materials is resisted by high modulus of elasticity of the matrix and high volume fraction of fibers in the loading direction. This is a soft material requirement but a lower limit will be determined by the limitation on the outer diameter.

- Corrosion that can be resisted either by selecting materials with inherently good corrosion resistance or by protective coating.
- Reliability of the component in service. A factor of safety of 1.5 is taken for the axial loading, that is, the working axial force will be taken as 230 kN to improve reliability.

In addition to the mentioned requirements, the limitations set on dimensions and weight should be observed.

Initial Screening of Materials

The requirement for fracture toughness of the material is used to eliminate ceramic materials. Because of the limitations set on the outer and inner diameters, the maximum possible cross section of the component is about 2300 mm^2. To avoid yielding under the axial working load, the yield strength of the material should be more than 100 MPa, which excludes engineering polymers, woods, and some of the lower strength engineering alloys (see Figure 9.2). Corrosion resistance is desirable but will not be considered a factor for screening, since the possibility of protection for less corrosion materials exists, but will be considered as a soft requirement.

Comparing and Ranking Alternative Solutions

Table 9.17 shows a sample of materials that satisfy the conditions set in the initial screening stage. In a real-life situation the list in the table could be much longer, but the intent here is to illustrate the procedure. The yield strength, elastic modulus, specific gravity, corrosion resistance, and cost category are given for each of the materials. At this stage, it is sufficient to classify materials into very inexpensive, inexpensive, etc. Better estimate of the material and manufacturing cost will be needed in making the final decision in selection. Because the weight of the component is important in this application, specific strength and specific modulus would be better indicators of the suitability of the material (Table 9.18). The relative importance of the material properties is given in Table 9.19 and the performance indices of the different materials, as determined by the weighted properties method, are given in Table 9.20. The seven candidate materials with high-performance indices ($\gamma > 45$) are selected for making actual component designs.

Selecting the Optimum Solution

As shown earlier, the possible modes of failure of a hollow cylinder include yielding, local buckling and global buckling, and internal fiber buckling. These four failure modes are used to develop the design formulas for the mast component. For more details on the design and optimization procedure or more details about Equations 9.11 through 9.14, please refer to the paper by Farag and El-Magd (1992).

TABLE 9.17
Properties of Sample Candidate Materials

Material	Yield Strength (MPa)	Elastic Modulus (GPa)	Specific Gravity	Corrosion Resistance[a]	Cost Category[b]
AISI 1020 (UNS G10200)	280	210	7.8	1	5
AISI 1040 (UNS G10400)	400	210	7.8	1	5
ASTM-A242 Type1 (UNS K11510)	330	212	7.8	1	5
AISI 4130 (UNS G41300)	1520	212	7.8	4	3
AISI 316 (UNS S31600)	205	200	7.98	4	3
AISI 416 Ht. Treated (UNS S41600)	440	216	7.7	4	3
AISI 431 Ht. Treated (UNS S43100)	550	216	7.7	4	3
AA 6061 T6 (UNS A96061)	275	69.7	2.7	3	4
AA 2024 T6 (UNS A92024)	393	72.4	2.77	3	4
AA 2014 T6 (UNS A92014)	415	72.1	2.8	3	4
AA 7075 T6 (UNS A97075)	505	72.4	2.8	3	4
Ti-6Al-4V	939	124	4.5	5	1
Epoxy–70% glass fabric	1270	28	2.1	4	2
Epoxy–63% carbon fabric	670	107	1.61	4	1
Epoxy–62% aramid fabric	880	38	1.38	4	1

Note: Based on Farag, M.M. and El-Magd, E., *Mater. Design*, 13, 323–327, 1992.

[a] 5, Excellent; 4, very good; 3, good; 2, fair; 1, poor.

[b] 5, Very inexpensive; 4, inexpensive; 3, moderate price; 2, expensive; 1, very expensive.

Condition for yielding:

$$\frac{F}{A} < \sigma_y \tag{9.11}$$

where

σ_y = Yield strength of the material
F = External working axial force
A = Cross-sectional area

TABLE 9.18
Properties of Sample Candidate Materials

Material	Specific Strength (MPa)	Specific Modulus (GPa)	Corrosion Resistance[a]	Cost Category[b]
AISI 1020 (UNS G10200)	35.9	26.9	1	5
AISI 1040 (UNS G10400)	51.3	26.9	1	5
ASTM-A242 Type1 (UNS K11510)	42.3	27.2	1	5
AISI 4130 (UNS G41300)	194.9	27.2	4	3
AISI 316 (UNS S31600)	25.6	25.1	4	3
AISI 416 Ht. Treated (UNS S41600)	57.1	28.1	4	3
AISI 431 Ht. Treated (UNS S43100)	71.4	28.1	4	3
AA 6061 T6 (UNS A96061)	101.9	25.8	3	4
AA 2024 T6 (UNS A92024)	141.9	26.1	3	4
AA 2014 T6 (UNS A92014)	148.2	25.8	3	4
AA 7075 T6 (UNS A97075)	180.4	25.9	3	4
Ti–6Al–4V	208.7	27.6	5	1
Epoxy–70% glass fabric	604.8	28	4	2
Epoxy–63% carbon fabric	416.2	66.5	4	1
Epoxy–62% aramid fabric	637.7	27.5	4	1

[a] 5, Excellent; 4, very good; 3, good; 2, fair, 1, poor.

[b] 5, Very inexpensive; 4, inexpensive; 3, moderate price; 2, expensive; 1, very expensive.

TABLE 9.19
Weighting Factors

Property	Specific Strength (MPa)	Specific Modulus (GPa)	Corrosion Resistance	Relative Cost
Weighting factor (α)	0.3	0.3	0.15	0.25

Condition for local buckling:

$$\frac{F}{A} < 0.121\,\frac{ES}{D} \tag{9.12}$$

where

D = Outer diameter of the cylinder
S = Wall thickness of the cylinder
E = Elastic modulus of the material

TABLE 9.20
Calculation of the Performance Index

Material	Scaled-Specific Strength × 0.3	Scaled-Specific Modulus × 0.3	Scaled Corrosion Resistance × 0.15	Scaled Relative Cost × 0.25	Performance Index
AISI 1020 (UNS G10200)	1.7	12.3	3	25	42
AISI 1040 (UNS G10400)	2.4	12.3	3	25	42.7
ASTM-A242 Type 1 (UNS K11510)	2	12.3	3	25	42.3
AISI 4130 (UNS G41300)	9.2	12.3	6	15	42.5
AISI 316 (UNS S31600)	1.2	11.3	12	15	39.5
AISI 416 Ht. Treated (UNS S41600)	2.7	12.7	12	15	42.4
AISI 431 Ht. Treated (UNS S43100)	3.4	12.7	12	15	43.1
AA 6061 T6 (UNS A96061)	4.8	11.6	9	20	45.4
AA 2024 T6 (UNS A92024)	6.7	11.8	9	20	47.5
AA 2014 T6 (UNS A92014)	7	11.6	9	20	47.6
AA 7075 T6 (UNS A97075)	8.5	11.7	9	20	49.2
Ti–6Al–4V	9.8	12.5	15	5	42.3
Epoxy–70% glass fabric	28.4	12.6	12	10	63
Epoxy–63% carbon fabric	19.6	30	12	5	66.6
Epoxy–62% aramid fabric	30	12.4	12	5	59.4

Condition for global buckling:

$$\sigma_y > \frac{F}{A}\left[1 + \left(\frac{LDA}{1000I}\right)\sec\left\{\left(\frac{F}{EI}\right)^{1/2}\frac{L}{2}\right\}\right] \tag{9.13}$$

where

I = Second moment of area
L = Length of the component

Condition for internal fiber buckling:

$$\frac{F}{A} < \left[\frac{E_m}{4(1 + v_m)(1 - V_f^{1/2})}\right] \tag{9.14}$$

where

E_m = Elastic modulus of the matrix material
v_m = Poisson's ratio of the matrix material
V_f = Volume fraction of the fibers parallel to the loading direction

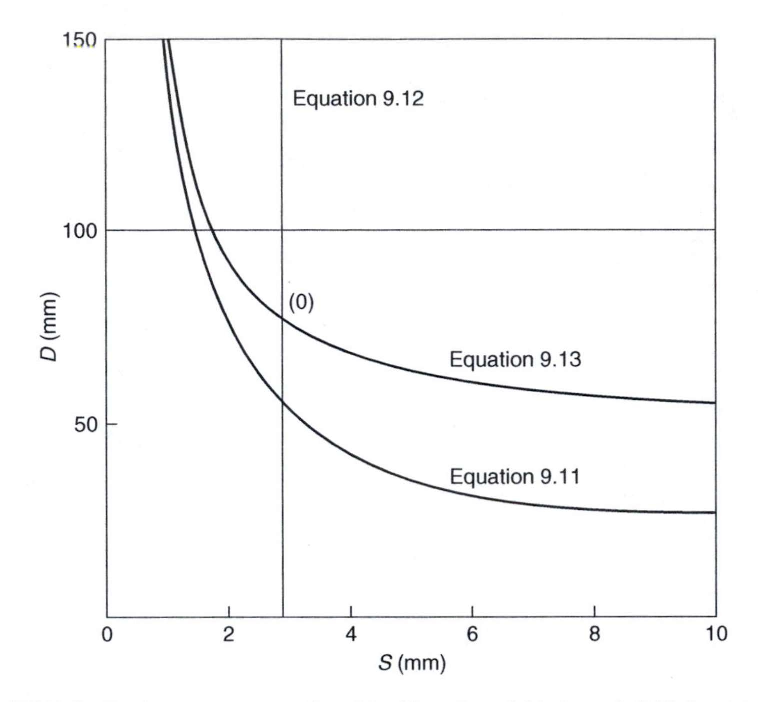

FIGURE 9.4 Design range as predicted by Equations 9.11 through 9.13 for AA 7075 aluminum alloy. (Based on Farag M.M. and El-Magd E., *Mater. Design*, 13, 323–327, 1992.)

TABLE 9.21
Designs Using Candidate Materials with Highest Performance Indices

Material	D_a (mm)	S (mm)	A (mm^2)	Mass (kg)	Cost/kg ($)	Cost of Component ($)
AA 6061 T6 (UNS A96061)	100	3.4	1065.7	2.88	8	23.2
AA 2024 T6 (UNS A92024)	88.3	2.89	801.1	2.22	8.3	18.4
AA 2014 T6 (UNS A92014)	85.6	2.89	776.6	2.17	9	19.6
AA 7075 T6 (UNS A97075)	78.1	2.89	709.1	1.99	10.1	20
Epoxy—70% glass fabric	78	4.64	1136.3	2.39	30.8	73.6
Epoxy—63% carbon fabric	73.4	2.37	546.1	0.88	99	87.1
Epoxy—62% aramid fabric	75.1	3.99	941.6	1.30	88	114.4

Note: Based on Farag, M.M. and El-Magd, E., *Mater. Design*, 13, 323–327, 1992.

Figure 9.4 shows the optimum design range of component diameter and wall thickness as predicted by Equations 9.11 through 9.14 for AA 7075 aluminum alloy. Point (O) represents the optimum design. Similar figures were developed for the different candidate materials to determine the mast component's optimum design dimensions when made of the materials and the results are shown in Table 9.21. Although all the materials in Table 9.21 can be used to make safe components that comply with the space and weight limitations, AA2024 T6 is selected since it gives the least expensive solution.

9.7 COMPUTER ASSISTANCE IN MAKING FINAL SELECTION

9.7.1 CAD/CAM Systems

Integrating material property database with design algorithms and CAD/CAM programs has many benefits including homogenization and sharing of data in the different departments, decreased redundancy of effort, and decreased cost of information storage and retrieval. Several such systems have been cited by Boardman and Kaufman (2000), the following is a sample:

- The computerized application and reference system (CARS) is developed by the AISI *Automotive Steel Design Manual* and performs first-order analysis of design using different steels.
- Aluminum design system (ADS) is developed by the Aluminum Association (United States) and performs design calculations and conformance checks of aluminum structural members with the design specifications for aluminum and its alloys.
- Material selection and design for fatigue life predictions is developed by ASM International and aids in the design of machinery and engineering structures using different engineering materials.
- Machine design's materials selection, developed by Penton Media (United States), combines the properties for a wide range of materials and the data set for design analysis.

9.7.2 Expert Systems

Expert systems, also called knowledge-based systems, are computer programs, which simulate the reasoning of a human expert in a given field of knowledge. Expert systems rely on heuristics, or rules of thumb, to extract information from a large knowledge base. Expert systems typically consist of three main components:

- Knowledge base, which contains facts and expert level heuristic rules for solving problems in a given domain. The rules are normally introduced to the system by domain experts through a knowledge engineer.
- Inference engine, which provides an organized procedure for sifting through the knowledge base and choosing applicable rules to reach the recommended

solutions. The inference engine also provides a link between the knowledge base and the user interface.
- User interface, which allows the user to input the main parameters of the problem under consideration. It also provides recommendations and explanations of how such recommendations were reached.

A commonly used format for the rules in the knowledge base is in the form:

IF (condition 1) and/or (condition 2)
THEN (conclusion 1)

For example, in the case of FRP selection:

IF: required elastic modulus, expressed in GPa, is more than 150 and specific gravity less than 1.7.
THEN: oriented carbon fibers at 60% by volume.

Expert systems are finding many applications in industry including the areas of design, troubleshooting, failure analysis, manufacturing, materials selection, and materials substitution. When used to assist in materials selection, expert systems provide impartial recommendations and are able to search large databases for optimum solutions. Another important advantage of expert systems is their ability to capture valuable expertise and make it available to a wider circle of users. An example is the Chemical corrosion expert system, which is produced by the National Association of Corrosion Engineers (NACE) in the United States, as reported by Boardman and Kaufman (2000). The system prompts the user for information on the environmental conditions and configuration of the component and then recommends candidate materials.

9.8 SOURCES OF INFORMATION FOR MATERIALS SELECTION

9.8.1 Locating Materials Properties Data

One essential requisite to successful materials selection is a source of reliable and consistent data on materials properties. There are many sources of information that include governmental agencies, trade associations, engineering societies, textbooks, research institutes, and materials producers. Locating the appropriate type of information is not easy and may require several cycles of iteration until the information needed is gathered. According to Kirkwood, the steps involved in each cycle are: define the question, set up a strategy to locate the needed information, use the resources you know best, go to less-known sources when necessary, evaluate the quality of data/information resources, start again if needed using the new information to help define the question better. To locate useful sources of materials data, it is important to identify the intended use of data, the type of data required, and the quality of data required. In general, progressively higher quality of data is needed as the selection process progresses from initial screening, to ranking, and finally to selecting the optimum solution.

The ASM International has recently published a directory of materials property databases (Boardman and Kaufman, 2000), which contains more than 500 data sources, including both specific databases and data centers. For each source, the directory gives a brief description of the available information, address, telephone number, e-mail, Web site, and approximate cost if applicable. The directory also has indices by material and property to help the user in locating the most appropriate source of material information. Much of the information is available on CD-ROM or PC disk, which makes it possible to integrate the data source in computer-assisted selection systems. Other useful reviews of the sources of materials property data and information are also given in Westbrook (1997) and Price (1993).

9.8.2 Types of Material Information

According to Cebon and Ashby, materials information can be classified into structured information either from reference sources or developed in-house and unstructured information either from reference sources or developed in-house. Structured information is normally in the form of databases of properties and is most suited for initial screening and for comparing and ranking of materials. Examples of structured reference sources of information material properties include ASM, Materials Universe, and Matweb databases. These databases, in addition to several others, are collected online under the Material Data Network (www.matdata.net).

Unstructured information gives details about performance of specific materials and is normally found in handbooks, publications of trade organizations, and technical reports in the form of text, pdf files, tables, graphs, photographs, etc. Such information is most suited for detailed consideration of the top-ranking candidates who were selected in the initial screening and the ranking stages, as discussed in Section 9.6.

9.8.3 Computerized Materials Databases

Computerized materials databases are an important part of any computer-aided system for selection. With an interactive database, as in the case of ASM metal selector, the user can define and redefine the selection criteria to gradually sift the materials and isolate the candidates that meet the requirements. In many cases, sifting can be carried out according to different criteria such as

1. Specified numeric values of a set of material properties
2. Specified level of processability such as machinability, weldability, formability, availability, and processing cost
3. Class of material, for example, fatigue-resistant, corrosion-resistant, heat-resistant, and electrical materials
4. Forms like rod, wire, sheet, tube, cast, forged, and welded
5. Designations such as UNS numbers, AISI numbers, common names, material group, or country of origin
6. Specifications, which allow the operator to select the materials that are acceptable to organizations such as ASTM and SAE
7. Composition, which allows the operator to select the materials that have certain minimum or maximum values of alloying elements

More than one of these sifting criteria mentioned can be used to identify suitable materials. Sifting can be performed in the AND or OR modes. The AND mode narrows the search since the material has to conform to all the specified criteria. The OR mode broadens the search since materials that satisfy any of the requirements are selected.

The number of materials that survive the sifting process depends on the severity of the criteria used. At the start of sifting, the number of materials shown on the screen is the total in the database. As more restrictions are placed on the materials the number of surviving materials gets smaller and could reach 0, that is, no materials qualify. In such cases, some of the restrictions have to be relaxed and the sifting restarted.

The Material Data Network (www.matdata.net) is an online search engine for materials information and is sponsored by ASM International and Granta Design Limited. Souces of information that are linked to the network, member sites, include ASM International, The Welding Institute, National Physical Laboratory (United Kingdom), National Institute of Materials Science (Japan), U.K. Steel Association "Steel Specifications," Cambridge Engineering Selector, MatWeb, and IDES plastics data. Member sites can be searched simultaneously with one search string for all classes of materials. The information provided is both quantitative and qualitative, with tables, graphs, micrographs, etc. Some of the sites are freely available and do not require registration, whereas others require registration to access the information.

9.9 SUMMARY

1. Experience has shown that it is desirable for product development teams to adopt the concurrent engineering approach, where materials and manufacturing processes are considered in the early stages of design and are more precisely defined as the design progresses from the concept to the embodiment and finally the detail stage.
2. Stages of the selection process can be summarized into analysis of the performance requirements and creating alternative solutions, initial screening of solutions, comparing and ranking alternative solutions, and selecting the optimum solution.
3. Cost per unit property method can be used for the initial screening of alternative solutions. In its simplest form, this is equal to the cost per unit weight multiplied times the density of the material and divided by the property that is considered to be most important to the application. Ashby's selection charts, Dargie's method, and Esawi and Ashby's method are computer-based alternative ways of initial screening.
4. Weighted property method of selection can be used for comparing and ranking alternative solutions. Each material requirement is assigned a certain weight depending on its importance to the function of the product. The individual weighted properties of each material are summed to give a material performance index. The material with the highest performance index is considered as the optimum for the application. The limits on properties method and the AHP are two additional methods that can be used for comparing and ranking alternatives.

5. After ranking of alternatives, candidates that have the most promising performance indices can each now be used to develop a detail design. Each detail design will exploit the points of strength of the material, avoid the weak points, and reflect the requirements of the manufacturing processes needed for the material. After completing the different designs, solutions are then compared, taking the cost elements into consideration to arrive at the optimum design–material–process combination.
6. Computers can be used to assist in materials selection. Several CAD/CAM programs and expert systems are available for such purposes.
7. Reliable and consistent sources of materials information are essential for successful materials selection. More detail and higher accuracy of information are needed as the selection process progresses from the initial screening to the final selection stage. Several databases and Internet sources are cited for these purposes.

9.10 REVIEW QUESTIONS

9.1 What are the main functional requirements and corresponding material properties for the following products: (a) milk containers, (b) gas turbine blades, (c) sleeve for sliding journal bearing, (d) piston for an internal combustion engine, (e) airplane wing structure?

9.2 What are the main reasons why graphite-reinforced epoxy is now widely used in sports equipment?

9.3 (a) What are the main design requirements for an elevator hoisting cable (the cable from which the elevator is pulled up and let down)? (b) Select the optimum material out of the following candidates for use in making the elevator hoisting cable:

Material	UTS (MPa)	Yield (MPa)	Elongation (%)	Relative Cost
ASTM-A675, 45	350	155	33	1
ASTM-A675,70	540	240	18	1.5
ASTM-A242 Type 1	450	320	21	2.1
ASTM-A717 Grade 70	550	485	19	5.0

9.4 The following three materials are being considered for manufacturing a welded structure in an industrial atmosphere. The structure is expected to be subjected to alternating stresses in addition to the static loading.

Material	A	B	C
Relative weldability (0.15)	5	1	3
Relative tensile strength (0.15)	3	5	2
Relative fatigue strength (0.25)	5	3	3
Relative corrosion resistance (0.20)	3	5	3
Relative cost (0.25)	2	5	3

Taking the weighting factors for the different properties as shown in parentheses after each property, what would be the best material? (Answer: material B)

9.5 If the weighting factors used in question 9.4 are changed to 0.25, 0.10, 0.2, 0.25, and 0.2 for weldability, tensile strength, fatigue strength, corrosion resistance, and cost, respectively, find the new optimum material. (Answer: material A)

9.6 The following three materials are being considered for making the frame of a racing car. Give appropriate weighting factors for each of the properties and apply the weighted properties method to select the appropriate material. If the cost is to be considered, what would be the appropriate way of doing so?

Material	Al-2014 T6	Steel AISI 1015	Epoxy–70% Glass Fabric
Yield strength (MPa)	248	329	680
Young's modulus (GPa)	70	207	22
Weldability index (5 = excellent, 4 = very good, 3 = good)	3	5	4
Specific gravity	2.8	7.8	2.1

9.7 The boiler shell in a power-generating plant is made of welded sheets. It is required to select materials for the shell. (a) What are the main functional requirements of the material used in making the shell? (b) Translate the functional requirements into material properties. (c) Suggest the weighting factors for the different properties. (d) What are the possible modes of failure of the shell? Suggest possible remedies for the different modes of failure. (e) Suggest possible materials for the shell.

9.8 The following three materials are being considered for making the sleeves for the shaft of a centrifugal pump. Taking the weighting factors for the different properties as shown in parentheses after each property, what would be the best material?

Material	Al-770	Bronze ASTM B22	Tin Alloy ASTM B23
Yield strength (MPa) (0.2)	173	168	40
Wear resistance (0.11)	2	5	2
Fatigue strength (MPa) (0.14)	150	120	32
Corrosion resistance (0.11)	3	2	5
Thermal conductivity (W/mk) (0.2)	167	42	50

9.9 It is required to produce a pair of scissors for kitchen use in household applications. Draw neat sketches showing the different elements and the main dimensions of the pair of scissors. Give the functional requirements and corresponding material requirements for each element. Suggest candidate materials and matching manufacturing processes for each of the elements as well as method of assembly knowing that the volume of production is 10,000 units per year.

9.10 It is required to design a screwdriver set for household use. The set is composed of five screwdrivers of different sizes for various applications in the household. Draw a neat sketch of a screwdriver and give the possible range of dimensions for the different screwdrivers in the set. Give the functional requirements and the corresponding material requirements for the different parts of the screwdriver. Suggest the different materials, manufacturing processes, and methods of assembly if the required number of sets is 10,000 per year.

9.11 It is required to design and select materials for a suitcase for air travel. (a) What are the main structural elements of the suitcase? (b) What are the main functional requirements of each structural element? (c) Translate the functional requirements into material properties. (d) Give weighting factors to the different properties. (e) Suggest possible materials for each structural element of the suitcase.

9.12 It is required to design and select materials for an overhead pedestrian crossing to connect two parts of a company. (a) What are the main design features and structural elements? (b) What are the main functional requirements of each structural element? (c) What are the corresponding material requirements for each element? (d) Use the digital logic method to determine the relative importance of each property for the different structural elements. (e) Recommend the possible materials that may be used as candidates for the final selection of the optimum material for each structural element.

9.13 The following three materials are being considered for making the outer body of a freshwater valve in a power-generation plant. Select the optimum material.

Material	AA-770 Aluminum Alloy	Bronze ASTM B22	AISI 302 Stainless Steel
Yield strength (MPa)	173	168	280
Wear resistance	2	5	5
Fatigue strength (MPa)	150	120	315
Corrosion resistance	2	4	4
Processability	3	5	3
Cost relative to aluminum	1	2	3

9.14 Draw a neat sketch of a bicycle for use by children aged 3–5. It is required to produce the frame of this bicycle (the frame is part of the bicycle to which the front- and back-wheel pedals and seat are attached). Select a reasonable shape for the frame and give the possible dimensions of each element of the frame. Give the functional requirements and corresponding material requirements for the frame. Suggest candidate materials and possible matching manufacturing processes for the elements of the frame as well as possible methods of assembling the elements to make the frame. Expected volume of production is 100,000 frames per year.

9.15 It is required to select the material for a structural member in a tensile testing machine. The member is 2 m long and will carry a maximum tensile load of 50 kN. The maximum extension should not exceed 0.5 mm. Select the optimum material out of the following candidates:

Material	UTS (MPa)	Yield (MPa)	Young's Modulus (GPa)	Elongation (%)	Relative Cost
ASTM-A675, 45	350	155	212	33	1
ASTM-A675,70	540	240	212	18	1.5
ASTM-A242 Type 1	450	320	212	21	2.1
ASTM-A717 Grade 70	550	485	212	19	5.0

9.16 The following three materials are being considered for making the frame of a racing car.

Material	Al-2014 T6	Steel AISI 1015	Epoxy–70% Glass Fabric
Yield strength (MPa)	248	329	680
Young's modulus (GPa)	70	207	22
Weldability index (5 = excellent, 4 = very good, 3 = good)	3	5	4
Specific gravity	2.8	7.8	2.1

Give appropriate weighting factors for each of the properties and apply the weighted properties method to select the appropriate material. If the cost is to be considered, what would be the appropriate way of doing so?

9.17 Case study: Packaging materials, types, and selection.

Background information: Packaging is an important and fast growing industry, which utilizes modern design, materials, and manufacturing technology. The shape of the package can range from a simple box, as in the case of industrial packages, to a complex design, as in the case of cosmetics. The materials used in making the package cover a wide variety of engineering materials, which include paper, wood, glass, metal, plastic, and composites of various materials. Manufacturing techniques used in making the package range from manual cutting and assembly to fully automatic processing and filling. The cost of the package can represent a considerable portion of the total cost of the product, especially in the case of consumer products. For example, the cost of packaging is about 30% of the selling price of cosmetics, 25% for drugs and pharmaceuticals, 20% for foods, and 10% for toys.

A well-designed package in the consumer industry should satisfy the following requirements:

1. Does not adulterate the contents, especially in the case of food and pharmaceutical packages
2. Maintains quality of the contents after it has been opened and until the consumer finishes the contents
3. Protects contents against environment and handling during shipping from manufacturer to consumer
4. Provides a convenient and efficient means of storage and handling at the wholesaler's warehouse, retailer's store room, and consumer's home
5. Conforms to the specifications of the transportation company or post office
6. Allows clear labeling and identification of the type, composition, and amount of the contents
7. Provides an attractive visual appearance and a high value as a sales tool
8. Does not endanger public safety at any stage of its life
9. Is easy to dispose of and recycle after the contents have been consumed
10. Has reasonable cost

The relative importance of the requirements mentioned depends on the contents of the package, expected shelf life, distance between manufacturer and consumer and method of delivery, as well as local and international laws.

In the present case study, a large manufacturer of instant coffee is in the process of reviewing the packaging policy. It is required to analyze the requirements, design the package, select the materials, and propose the method of manufacturing of coffee packages for the following cases:

1. Package for a single cup of coffee, 2 g (0.07 oz)
2. Small container for household use, 50 g (1.75 oz)
3. Medium-size container for household use, 100 g (3.5 oz)
4. Large-size container for household use, 200 g (7 oz)
5. Commercial-size container for cafeteria and restaurant use, 1000 g (2.2 lb).

BIBLIOGRAPHY AND FURTHER READING

Ashby, M.F., Materials selection charts, in *ASM Metals Handbook*, Vol. 20, Dieter, G., Volume Chair. ASM International, Materials Park, OH, 1997a, pp. 266–280.

Ashby, M.F., Performance indices, in *ASM Metals Handbook*, Vol. 20, Dieter, G., Volume Chair. ASM International, Materials Park, OH, 1997b, pp. 281–290.

Ashby, M.F., *Materials Selection in Mechanical Design*, 3rd Ed., Elsevier, London, 2005.

Boardman, B.E. and Kaufman, J.G., *Directory of Materials Properties Databases, Special Supplement to Advanced Materials & Processes*, ASM, New York, August 2000.

Bourell, D., Decision matrices in materials selection, in *ASM Metals Handbook*, Vol. 20, Dieter, G., Volume Chair. ASM International, Materials Park, OH, 1997, pp. 291–296.

Boyer, H.E. and Gall, T.L., *Metals Handbook Desk Ed.*, ASM, Materials Park, OH, 1985.

Catalyze, www.catalyze.co.uk.

Cebon, D. and Ashby, M.F., Data systems for optimal materials selection, *Adv. Mater. Processes*, **161**, 51–54, 2003.

CES 4 Software. Granta Design Limited Cambridge, UK, 2002, www.Grantadesign.com.

Clark, J., Roth, R., and Field III, F., Techno-economic issues in materials selection, in *ASM Metals Handbook*, Vol. 20, Dieter, G., Volume Chair. ASM International, Materials Park, OH, 1997, pp. 255–265.

Crane, F.A. and Charles, J.A., *Selection and Use of Engineering Materials*, Butterworths, London, 1984.

Dargie, P.P., Parmeshwar, K., and Wilson, W.R.D., MAPS 1: computer aided design system for preliminary material and manufacturing process selection, *Trans. ASME J. Mech. Design*, **104**, 126–136, 1982.

Dieter, G., Overview of the materials selection process, in materials selection and design, in *ASM Metals Handbook*, Vol. 20, Dieter, G., Volume Chair. ASM International, Materials Park, OH, 1997, pp. 243–254.

Easymind, www.easymind.info/fun/index_en.php.

Esawi, A.M.K. and Ashby, M.F., Cost estimates to guide pre-selection of processes, *Mater. Design*, **24**, 605–616, 2003.

Farag, M.M., *Materials and Process Selection in Engineering*, Applied Science Publishers, London, 1979.

Farag, M.M., *Selection of Materials and Manufacturing Processes for Engineering Design*, Prentice-Hall, New York, 1989.

Farag, M.M., Properties needed for the design of static structures, in ASM metals handbook, in *Materials Selection and Design*, Vol. 20, Dieter, G., Volume Chair. ASM International, Materials Park, London, OH, 1997, pp. 509–515.

Farag, M.M., *Materials Selection for Engineering Design*, Prentice-Hall, Europe, 1997.

Farag, M.M., Quantitative methods of materials selection, in *Handbook of Materials Selection*, Kutz, M., Editor. Wiley, New York, NJ, 2002, pp. 3–26.

Farag, M.M., Quantitative methods of materials selection, in *Mechanical Engineers Handbook: Materials and Mechanical Design*, 3rd Ed., Kutz, M., Editor. Wiley, Hoboken, NJ, 2006, pp. 466–488.

Farag, M.M. and El-Magd, E., An integrated approach to product design, materials selection, and cost estimation, *Mater. Design*, **13**, 323–327, 1992.

Fowler, T., Value analysis in materials selection and design, in *ASM Metals Handbook*, Vol. 20, Dieter, G., Volume Chair. ASM International, Materials Park, OH, 1997, pp. 315–321.

InfoHarvest, Inc., www.infoHarvest.com, 2005.

Kaufman, J.G., Sources of materials data, in *Handbook of Materials Selection*, Kutz, M., Editor. Wiley, New York, 2002, pp. 457–473.

Kirkwood, P.E., How to find materials properties data, in *Handbook of Materials Selection*, Kutz, M., Editor. Wiley, New York, 2002, pp. 441–456.

Metal Selector, Heller, M.E., ASM, Materials Park, OH, 1985. (www.asminternational.org).

Price, D., A guide to materials databases, *Mater. World*, July, 418–421, 1993.

Ragsdale, C., *Spreadsheet Modeling and Decision Analysis*, 4th Ed., Thomson, South-Western, Mason, OH, 2004.

Saaty, T.L., *The Analytic Hierarchy Process*, McGraw-Hill, New York, 1980.

Weiss, V., Computer-aided materials selection, in *ASM Metals Handbook*, Vol. 20, Dieter, G., Volume Chair. ASM International, Materials Park, OH, 1997, pp. 309–314.

Westbrook, J.H., Sources of materials property data and information, in *ASM Metals Handbook*, Vol. 20, Dieter, G., Volume Chair. ASM International, Materials Park, OH, 1997, pp. 491–506.

www.matdata.net.

10 Materials Substitution

10.1 INTRODUCTION

Manufacturers and engineers are always on the look out for new materials and improved processes to use in manufacturing better products, and thus maintain their competitive edge and increase their profit margin. For example, in the automotive industry, the effort to reduce weight to improve fuel economy and to comply with tighter governmental regulations on safety and emission has led to the introduction of increasing amounts of plastics and composite materials in place of the traditionally used steels. However, suppliers of the traditional materials are trying to regain some of the lost ground by introducing higher-strength steels and prefinished sheets. This competition among materials makes it necessary to periodically perform a materials audit to determine whether the material used in making a given component should be substituted.

Reaching a decision on substitution can be complicated and may involve materials engineers, designers, manufacturing engineers, as well as marketing and purchasing personnel. The process can be further complicated if future investment plans for new plants and equipments are involved. Unless the substitution process is carried out methodically, it can lead to confusion, delays, unnecessary expense, or even the wrong decision being made.

The goal of this chapter is to analyze the various incentives and constraints involved in substituting one material for another in making an existing component.

The main objectives are to get a better understanding about

1. materials audit;
2. constraints and incentives in materials substitution;
3. life cycle energy impact of materials substitution;
4. financial implications of materials substitution; and
5. some quantitative methods of initial screening, comparing alternatives, and making final decision for materials substitution.

10.2 MATERIALS AUDIT

The audit process could start by asking the following questions

- When were the materials last selected and specified?
- Who initiated the last changes in materials? Was it company, personnel, or materials suppliers?
- Why was the material changed? Was it legislation, lack of availability, or increased cost of the established material, or the possibility of improved

performance and reduced cost? If it were performance, then which properties dictated the change?

- What feedback do you have on the performance of your product? Product failures and returns are an obvious source of information but user survey may be necessary.
- What progress has been made by the materials manufacturers since the last change?
- Have new manufacturing processes been introduced since the last change?
- Have the new materials been developed with your product in mind? When materials manufacturers develop new or improved grades, they normally have a number of specific targets in mind. Is your product among them? If not, how do you ascertain that the new material is optimized for your application? Have you consulted the materials supplier?
- Is the advantage of adopting a new and untried material worth the risk of abandoning the current and established material?
- Is the cost of conversion to the new material less than the benefits?
- Would new equipments and plants be needed?
- Assuming that the substitution has been made, what are the implicates of that substitution on the system at large?
- What are the institutional, legal, social, and environmental consequences?

Answers to the preceding questions provide the background against which the decision to explore materials substitution and the driving force required to overcome the resistance to substitution and the tendency to continue doing what has been done in the past. The degree of resistance to substitution of one material by another depends on the following:

1. Company policy
2. Availability of design guidelines and in-service experience for new materials
3. Cost of redesign and investment required for new production facilities and equipment
4. Increased inventory required for more than one spare replacement
5. Type of product and the extent to which the new material is being used When the market share of a new material is small, resistance to substitution is high. As market share increases, resistance to change decreases, therefore increasing the rate of substitution.

There are powerful arguments for not changing the *status quo* unless the benefits can be seen to be considerable. However, engineering products are subject to continual evolution to meet increased performance demands and to lower manufacturing costs. To stand still is to invite the competition to overtake. New and improved materials and processes can make a vital contribution to improved competitiveness, and the opportunities should be continuously assessed.

10.3 CONSIDERATIONS IN MATERIALS SUBSTITUTION

If a decision is taken to substitute a new material for an established one, care must be taken to ensure that all the characteristics of the new material are well understood. A large number of product failures have resulted from new materials being used before their long-term properties were fully known. Another source of failure results from substituting a new material without reviewing the design. As an example, consider the case where thinner high-strength low-alloy (HSLA) steel sheets are substituted for the currently used thicker steel sheets in motorcar bodies. Having overcome the processing problems, the substitution appears attractive in view of the weight saving. It should be remembered, however, that although the strength of HSLA steel is higher, corrosion resistance and elastic modulus are essentially the same as those for low-carbon steel. Thinner HSLA sheets will be damaged by corrosion in shorter time, get dented more easily, and undesirable vibrations can be more of a problem.

Generally, a simple substitution of one material for another (part-for-part substitution) does not provide optimum utilization of the new material. This is because it is not possible to realize the full potential of a new material unless the component is redesigned to exploit its properties and manufacturing characteristics. This is illustrated in Figure 10.1. Wood and steel wire are used in making the clothes hangers 1 and 2. Hanger 3 is mostly made of plastic but uses a steel hook and is styled in the same way as hangers 1 and 2. Hanger 4 is all plastic and introduces additional useful features that are only possible with injection molding.

In making a substitution, the new material should be mechanically, physically, and chemically compatible with the surrounding materials. For example, replacing steel with plastic may cause changes in deflection, thermal conduction, and thermal expansion due to differences in modulus of elasticity and thermal properties. Replacing steel with aluminum may cause galvanic corrosion with neighboring steel components. This means that changing the material of one component may entail several changes in the neighboring components. In some applications, the currently used material may have inherent characteristics that are not specified but are useful. For example, the high damping capacity of cast iron will be lost to the system if it is replaced by steel to save weight. System replacement provides the designer with the opportunity of integrated design to gain maximum benefit.

The main parameters that need to be examined for material substitution can be grouped as follows:

a. Technical performance advantage, as a result of introducing a stronger, stiffer, tougher, or lighter material.
b. Economic advantage over the total life cycle of the product. This can be achieved as a result of introducing cheaper material, more cost-effective use of material, lower cost of processing, better recyclability and lower cost of disposal, or lower running cost of the product.
c. Changing the character of the product by incorporating a material that is esthetically more attractive, with a different feel or that can provide more comfort to the user through sound or heat insulation, for example.

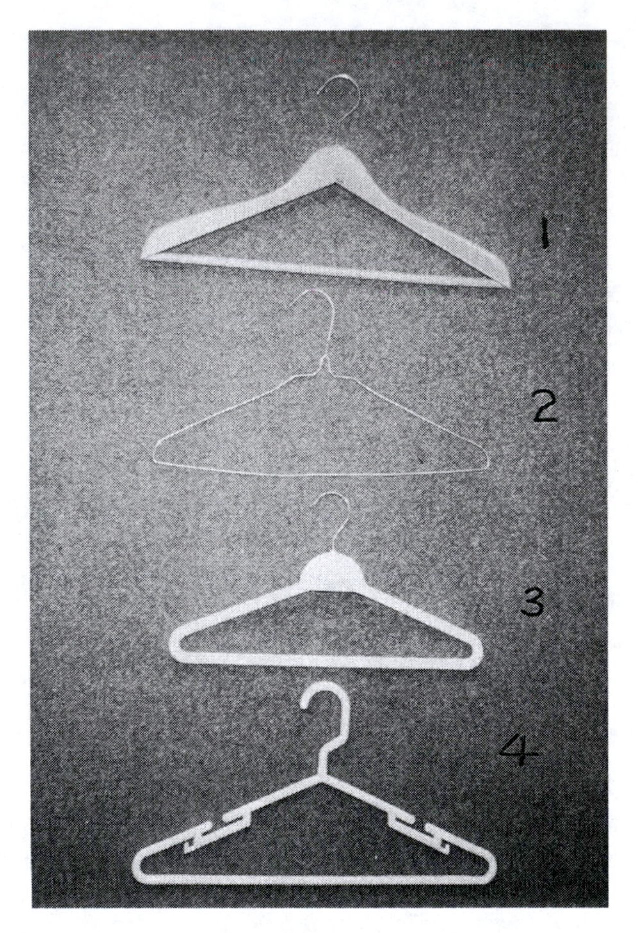

FIGURE 10.1 An example of how materials substitution affects the design and manufacturing processes of clothes hangers. Wood is shaped by cutting, the steel wire is shaped by bending, and the plastic is shaped by injection molding. Injection molding allows the flexibility of introducing useful features in the bottom hanger.

d. Environmental and legislative considerations including less damage to the environment over the life cycle of the product, better recycling or reuse, and compliance with environmental regulations.

10.4 SCREENING OF SUBSTITUTION ALTERNATIVES

In some cases, the need to search for a substitute material is recognized, but no definite alternative is presented. Under these conditions, it may be useful to start with a brainstorming session to identify a variety of alternatives that can then be screened to narrow the choices down to a few promising solutions. The method described by Pugh (1991) can be useful for initial screening of alternatives in the early stages of materials substitution. In this method, the performance requirements are first defined

TABLE 10.1
Example of the Use of the Pugh Decision Matrix for Materials Substitution

Property	Currently Used Material	New Material (1)	New Material (2)	New Material (3)
Property (1)	C1	−	+	+
Property (2)	C2	+	+	+
Property (3)	C3	+	+	−
Property (4)	C4	0	+	−
Property (5)	C5	−	0	−
Property (6)	C6	0	0	0
Property (7)	C7	−	−	0
Property (8)	C8	−	+	0
Property (9)	C9	−	0	0
Total (+)		2	5	2
Total (−)		5	1	3
Total (0)		2	3	4

and the corresponding material properties are identified as described in Section 9.3. A decision matrix is then constructed as shown in Table 10.1. Each of the properties of possible alternative new materials is compared with the corresponding property of the currently used material and the result is recorded in the decision matrix as (+) if more favorable, (−) if less favorable, and (0) if the same. The decision on whether a new material is better than the currently used material is based on the analysis of the result of comparison, that is, the total number of (+), (−), and (0). New materials with more favorable properties than drawbacks are selected as serious candidates for substitution and can be ranked using one of the two quantitative methods described in Section 10.5.

10.5 COMPARING AND RANKING OF ALTERNATIVE SUBSTITUTES

After narrowing down the field of possible substitute materials as described in Section 10.4, quantitative methods can be used to rank the most feasible alternatives. Following is a description of two methods of ranking alternative substitutes.

10.5.1 Cost of Performance Method of Substitution

In this method, the performance index, γ, and the total cost, C_t, of the candidate materials are separately compared against the currently used material. The performance index, γ, covers all requirements of the material except cost and is estimated using one of the ranking procedures outlined in Section 9.5. Various scenarios of substitution can be developed in this method by assigning different weighting factors to the performance requirements.

The total cost, C_t, is considered to consist of several cost elements as follows:

$$C_t = C_1 + C_2 + C_3 + C_4 \tag{10.1}$$

where

C_1 = Cost of material used in making the component
C_2 = Cost of manufacturing and finishing the component including cost of redesign and new tools
C_3 = Running cost over the entire life of the component
C_4 = Cost of disposal and recycling

If the purpose of substitution is to reduce the total cost of the component, an acceptable candidate must perform as well as the current material, that is, has similar performance index, γ, but at a lower total cost, C_t. If several candidates fulfill this condition, the one that gives the most cost reduction is selected.

If, however, the purpose is to improve performance, acceptable candidates must perform at a higher level than the currently used material. If cost is not the objective, the candidate with the highest performance index, γ, can be selected. In most situations, however, it is more realistic to calculate the percentage increase in performance ($\Delta\gamma\%$) and the corresponding percentage increase in cost ($\Delta C_t\%$) as follows:

$$\Delta\gamma\% = \frac{100(\gamma_n - \gamma_o)}{\gamma_o} \tag{10.2}$$

$$\Delta C_t\% = \frac{100(C_{tn} - C_{to})}{C_{to}} \tag{10.3}$$

where

γ_n and γ_o = Performance indices of the new and original materials, respectively
C_{tn} and C_{to} = Cost of the new and original materials, respectively

The substitute material that gives the highest ($\Delta\gamma\%/\Delta C_t\%$) can be selected if it has a clear advantage over the currently used material. Otherwise, further detailed analysis may be performed as discussed in Section 10.6.

The use of the cost of performance method is illustrated in Case study 10.1.

10.5.2 The Compound Performance Function Method

This method compares the compound performance function (CPF) of the candidate materials against the currently used material. CPF is defined as the weighted sum of all the normalized material performance requirements, including total cost (C_t), and is estimated using one of the ranking procedures outlined in Section 9.5. Various scenarios of substitution can be developed in this method by assigning different weighting factors to the performance requirements.

For a successful substitution, the candidate material must have a higher CPF than the currently used material. The process may stop at this stage if the top-ranking alternative has a clear advantage over the currently used material. Otherwise, a more detailed analysis may be necessary as described in Section 10.6. The use of the CPF method in materials substitution is illustrated in Case studies 10.1 and 10.2.

Case Study 10.1: Materials Substitution in a Tennis Racket

Introduction

In this case study, a manufacturer of tennis rackets is considering the introduction of a new, more powerful model. The main evaluation criteria for the racket can be grouped into power, damping, and cost. The power of a racket allows the delivery of faster balls with less effort. Damping is the ability of the racket material to reduce the vibrations in the strings after hitting the ball and thus reduce the possibility of the player developing tennis elbow.

Analysis

The current material is epoxy–50% CF (carbon fibers). Possible substitute materials to increase the power are shown in Table 10.2. For the purpose of this case study, power is taken as equal to (E/ρ), where E is the elastic modulus and ρ the density. Damping is taken as inversely proportional to E, and the material with the lowest E is given a damping of 10. The cost is taken as cost of the material per unit mass. The normalized values in Table 10.2 were obtained as described in the weighted properties method given in Section 9.5.

Cost of Performance Method

The performance index, γ, in Table 10.3 is calculated by giving weights of 0.7 for power and 0.3 for damping. $\Delta\gamma$ and $\Delta C\%$ are percentage increases in γ and cost relative to the base material, respectively. The table shows that epoxy + 65% CF is a preferable substitution material as it has the highest $(\Delta\gamma\%/\Delta C\%)$. Epoxy + 60% CF comes as a close second best.

Compound Performance Function Method

The CPF in Table 10.4 is calculated by giving the weights of 0.55 for power, 0.2 for damping, and 0.25 for cost. The table shows that epoxy + 65% CF is

TABLE 10.2
Characteristics of Tennis Racket Materials

Material	E (GPa)	Density ρ (g/cc)	Cost ($/kg)	Power	Damping	NP	ND	NC
Epoxy + 50% CF	136	1.87	93	73	10	82	100	100
Epoxy + 55% CF	146.4	1.873	101	78	9.3	88	93	92
Epoxy + 60% CF	156.8	1.876	109	84	8.7	94	87	85
Epoxy + 65% CF	167.2	1.879	117	89	8.1	100	81	80

Note: NP = normalized power, ND = normalized damping, NC = normalized cost.
Source: Data based on Esawi, A.M.K. and Farag, M.M., Carbon nanotube reinforced composites: potential and current challenges, *Mater. Design*, 28(9), 2394–2401, 2007.

TABLE 10.3
Cost of Performance Method

Material	γ	$\Delta\gamma$%	ΔC%	$\Delta\gamma$%/ΔC%
Epoxy + 50% CF	87.4	—	—	—
Epoxy + 55% CF	89.5	2.40	8.6	0.28
Epoxy + 60% CF	91.9	5.15	17.2	0.3
Epoxy + 65% CF	94.3	7.9	25.8	0.31

TABLE 10.4
Compound Performance Function Method

Material	0.55 NP	0.20 ND	0.25 NC	CPF
Epoxy + 50% CF	45.1	20	25	90.1
Epoxy + 55% CF	48.4	18.6	23	90.0
Epoxy + 60% CF	51.7	17.4	21.25	90.35
Epoxy + 65% CF	55	16.2	20	91.2

Note: NP = normalized power, ND = normalized damping, NC = normalized cost.

a preferable substitution material as it has the highest CPF. Epoxy + 60% CF comes as a close second best.

Conclusion

The results obtained using the cost of performance and CPF methods agree that epoxy + 65% CF is an optimum substitute material with epoxy + 60% CF as a close second best.

Case Study 10.2: Materials Substitution for a Cryogenic Tank

Problem

Consider the case of the cryogenic tank discussed in Case study 9.2 in Section 9.5. The results of the analysis show that SS 301-FH is the optimum material and is therefore used in making the tank. Suppose that at a later date a new fiber-reinforced material is available and it is proposed to manufacture the tank from the new material by the filament-winding technique. The properties of the new fiber-reinforced material are given in Table 10.5 together with the properties of SS 301-FH.

Analysis

Following the procedure in Section 9.5, the properties are first scaled. Using the same weighting factors as in Table 9.5, the performance index is calculated and

TABLE 10.5
Properties of Candidate Materials for Cryogenic Tank

	1	2	3	4	5	6	7
Material	**Toughness Index**	**Yield Strength**	**Young's Modulus**	**Specific Gravity**	**Thermal Expansion**	**Thermal Conductivity**	**Specific Heat**
SS 310-FH	770	1365	189	7.9	16.9	0.04	0.08
Composite	175	1500	200	2.0	12	0.005	0.1

TABLE 10.6
Scaled Values of Properties and Performance Index

	Scaled Properties							
Material	**1**	**2**	**3**	**4**	**5**	**6**	**7**	**Performance Index (y)**
SS 301-FH	100	91	95	25	71	12.5	100	70.9
Composite	23	100	100	100	100	100	80	77.4

TABLE 10.7
Relative Cost and Cost of Unit Strength for Candidate Materials

Material	**Relative Cost**	**Cost of Unit Strength × 100**	**Figure of Merit (y/Cost of Unit Strength)10^{-2}**
SS 301-FH	1.4	0.81	87.53
Composite	7	0.93	83.23

the results in Table 10.6 show that the composite material is technically better than the stainless steel.

Final comparison between the original and candidate materials will be carried out according to the CPF method. The basis of comparison is chosen as the figure of merit, as described in Section 9.5. Following the same procedure of Section 9.5, the cost of unit strength is calculated as shown in Table 10.7.

Conclusion

As the figure of merit of SS 301-FH is higher than that of the composite material, the basis material still gives better value than the new material and no substitution is required.

If, however, the increasing use of the new composite material causes its relative cost to decrease to 6.6 instead of 7 (Table 10.7), the cost of unit property becomes 0.837×100 instead of 0.93×100. In this case, the figure of merit of the composite material becomes 92.5×10^{-2}, which means that it gives better value and is, therefore, a viable substitute.

10.6 REACHING A FINAL DECISION

A final step in the material substitution process is to perform a detailed comparison of the technical and economic implications of adopting the substitute material. The following cost–benefit analysis procedure is a possible rational way of arriving at a final decision.

10.6.1 Cost–Benefit Analysis

The cost–benefit analysis is more suitable for the detailed analysis involved in making the final material substitution decision. In cases where the new material is technically better but more complex and requires closer control and new technologies for its processing, components made from it would have better performance but would also be more expensive. In such cases, if material substitution is to be economically feasible, the economic gain as a result of improved performance, $\Delta\gamma_e$, should be more than the additional cost, ΔC_t, incurred as a result of substitution.

$$\Delta\gamma_e - \Delta C_t > 1 \quad (10.4)$$

10.6.2 Economic Advantage of Improved Performance

The economic gain as a result of improved performance, $\Delta\gamma_e$, can be estimated based on the expected improved performance of the component, which can be related to the increase in performance index of the new material compared with the currently used material, γ_n and γ_o, respectively. The performance index, γ, can be calculated using one of the ranking procedures outlined in Section 9.5. The increase in performance can include the saving gained as a result of weight reduction, increased service life of the component, and reduced cost of disposal.

$$\Delta\gamma_e = A(\gamma_n - \gamma_o) \quad (10.5)$$

where

γ_n and γ_o = Performance indices of the new and original materials, respectively

A = Benefit of improved performance of the component expressed in dollars per unit increase in material performance index, γ

The use of the parameter A in the substitution process is illustrated in Case study 10.3.

Case Study 10.3: Reaching a Final Decision on Material Substitution for the Sailing-Boat Mast Component

Problem

In the Case study 9.5 that was discussed in Chapter 9, the aluminum alloy AA 2024 T6 was selected for the sailing-boat mast component since it gives the least-expensive solution. Of the seven materials in Table 9.21, AA 6061 T6, epoxy–70% glass fabric, and epoxy–62% aramid fabric result in components that are heavier and more expensive than those of the other four materials and will be rejected as they offer no advantage. Of the remaining four materials, AA 2024 T6 results in the least-expensive but the heaviest component. The other three materials—AA 2014 T6, AA 7075 T6, and epoxy–63% carbon fabric—result in progressively lighter components at progressively higher cost.

Analysis

For the cases where it is advantageous to have a lighter component, the cost–benefit analysis can be used in finding a suitable substitute for AA 2024 T6 alloy. For this purpose, Equation 10.5 is used with the performance index, γ, being considered as the weight of the component; ΔC the difference in cost of component; and A the benefit expressed in dollars, of reducing the mass by 1 kg. Comparing the materials in pairs shows that

For $A <$ \$7/kg saved,	AA 2024 T6 is the optimum material
For $A =$ \$7–\$60.5/kg saved,	AA 7075 T6 is a better substitute
For A > \$60.5/kg saved,	epoxy–63% carbon fabric is optimum

10.6.3 Total Cost of Substitution

The additional cost (ΔC_t) incurred as a result of substitution can be divided into

- *Cost of redesign and testing.* Using new materials usually involves design changes and testing of components to ensure that their performance meets the requirements. The cost of redesign and testing can be considerable in the case of critical components.
- *Cost differences in materials used.* When smaller amounts of a new, more expensive material is used to make the product, the increase in direct material cost may not be as great as it would appear at first.
- *Cost differences in labor.* This may not be an important factor in substitution if the new materials do not require new processing techniques and assembly procedures. This element of cost can be a source of cost saving if the new material does not require the same complex treatment or finishing processes used for the original material. If, however, new processes are needed, new cycle times may result and the difference in productivity has to be carefully assessed.
- *Cost of new tools and equipment.* Changing materials can have considerable effect on the life and cost of tools, and it may influence the heat treatment and finishing processes. The cost of equipment needed to process new materials can be considerable if the new materials require new production facilities as in the case of replacing metals with plastics.

Based on this analysis, the total cost, ΔC_t, of substituting a new material, n, in place of an original material, o, in a given part is

$$\Delta C_t = (P_n M_n - P_o M_o) + f(C_1/N) + (C_2/N) + (T_n - T_o) + (L_n - L_o) \quad (10.6)$$

where

P_n and P_o = Price/unit mass of new and original materials used in the part
M_n and M_o = Mass of new and original materials used in the part
f = Capital recovery factor; it can be taken as 15% in the absence of information
C_1 = Cost of transition from original to new materials including cost of new equipment
C_2 = Cost of redesign and testing
N = Total number of new parts produced
T_n and T_o = Tooling cost per part for new and original materials
L_n and L_o = Labor cost per part using new and old materials

The use of cost analysis in substitution is illustrated in Case study 10.4.

Case Study 10.4: Materials Substitution of a Panel in Aerospace Industry

Introduction

The main driving force for materials substitution in aerospace industry is weight reduction at a reasonable cost while maintaining reliability and safety standards. Reducing the weight of the structure allows lifting a greater payload and traveling longer distances without refueling. In addition, weight saving could allow reduction of structures such as wing area, which would lead to further weight reduction. The outstanding values of strength/weight and stiffness/weight of FRPs make them prime challengers to the traditionally used aluminum alloys. As carbon fiber–reinforced plastics (CFRP) offer the highest potential for weight saving, their use for panels in the upper wing surface of a civilian aircraft is discussed here.

This case study gives an analysis of the different factors involved in materials substitution in aerospace industry. The merits and drawbacks of substituting CFRP for the traditionally used aluminum alloys are examined. Body panels are used to illustrate the procedure, but similar analysis may be used for other parts of the structure of the aircraft. This case study was prepared when the author visited MIT. The help extended by Professor Thomas Eagar is acknowledged.

Analysis of Candidate Materials

Aluminum Alloys

Aluminum alloys are the traditional materials for civilian aircraft panels. Aluminum alloys in the 5xxx, 2xxx, and 7xxx series can be used for panel applications in the aerospace industry.

Fiber-Reinforced Plastics

FRPs are being increasingly used in aerospace industry as a result of their superior strength/weight and stiffness/weight. In low production volumes, composite panels containing continuous fibers can be made by stacking the required number of layers of preimpregnated fibers, prepregs, in the form of tapes or fabrics and then shaping them into matched dies. Stacking of the prepregs can be done manually or by using tape laying machines. Composite structures are easily joined using structural adhesives, but machining and drilling are difficult as a result of the widely different properties of their constituents.

Required Mechanical Properties

Body panels of an aircraft can be subjected to a variety of loading conditions depending on their position and function. For example, the loading conditions on the wing of an airplane in flight can be approximately represented by a uniformly distributed load acting in the upward direction on a cantilever beam. In such a case, the load on a panel in the upper wing surface can be approximated to uniform in-plane compression. A major requirement for such a panel is resistance to buckling. However, a major requirement for a panel in the lower wing surface is resistance to static and fatigue tensile stresses. A panel in the control surfaces, for example, rudder, spoilers, and elevators, is not highly stressed but must be provided with adequate stiffness to keep its shape under wind forces. For such panels, it can be shown that $E^{1/3}/\rho$ is the major design parameter for comparing the candidate materials, where E is the modulus of elasticity and ρ is the density (Farag, 1997, 2007).

Comparison of Candidate Materials

In the present case study, the upper wing panel is considered to be a flat rectangle of width, $b = 50$ cm and length, $l = 100$ cm. Although the main load on such a panel is in-plane compression, some transverse and torsional loading may occur in maneuvering the aircraft or as a result of unfavorable weather conditions. Such secondary loads are not serious when isotropic materials, such as aluminum alloys, are used for the panel. CFRP, however, is not isotropic, and if all the fibers are oriented in the direction of the compressive load, the panel could easily fail under relatively small loads at right angles to the fibers. This can be avoided by placing some of the fibers at 90° or arranging them in the +45°/−45° directions, depending on the required degree of isotropy. In the present case study, the epoxy matrix will be strengthened using 33% carbon fabric + 30% carbon fibers. Composites with this arrangement of fibers are not as strong as those where all the carbon fibers (63% by volume) are oriented in the same direction, but it provides the required strength in the 90° direction. Table 10.8 shows the superior structural efficiency of CFRP in comparison with the aluminum alloys. For a simply supported thin rectangular panel of thickness (t) and width (b) under in-plane compressive load (P), buckling will occur when

$$P = S_B tb = \frac{\pi^2 E}{3(1 - \nu^2)}\left(\frac{t}{b}\right)^2 tb \tag{10.7}$$

where

S_B = Buckling stress

ν = Poisson's ratio (about 0.3 for most materials)

TABLE 10.8
Properties of Candidate Materials for Aircraft Body Panels

Material	Modulus of Elasticity (GPa)	(ksi)	Density (Mg/m^3)	(lb/in.^{3})	$E^{1/3}/\rho$ (SI units)	Cost ($/kg)	($/lb)
Aluminum alloy (average of 2xxx and 7xx series)	71	10,000	2.7	0.097	71.2	4.3	1.95[a]
Epoxy–33% carbon fabric + 30% carbon fibers	100	14,286	1.61	0.058	134.65	110	50[b]

[a] Aluminum alloys' cost is based on the average of 2024 and 7075 alloys, 1987 prices. (Adapted from Charles, J.A. and Crane, F.A.A., *Selection and Use of Engineering Materials*, Butterworths, London, 1989.)

[b] Adapted from McAffee, A.P., *On the appropriate level of automation for advanced structural composites manufacturing for commercial aerospace applications*, M.Sc. thesis, Department of Mechanical Engineering and Sloan School of Management, MIT, Cambridge, Massachusetts, USA, 1990.

From Equation 10.7, the thickness is written as

$$t = \left(\frac{Pb}{3.62E}\right)^{1/3} \tag{10.8}$$

The mass of a panel (M) of length (l) is

$$M = \rho t b l$$

According to Charles and Crane (1989), a typical compressive end load for an aluminum alloy panel is about 3.5 MN/m, that is, $P/b = 3.5$ MN/m. This value is used to calculate the thicknesses of aluminum and CFRP panels according to Equation 10.8. The masses and costs of materials are then calculated using the information in Table 10.8 and the results are given in Table 10.9.

Shipp (1990) analyzed the nonrecurring cost of transition from aluminum of CFRP for the spoilers of model 737-200/300 at Boeing Commercial Airplanes. The cost was $948,198 for 150 shipsets, weighing 151 lb (68.6 kg) each. Dividing this sum by the total weight of the spoilers, gives the cost of transition as $41.86/lb ($92.1/kg) of CFRP. This relatively high value reflects the man-hours needed to design the unfamiliar material and the extensive amount of testing required to verify its reliability. This value is used to calculate the cost of transition from aluminum to CFRP in the present case study, as shown in Table 10.9.

According to Shipp (1990), American Airlines literature shows that a DC 10-10 airplane cuts its operating costs by about $29/year for a 1 lb (450 g) decrease in weight at the fuel prices of 1989. Assuming an expected life of 20 years for the airplane and a 10% cost of capital, Shipp estimated the present total value of the cost savings as $272/lb ($598/kg) decrease in weight. Cost savings are expected to depend on the size and function of the aircraft, current fuel prices,

TABLE 10.9
Estimates for Aircraft Panel Substitution

	Aluminum	CFRP
Thickness for equal buckling resistance, in. (mm)	0.59 (15)	0.53 (13.4)
Mass of panel, lb (kg)	44.64 (20.25)	23.79 (10.79)
Cost of material in panel ($)	87.08	1186.90
Cost of transition per panel ($)	—	1002.51[a]
Cost of labor per pound of panel material ($)	10–50[b]	50–300[a]
Cost of labor per panel ($)	446.4–2232	1189.5–7137
Cost savings per panel due to less weight ($)	—	5671.20

[a] Adapted from Shipp, C.T., *Cost-effective use of advanced composite materials in commercial aircraft manufacture*, M.Sc. thesis, Department of Mechanical Engineering and Sloan School of Management, MIT, Cambridge, Massachusetts, USA, 1990.

[b] Estimated.

and the general economic conditions; they can range from $100/lb to $500/lb ($220/kg–$1100/kg). Similar values of cost savings were reported by Charles and Crane (1989). In the present case study, a cost saving of $272/lb ($598/kg) is used to calculate the cost saving per panel due to less mass as shown in Table 10.9.

In a survey of several aerospace firms involved in manufacturing with advanced composites, Shipp found that the cost of labor involved in making a given part depends on its weight. Smaller parts are relatively more labor-intensive. Components weighing more than 10 lb (4.54 kg) are found to need 0.8–4.6 h/lb (1.76–10.1 h/kg) of direct labor. Using an industry average of $65/h for fully burdened labor rate, Shipp estimated the cost of labor as $50/lb–$300/lb ($110/kg–$660/kg) of advanced composite materials. Aluminum labor cost is expected to be lower than the figures mentioned, as its manufacture is less labor-intensive. For the present case study, a labor cost in the range $10/lb–$50/lb ($22/kg–$110/kg) is assumed.

Because the labor cost represents a large proportion of the cost (see Table 10.9), relatively small variations in the labor rate can affect the economic feasibility of substitution. The effect of labor rate variations on the total cost of substitution (ΔC_t) for a panel in the upper wing surface is shown in Figure 10.2. As would be expected, lowering the labor rate for one material makes it more attractive economically. For example, at a labor rate of $200/lb ($440/kg) for CFRP, aluminum is more attractive if its labor rate is $20/lb ($44/kg), but not attractive if its labor rate is $40/lb ($88/kg). Similarly, at a labor rate of $20/lb ($44/kg) for aluminum, CFRP is more attractive if its labor rate is $150/lb ($330/kg), but not attractive if its labor rate is $200/lb ($440/kg).

Conclusion

As the long-range behavior of the new materials is not well established, the present design codes require higher factors of safety in design and extensive testing programs when adopting FRP for critical components. This adds to the economic disadvantage of FRP. Such difficulty can only be solved gradually

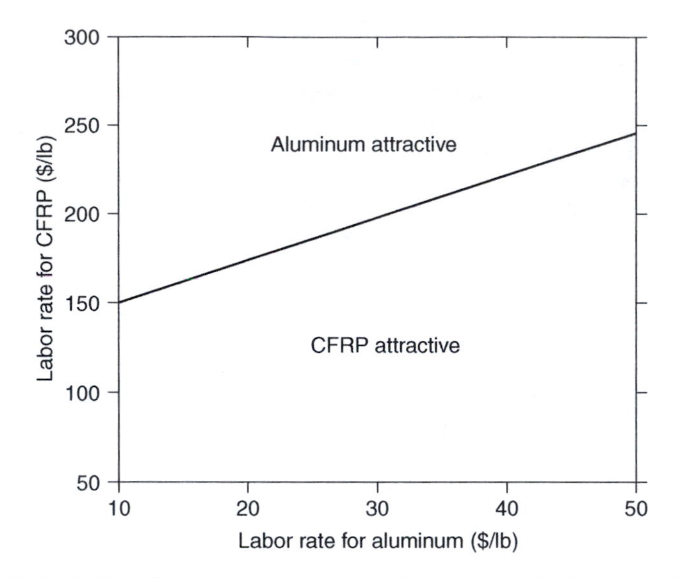

FIGURE 10.2 Effect of labor rate variations on the total cost of substitution for a panel in the upper wing surface of an aircraft.

because engineers need to be more familiar with the unusual behavior of the new materials and need to gain more confidence in their long-range performance.

10.7 SUMMARY

1. Materials substitution is an ongoing process and materials used for a given product should be reviewed on a regular basis through a materials audit process. The audit process answers questions related to when the material was last selected, feedback on performance, and progress made in the area of materials and manufacturing.
2. If a decision is taken to substitute a new material for an established one, care must be taken to ensure that the characteristics of the new material are well understood and that advantages outweigh drawbacks of adopting it. Risk, cost of conversion, equipment needed, as well as the environmental impact need to be carefully evaluated.
3. The economic parameters involved in material substitution include direct material and labor, cost of redesign and testing, cost of new tools and equipment, cost of change in performance, and overheads.
4. The major stages of materials substitution include screening of alternatives, comparing and ranking alternative substitutes, and reaching a final decision. The initial stages involve only rough estimates, which become more elaborate as the substitution process progresses to the screening and then the final selection stages.
5. Several quantitative methods of substitution are described in this chapter. The use of quantitative methods ensures that decisions are made rationally and

that no viable alternative is ignored. These methods include Pugh's method for initial screening, cost of performance and CPF methods for ranking alternative solutions, and cost–benefit analysis for reaching a final decision.

10.8 REVIEW QUESTIONS

10.1 Compare the following materials as possible replacement for glass in containers for apple juice drink: plastic and carton laminate. Consider the following factors in your discussion: whether the container is disposable or returnable, shelf life, weight, cost, environmental impact, and sales appeal.

10.2 What are the main materials that can be used to replace glass for packaging fresh milk? Compare the advantages and disadvantages of each material and give the milk distribution system that is most suitable for each material.

10.3 Materials B and C are being considered as replacement for material A in manufacturing a welded structure to serve in an industrial atmosphere. The structure is expected to be subjected to alternating stresses in addition to the static loading.

Material	A	B	C
Relative weldability (0.15)	5	1	3
Tensile strength (MPa) (0.15)	300	500	200
Fatigue strength (MPa) (0.25)	150	90	90
Relative corrosion resistance (0.20)	3	5	3
Relative cost (0.25)	2	5	3

Taking the weighing factors for the different properties as shown in parentheses after each property, what would be the best substitute material?

10.4 Aluminum alloy and GFRP are being considered for replacing steel in making the frame of a firefighting ladder to be fixed to a fire engine.

Material	Steel AISI 1015	Al-2014 T6	Epoxy–70% Glass Fabric
Yield strength (MPa)	329	248	680
Young's modulus (GPa)	207	70	22
Weldability index (5 = excellent, 4 = very good, 3 = good)	5	3	4
Specific gravity	7.8	2.8	2.1
Relative cost	1	4	5

Give appropriate weighing factors for each of the properties and decide whether any of the alternative materials can serve as a successful substitute.

10.5 Develop various scenarios of the solution to the problem in Case study 10.1 by changing the weights given to the power, damping, and cost. Comment on the results.

10.6 Currently a company is making the frame of a racing bicycle out of medium-carbon steel (the frame is part of the bicycle to which the front- and back-wheel pedals and seat are attached). Select a reasonable shape for the frame and give the possible dimensions of each element of the frame. Give the functional requirements and corresponding material requirements for the frame. Suggest candidate materials and possible matching manufacturing processes for replacing steel in making the frame. Expected volume of production is 10,000 frames/year.

10.7 Develop various scenarios of the solution in the Case study 10.2 by changing the weights allocated to the power, damping, and cost. Comment on the results and the relative merits of the two methods, cost of performance, and CPF.

10.8 Term Project

Objective. The objectives of the term project are to train students to work in teams and to present an integrated study of the design, materials, and process selection of an engineering product.

Project teams. Project teams consist of about five members who will organize themselves so that each member will carry out a fair share of the work. The team will submit one final report but all members of the team will participate in an oral presentation of the project.

Guidelines for project work.

1. Select a product or an engineering system in consultation with the course instructor.
2. Define the uses and function of the product or system.
3. Identify the different components, assemblies, or subassemblies. Define their function, operating conditions, and expected performance level.
4. Select one component for detailed analysis.
5. Define the principles of design calculations.
6. Define the material and manufacturing processes that are currently used in making the component.
7. Deduce the mechanical, physical, and chemical properties of the materials that could be used in making the component.
8. Classify possible materials and manufacturing processes.
9. Rank the alternative materials and processes based on the required characteristics.
10. Make final design for the component using the optimum candidate material and matching processes.
11. Compare expected performance of alternative component with that of the current component.
12. Submit a written report and give an oral presentation if required.

Note: These steps are only meant as guidelines. The team may change them to suite the project.

BIBLIOGRAPHY AND FURTHER READING

Arnold, S.A., Economic modelling of multi-sequential manufacturing process: a case study analysis of the automotive door, Ph.D. thesis, MSL, MIT, Cambridge, Massachusetts, USA, 1989.

Ashby, M.F., *Materials Selection Charts, Vol. 20 of the ASM Metals Handbook*, Dieter, G., Volume Chair. ASM International, Materials Park, OH, 1997a, pp. 266–280.

Ashby, M.F., *Performance Indices, Vol. 20 of the ASM Metals Handbook*, Dieter, G., Volume Chair. ASM International, Materials Park, OH, 1997b, pp. 281–290.

Ashby, M.F., *Materials Selection in Mechanical Design*, 3rd Ed., Elsevier, Amsterdam, 2005.

Ashby, M.F. and Johnson, C., *Materials and Design: The Art and Science of Material Selection in Product Design*, Butterworth-Heinemann, Oxford, 2002.

Bittence, J.C., Metals in Power train and chassis, *Adv. Mater. Proc.*, **5**, 40–63, 1987.

Bourell, D., *Decision Matrices in Materials Selection, Vol. 20 of the ASM Metals Handbook*, Dieter, G., Volume Chair. ASM International, Materials Park, OH, 1997, pp. 291–296.

CES 4 Software Granta Design Limited, Cambridge, U.K., 2002, www.Grantadesign.com.

Charles, J.A. and Crane, F.A.A., *Selection and Use of Engineering Materials*, 2nd Ed., Butterworths, London, 1989.

Clark, J., Roth, R., and Field III, F., *Techno-Economic Issues in Materials Selection, Vol. 20 of the ASM Metals Handbook*, Dieter, G., Volume Chair. ASM International, Materials Park, OH, 1997, pp. 255–265.

Destefani, J.D., *Advanced Materials and Processes*, No. 4, pp. 28–31, 1989.

Dieter, G., Overview of the Materials Selection Process, in *Materials Selection and Design, Vol. 20 of the ASM Metals Handbook*, Dieter, G., Volume Chair. ASM International, Materials Park, OH, 1997, pp. 243–254.

Edwards, K.L., Strategic substitution of new materials for old: applications in automotive product development. *Mater. Design*, **25**, 529–533, 2004.

Ermolaeva, N.S., Kaveline, K.G., and Spoormaker, J.L., Materials selection combined with optimal structural design: concept and some results. *Mater. Design*, **23**, 459–470, 2002.

Esawi, A.M.K. and Ashby, M.F., Cost estimates to guide pre-selection of processes, *Mater. Design,* **24**, 605–616, 2003.

Esawi, A.M.K. and Farag, M.M., Carbon nanotube reinforced composites: potential and current challenges, *Mater. Design*, **28**(9), 2394–2401, 2007.

European Directive 2000/53/EC—End of Life Vehicle, September 2000.

Farag, M.M., *Materials Selection for Engineering Design*, Prentice-Hall, Europe, 1997.

Farag, M.M., Quantitative methods of materials selection, in *Mechanical Engineers Handbook: Materials and Mechanical Design*, 3rd Ed., Kutz, M., Editor. Wiley, New York, 2006, pp. 466–488.

Farag, M.M., Quantitative methods of materials substitution: application to automotive components, *Mater. Design*, 2007. Available at www.sciencedirect.com.

Fowler, T., *Value Analysis in Materials Selection and Design, Vol. 20 of the ASM Metals Handbook*, Dieter, G., Volume Chair. ASM International, Materials Park, OH, 1997, pp. 315–321.

Gauthier, M.M., *Advanced Materials and Processes*, No. 7, pp. 26–35, 1990.

Giudice, F., La Rosa, G., and Risitano, A., Materials selection for life-cycle design process: a method to integrate mechanical and environmental performances in optimal choice, *Mater. Design*, **26**, 9–20, 2005.

Gordon, R.B., Analytical techniques for studying substitution among materials, Publication NMAB-385, National Academy Press, Washington, D.C., 1982.

Gray, G. and Savage, G.M., Advanced thermoplastic composite materials, *Metals Mater.*, 9, 513–517, 1989.

ISO 1401: 1998—Environmental management—life cycle assessment—goal and scope definition and inventory analysis. International Organization for Standardization, 2000.

Kubel Jr., E.J., *Advanced Materials and Processes*, No. 4, pp. 17–27, 1989.

Lewis, G., *Selection of Engineering Materials*, Prentice-Hall, Englewood Cliffs, NJ, 1990.

Matos, M.J. and Simplicio, M.H., Innovation and sustainability in mechanical design through materials selection. *Mater. Design*, **27**, 74–78, 2006.

McAffee, A.P., On the appropriate level of automation for advanced structural composites manufacturing for commercial aerospace applications, M.Sc. thesis, Department of Mechanical Engineering and Sloan School of Management, MIT, Cambridge, Massachusetts, USA, 1990.

Pugh, S., *Total Design: Integrated Methods for Successful Product Development*, Addison-Wesley, Reading, MA, 1991.

Shipp, C.T., Cost effective use of advanced composite materials in commercial aircraft manufacture, M.Sc. thesis, Department of Mechanical Engineering and Sloan School of Management, MIT, Cambridge, Massachusetts, USA, 1990.

11 Case Studies in Materials Selection and Substitution

11.1 INTRODUCTION

Different chapters of this book have discussed various issues related to materials selection and substitution, and how they fit with customer needs and design limitations. Illustrative examples in the form of design examples and case studies were used to explain the points of discussion, whenever possible. However, such examples had to be kept simple and to just address the point under discussion in order not to disrupt the continuity of the subject matter. This chapter tries to address this limitation by presenting more detailed case studies that are hopefully more representative of real-world material selection and substitution problems.

Five case studies are presented in this chapter:

Design and selection of materials for a turnbuckle (Section 11.2)
Design and selection of materials for surgical implants (Section 11.3)
Design and materials selection for lubricated journal bearings (Section 11.4)
Analysis of the requirements and substitution of materials for tennis racket (Section 11.5)
Materials substitution in automotive industry (Section 11.6)

Each of the case studies starts by an introduction that provides background information about the product under consideration, followed by an analysis of its functional requirements and design. Possible materials and manufacturing processes are then briefly presented and compared using an appropriate quantitative method. Final conclusions are then drawn based on the results of the comparison.

11.2 DESIGN AND SELECTION OF MATERIALS FOR A TURNBUCKLE

11.2.1 INTRODUCTION

A turnbuckle is a loop with opposite internal threads in each end for the threaded end of two ringbolts, forming a coupling that can be turned to tighten or loosen the tension in the members attached to the ringbolts. Figure 11.1 shows an assembly of a typical turnbuckle. The turnbuckle is used in different applications involving widely different requirements of forces, reliability, and service conditions. Examples

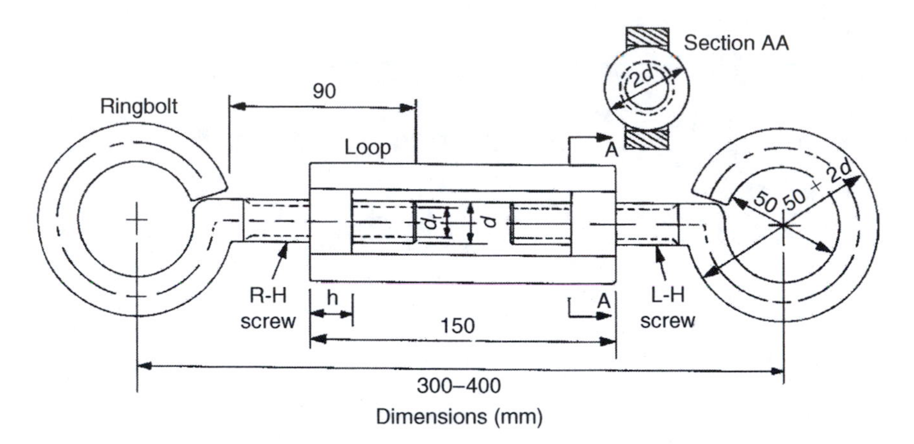

FIGURE 11.1 Assembly and material-independent dimensions of the turnbuckle.

include guy wires for telegraph poles, ship rigs, sports equipment, and camping gear. The main functional requirements of a turnbuckle are to apply and maintain tensile forces to the members attached to the ringbolts. It should be possible for an operator to release and reapply the tensile forces when needed.

11.2.2 Factors Affecting Performance in Service

The forces acting on the turnbuckle are usually tensile, although fatigue and impact loading can be encountered. Corrosion becomes a problem in aggressive environments, especially if the loop is made from a different material other than the ringbolt material. The possible modes of service failure and their effect on the performance of the turnbuckle are the following:

1. Yielding of the loop or one of the ringbolts. This will release the tensile forces in the system and could make operation unsafe.
2. Shearing, or stripping, of threads on the loop or on one of the ringbolts. This will release the tensile forces in the system and would make it impossible to reapply the required forces.
3. Fatigue fracture of the loop or one of the ringbolts. Fatigue fracture could start at any of the points of stress concentration in the turnbuckle assembly.
4. Creep strain in the loop or one of the ringbolts. This will relax the tensile forces in the system and could make operation unsafe.
5. Fracture of the loop or one of the ringbolts. This could take place as a result of excessive loading of the system, or as a result of impact loading if materials lose their toughness in service.
6. Corrosion as a result of environmental attack and galvanic action between ringbolt and loop if they are made of widely different materials. Excessive corrosion will make it difficult to apply and release the tension in the system and could reduce the cross-sectional area to dangerous limits. SCC can also occur in this system.

One or more of the mentioned failure modes could prove to be critical depending on the materials used in making the turnbuckle components, type of loading, and service environment. For example, fatigue is expected to be critical if the load is fluctuating, whereas creep should be considered for high-temperature service.

General specifications:

- The tensile force to be applied by the turnbuckle consists of a static component, $L_m = 20$ kN and an alternating component, $L_a = 5$ kN.
- Inner diameter of the rings at the end of ringbolts = 50 mm.
- Shortest distance between centers of rings on ringbolts = 300 mm.
- Longest distance between centers of rings on ringbolts = 400 mm.
- Other dimensions of the turnbuckle, which are material-independent, are shown in Figure 11.1.
- Service environment is industrial atmosphere.

11.2.3 Design Calculations

Figure 11.1 shows that the threaded length of the ringbolt is critical in view of the reduction in diameter involved in manufacturing the thread and the stress concentration at the roots of the teeth. It will, therefore, be assumed that if the threaded part can carry the service loads, the rest of the ringbolt will be safe.

In calculating the axial stress carried by a threaded bolt, an effective cross-sectional area, called tensile stress area (A_s), is used. The tensile stress area for standard metric threads is given by

$$A_s = 0.25\pi\,(d - 0.9382p)^2 \tag{11.1}$$

where

d = Major diameter of the bolt (see Figure 11.1)
p = Pitch of the thread

Table 11.1 gives the values of A_s for some standard metric threads.

11.2.4 Design for Static Loading

The tensile stress on the threaded part of the ringbolt (S_t) is given by

$$S_t = \frac{LK_t}{A_s} = \frac{\mathrm{YS_b}}{n_b} \tag{11.2}$$

where

L = Applied load
$\mathrm{YS_b}$ = Yield strength of the ringbolt material
n_b = Factor of safety used for the ringbolt calculations
K_t = Stress concentration factor

If a ductile material is used for making the ringbolt, there will be no need to introduce a stress concentration factor in Equation 11.2 and K_t can be ignored. However,

TABLE 11.1
Stress Areas for Some Standard Metric Threads

Major Diameter and Pitch (mm)		Tensile Stress Area (A_s) (mm²)	Thread Shear Area per Millimeter of Engaged Threads	
			AS_b (mm²)	AS_l (mm²)
M4	0.7	8.78	5.47	7.77
M5	0.8	14.20	7.08	9.99
M6	1.0	20.10	8.65	12.20
M8	1.25	36.60	12.20	16.80
M10	1.5	58.00	15.60	21.50
M12	1.75	84.30	19.00	26.10
M14	2.0	115	22.40	31.00
M16	2.0	157	26.10	35.60
M18	2.5	192	29.70	40.50
M20	2.5	245	33.30	45.40
M22	2.5	303	37.00	50.00
M24	3.0	353	40.50	55.00
M27	3.0	459	46.20	62.00
M30	3.5	561	51.60	69.60
M36	4.0	817	61.30	84.10
M42	4.5	1120	74.30	99.20

with brittle materials, such as cast iron, the stress concentration at the roots of the threads should be taken into consideration.

In the case of the loop, the combined area of the two sides ($2 \times A_l$) should withstand the applied load L without failure. Thus,

$$\frac{L}{2A_l} = \frac{YS_l}{n_l} \tag{11.3}$$

where

YS_l = Yield strength of the loop material
n_l = Factor of safety used for the loop calculations

Stripping of threads on the ringbolt will occur when its threads fail in shear at the minor diameter (d_r) (Figure 11.1). The shear stress in the ringbolt threads (τ) is given by

$$\tau = \frac{L}{AS_b \times h} \tag{11.4}$$

where

AS_b = Shear stress area per unit length of the ringbolt
h = Length of engagement between the ringbolt and loop

The values of AS_b for some standard metric threads are given in Table 11.1.

Similarly, stripping of the internal threads in the loop will occur when its threads fail in shear at the major diameter of the ringbolt (d). The shear stress in the loop threads (τ) is given by

$$\tau = \frac{L}{\mathrm{AS_l} \times h} \tag{11.5}$$

where $\mathrm{AS_l}$ is the shear stress area per unit length of the loop thread.

The values of $\mathrm{AS_l}$ for some standard metric threads are given in Table 11.1.

As Table 11.1 shows, the shear area of internal threads on the loop is greater than that of the external threads on the ringbolt. This means that, if the ringbolt and loop materials have the same shear strengths, stripping of the ringbolt threads will normally occur before stripping of the internal threads on the loop. For this reason, the material selected for the loop can be weaker than that of the ringbolt.

11.2.5 Design for Fatigue Loading

When the turnbuckle is subjected to fatigue loading, the procedure outlined in Section 6.5 can be used. Equation 6.12 can be used to calculate the ringbolt tensile stress area A_s. Thus,

$$\frac{n_m K_t L_m}{\mathrm{UTS} A_s} = \frac{n_a K_f L_a}{S_e A_s} = 1 \tag{11.6}$$

where

n_m and n_a = Factors of safety for static and fatigue strengths, respectively
L_m and L_a = Static and alternating loads, respectively
K_t and K_f = Static and fatigue stress concentration factors, respectively (K_t can be ignored for ductile materials)
UTS and S_e = Tensile strength and modified endurance limit, respectively

According to Equation 6.10, S_e can be calculated by multiplying the endurance limit of the material by a set of modifying factors. For the present case study, all the modifying factors will be grouped as one modifying factor k_i.

A similar procedure can be used to calculate the length of engagement between the ringbolt and loop h.

For the ringbolt:

$$\frac{n_m K_t L_m}{\tau_{bu} \mathrm{AS_b} h} = \frac{n_a K_f L_a}{\tau_{be} \mathrm{AS_b} h} = 1 \tag{11.7}$$

where τ_{bu} and τ_{be} are the static and fatigue shear strengths of the ringbolt material, respectively.

K_t can be ignored for ductile materials, as in the case of tensile stress calculations.

For the loop:

$$\frac{n_m K_t L_m}{\tau_{lu} \mathrm{AS_l} h} = \frac{n_a K_f L_a}{\tau_{le} \mathrm{AS_l} h} = 1 \tag{11.8}$$

where τ_{lu} and τ_{le} are the static and fatigue shear strengths of the loop material, respectively.

The values of h calculated from Equations 11.7 and 11.8 are compared and the larger one is selected.

For the present case study, the maximum-shear-stress theory will be used to predict the shear strengths of the ringbolt and loop materials. Thus, $\tau_u = 0.5\text{UTS}$ and $\tau_e = 0.5S_e$.

11.2.6 Candidate Materials and Manufacturing Processes

As shown in the design calculations, the strength of the loop material need not be as high as the ringbolt material. At the same time, the two materials should not be far apart in the galvanic series to avoid failure due to galvanic corrosion, as discussed in Section 6.7. For the present case, the ringbolt and loop materials will be selected with guidance from Table 3.1.

The ringbolt can be manufactured by the following methods:

1. From bar stock by first threading and then forming the ring by bending. Threading can be done by cutting or rolling
2. From bar stock by upset forging to form a head, flattening the head and forming the ring by forging, and then threading as mentioned
3. Sand-casting and then thread cutting
4. Shell molding and then thread cutting
5. Die-casting and then thread cutting

For the present case study, it is assumed that the available facilities favor the first method of manufacturing the ringbolt, where a bar stock is threaded by rolling and then bent round a die at room temperature to form the ring. The main processing requirement in this case is ductility. A minimum elongation of 15% is assumed to be necessary. This figure can be arrived at from experience with similar products or by performing development experiments.

The loop can be manufactured by the following methods:

1. Sand-casting using wooden or metal pattern and then thread cutting
2. Shell molding using metal pattern and then thread cutting
3. Die-casting and then thread cutting
4. Die forging of a bar stock and then thread cutting
5. Machining from a bar stock and then thread cutting
6. Welding of the threaded ends to round or square bars

For the present case study, it is assumed that the available facilities favor the first method of manufacturing the loop.

Table 11.2 lists the properties of some candidate wrought alloys for the ringbolt and cast alloys for the loop materials. The values of K_f in the table are calculated according to Equation 4.9 using the given values of q and assuming that $K_t = 2.5$, which is reasonable for coarse threads.

TABLE 11.2
Candidate Materials for the Ringbolt and Loop Materials

Material	UTS (MPa)	YS (MPa)	$S_{e'}$ (MPa)	k_i	q	K_f	ρ	C	Relative Cost[a]	
									Mat.	Mfr.
Ringbolt Materials (Minimum Elongation 15%)										
Steels										
AISI 1015	430	329	195	0.7	0.1	1.15	7.8	1	1	1
AISI 1040	599	380	270	0.6	0.2	1.30	7.8	1	1.1	4
AISI 1340	849	567	420	0.5	0.7	2.05	7.8	3	2.2	4
AISI 4820	767	492	385	0.55	0.6	1.90	7.8	3	2.5	8
Aluminum alloys										
AA 3003 O	112	42	56	0.7	0.1	1.15	2.73	4	5	1
AA 5052 O	196	91	90	0.65	0.4	1.60	2.68	4	7	1
AA 6061 O	126	56	55	0.6	0.3	1.45	2.7	3	7	1
Copper-base alloys										
Al bronze	420	175	147	0.7	0.4	1.60	8.1	4	12	6
Si bronze	441	210	175	0.6	0.5	1.75	8.5	4	12	6
70/30 Brass	357	133	145	0.75	0.3	1.45	8.5	4	10	4
Loop Materials (Cast Alloys)										
Gray cast irons ASTM A48-74										
Grade 20	140	140	70	0.5	0.2	1.30	7.5	4	1.2	1
Grade 40	280	280	130	0.45	0.2	1.30	7.5	4	1.25	1
Grade 60	420	420	168	0.4	0.2	1.30	7.5	4	1.3	1
Nodular cast irons ASTM A536										
60-40-18	420	280	210	0.6	0.2	1.30	7.5	4	1.9	4
80-55-06	560	385	280	0.55	0.2	1.30	7.5	4	2	4
120-90-02	840	630	420	0.5	0.2	1.30	7.5	4	2.1	4
Aluminum alloys										
AA 208.0	147	98	44	0.6	0.4	1.60	2.8	3	5	1
AA 356.0 T6	231	168	79	0.5	0.5	1.75	2.68	4	5	1
AA B443.0	133	56	40	0.6	0.7	2.05	2.7	4	4	1
Copper-base alloys										
Al bronze	590	120	200	0.55	0.4	1.60	8.1	4	12	6
Si bronze	420	125	120	0.6	0.4	1.60	8.3	4	12	6
Mn bronze	640	340	300	0.5	0.5	1.75	8.3	4	11	6

Note: k_i = endurance limit modifying factor; ρ = specific gravity; C = corrosion resistance; Mat. = material; Mfr. = manufacturing: 1 = poor, 2 = fair, 3 = good, 4 = very good.

[a] Relative materials and processing costs are based on the cost of steel AISI 1015, which is taken as unity.

11.2.7 Sample Calculations

The procedure for calculating the dimensions, weight, and cost of the turnbuckle will be illustrated here. The calculations are based on steel AISI 1015 as the ringbolt material and nodular cast iron ASTM A536 60-40-18 as the loop material. These two materials are compatible from the galvanic corrosion point of view.

As the turnbuckle is subjected to combined static and fatigue loading, Equation 11.6 will be used to calculate a preliminary value for A_s. According to Shigley and Mitchell (1983), the factors of safety can be taken as $n_m = 1.5$ for static loading, and $n_a = 3.0$ for fatigue loading. As the ringbolt material is ductile, the static stress concentration factor, K_t, can be taken as unity. Using the values in Table 11.2 for steel AISI 1015, Equation 11.6 can be written as

$$\frac{1.5 \times 1 \times 20000}{430 \times A_s} + \frac{3.0 \times 1.15 \times 5000}{195 \times 0.7 \times A_s} = 1 \quad (11.9)$$

which gives $A_s = 196.14$ mm^2.

From Table 11.1, it can be seen that the M18 standard metric thread has an A_s of 192 mm^2, which is close to the calculated A_s. This means that the major diameter of the bolt is 18 mm.

The mass of the ringbolt,

$$w_b = \text{density (volume of the straight part + volume of the ring)} = 603 \text{ g}$$

The next step is to calculate the length of engagement between the ringbolt and the loop using Equations 11.7 and 11.8. According to Equation 11.7, and using the same values n_m, n_a, K_t, and K_f as before

$$\frac{1.5 \times 1 \times 20000 \times 2}{430 \times 29.7 \times h} + \frac{3.0 \times 1.15 \times 5000 \times 2}{195 \times 0.7 \times 29.7 \times h} = 1 \quad (11.10)$$

This gives $h = 13.2$ mm.

In the case of the loop, K_t will be taken as unity since the nodular cast iron used is ductile. Factors of safety similar to those used for the ringbolt calculations will be used for the loop. According to Equation 11.8:

$$\frac{1.5 \times 1 \times 20000 \times 2}{420 \times 40.5 \times h} + \frac{3.0 \times 1.3 \times 5000 \times 2}{210 \times 0.6 \times 40.5 \times h} = 1 \quad (11.11)$$

This gives $h = 11.14$ mm, which is smaller than the value given by Equation 11.10.

The larger value of h is taken as the design value.

The two webs that connect the threaded ends of the loop have to resist the combined effects of the static and alternating loads. Equation 11.6 can be used to calculate the total cross-sectional area of the webs, A_w. Using the same factors of safety as mentioned,

$$\frac{1.5 \times 1 \times 20}{420 \times A_w} + \frac{3.0 \times 1.3 \times 5}{210 \times A_w} = 1 \quad (11.12)$$

This gives $A_w = 164.3$ mm^2.

The mass of the loop w_l = density (volume of threaded ends + volume of webs) = 336 g

$$\text{Total mass of the turnbuckle} = w_{tb} = 2 \times \text{weight of ringbolt} + \text{weight of loop}$$
$$= 1542 \text{ g} = 1.542 \text{ kg}$$

Relative cost of materials and processing

$$= C_m = 2 \times \text{relative cost of each ringbolt} + \text{relative cost of loop}$$
$$= 2 \times 0.603\,(1 + 1) + 0.336\,(1.9 + 4) = 4.39 \text{ units of relative cost}$$

To arrive at the optimum pair of materials for the turnbuckle, the calculations mentioned should be repeated for all possible combinations of ringbolt and loop materials given in Table 11.2. For example, each one of the three aluminum ringbolt alloys represents a possible candidate for each one of the three aluminum loop alloys, which involves nine combinations. However, aluminum alloys do not represent possible candidates for copper-base or ferrous alloys, as they are too far apart in the galvanic series, which could cause galvanic corrosion. From Table 11.2, it can be shown that there are 42 possible material combinations (24 combinations for ferrous alloys, 9 combinations for aluminum alloys, and 9 combinations for copper alloys).

It would be tedious and time consuming to perform all these required calculations manually. One of the screening methods discussed in Chapter 9 can be used to reduce the number of candidate materials in Table 11.2. Another approach, which is adopted here, is to write a computer program in a programming language, C++, for example, to perform the calculations. For each possible pair of materials, the program calculates the total weight of the turnbuckle and its relative cost.

To select the optimum pair of materials for the turnbuckle under consideration, the weighted properties method, which was discussed in Section 9.5, is used. When the corrosion resistances of the two materials in a turnbuckle are different, the lower value is taken to represent the corrosion resistance of the turnbuckle.

For the present case study, the weighting factors are taken as 0.5, 0.3, and 0.2 for the cost, corrosion resistance, and weight, respectively. The performance index (γ) of a turnbuckle made of a pair of materials is calculated as follows:

$$\gamma = 0.5\,(\text{scaled relative cost}) + 0.3\,(\text{scaled corrosion resistance})$$
$$+ 0.2\,(\text{scaled total weight})$$

Scaling was performed such that a turnbuckle with lower weight and cost was given a lower scaled value. To be consistent, material combinations with higher corrosion resistance were given a lower scaled value. With this method of scaling, turnbuckles with lower numerical value of the performance index (γ) are preferable to those with higher numerical value of the performance index.

The calculated relative total weight, relative cost, and performance index for 10 turnbuckles with the lowest numerical values of the performance index (γ) are given in Table 11.3. The results show that, with the present selection criteria, ferrous alloys are preferable. The main reasons for this are their lower cost and higher

TABLE 11.3
Comparison of Turnbuckle Materials

Material Pair		Relative	Relative	Merit
Ringbolt	Loop	Total Weight	Cost	Value
AISI 1340	Nodular CI 120-90-02	1.082	2.264	1.748
AISI 1340	Gray CI grade 60	1.221	2.504	1.896
AISI 1015	Gray CI grade 60	1.220	1.000	1.944
AISI 1340	Gray CI grade 40	1.313	2.580	1.952
AISI 1340	Nodular CI 80-55-06	1.117	2.717	1.982
AISI 1015	Nodular CI 120-90-02	1.091	1.157	1.997
AISI 1015	Gray CI grade 40	1.314	1.076	2.001
AISI 1340	Nodular CI 60-40-18	1.152	2.790	2.025
AISI 1015	Nodular CI 80-55-06	1.126	1.232	2.041
AISI 1015	Nodular CI 60-40-18	1.161	1.305	2.085

strengths. The higher strengths are reflected in the total weight of the turnbuckle as shown in Table 11.3. If the weighting factor given to corrosion resistance was increased at the expense of that given to cost, turnbuckles made of nonferrous alloys would be preferable.

11.3 DESIGN AND SELECTION OF MATERIALS FOR SURGICAL IMPLANTS

11.3.1 Introduction

Surgical implant materials are used in repairing many parts of the human body. The number of materials in current use as implant materials is large and includes metallic, polymeric, ceramic, and composite materials. Both hard tissues such as bones, and soft tissues such as skin can be restored or replaced with implants of similar mechanical properties, texture and color. For example, rigid metallic, ceramic, and composite materials are used for fixing or replacing bones and joints, foams and gels are used for soft tissue supplementation, and elastic materials for replacement of skin and blood vessels. Although the mechanical requirements of implant materials are relatively simple, the biocompatibility requirements are stringent and more difficult to meet. Biocompatibility means that the material and its possible degradation products must be tolerated and cause no tissue dysfunction at any time.

This case study discusses the design and selection of materials for a hip joint prosthesis. Figure 11.2 shows the components used for a complete hip joint. In this case, the femoral head is replaced by a rigid pin that is installed in the shaft of the femur, whereas the pelvic socket (acetabulum) is replaced by a rigid or soft cup that is fixed to the ilium. Both the pin and cup can be fixed to the surrounding bone with an adhesive. In recent designs, natural bone growth is used to provide the bond.

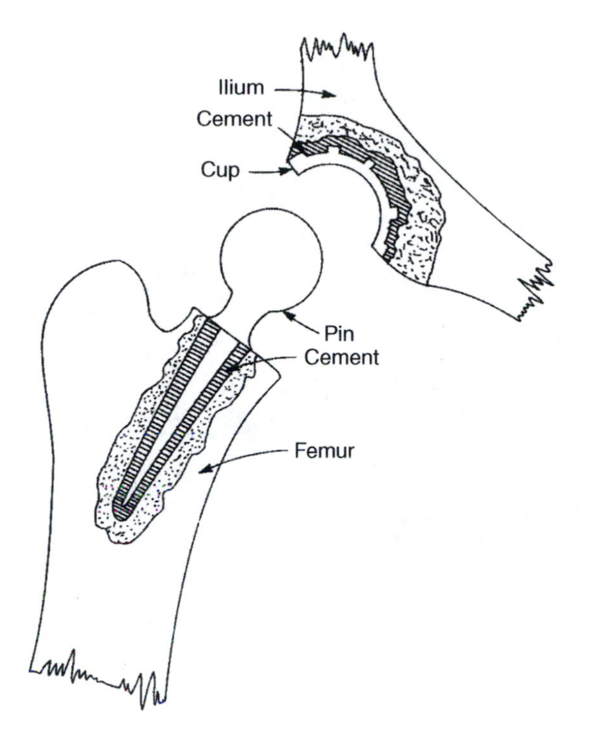

FIGURE 11.2 Components of a complete hip joint prosthesis.

11.3.2 Main Dimensions and External Forces

As the prosthesis is intended to replace the bone structure of the hip joint, it is important to study the bone structure and properties. Bone is a living tissue composed of inorganic and organic materials in dynamic equilibrium with the body fluids. Although the proportions vary from one part of the skeleton to another, water-free bone contains about two-thirds of inorganic and one-third of organic materials. The inorganic phase, primarily hydroxyapatite crystals, is hard and brittle and represents the main load-bearing component of the bone structure. The organic phase, primarily collagen fibers, is gelatinlike protein and its presence makes the bone tough, in addition to its biological functions. These phases are generally arranged in a complex structure to give maximum strength in the required direction. Figure 11.3 shows how the load-bearing phase is arranged in the direction of maximum stress in the head of the femur bone.

The compressive strength of compact bone is about 140 MPa, and the elastic modulus is about 14 GPa in the longitudinal direction and about one-third of that in the radial direction. These properties are modest in comparison with those of most engineering metallic and composite materials. However, live healthy bone is self-healing and has great resistance to fatigue loading. The implant material, however, does not have this ability to repair itself, and usually has a finite fatigue life. For this reason, the implant material must be stronger than bone especially under fatigue loading.

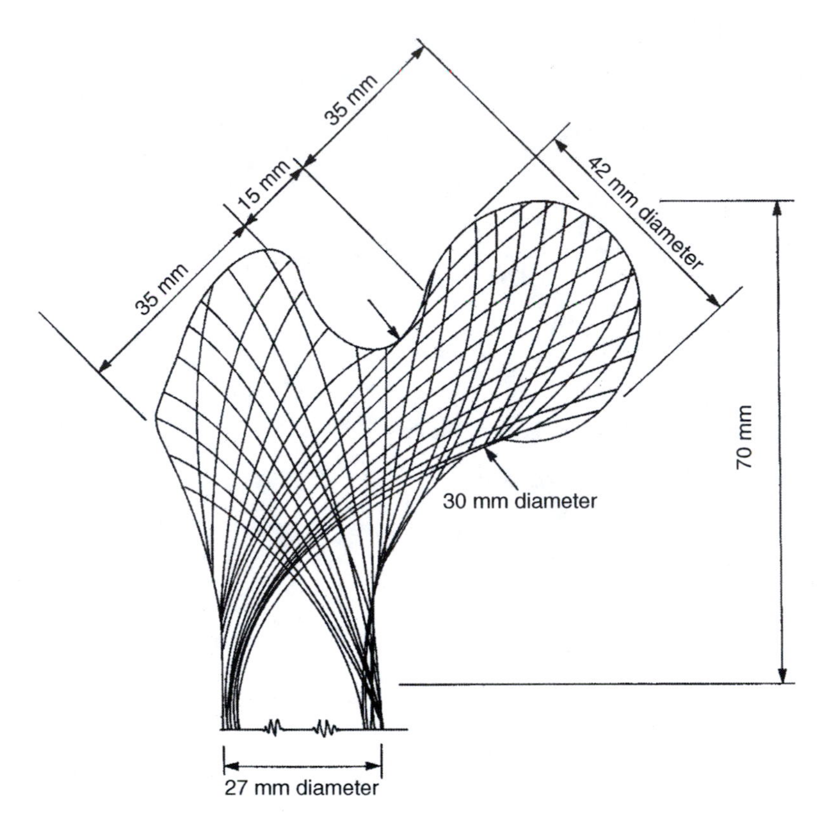

FIGURE 11.3 Main average dimensions and distribution of load-bearing phases in the head of the femur bone.

In the case of the hip joint, carrying the repeated loading caused by walking and similar activities is an important part of its function. The loading frequency ranges from 1 to 2.5 million cycles per year, depending on the activity and movement of the individual. Stress analysis of the forces acting on the hip joint shows that the fatigue loading is usually equal to 2.5–3 times the body weight and can possibly be higher depending on posture of the individual. In addition to the repeated loading, the joint is subjected to a static loading as a result of muscle action, which keeps the parts of the joint together. This static load is normally much smaller than the repeated loading. As the hip joint prosthesis is generally intended to be permanent, it is expected to resist fatigue fracture for years or decades.

11.3.3 Fatigue-Loading Considerations

In designing the hip joint prosthesis, it should be borne in mind that, regardless of the strength of the implant material properties, the dimensions of the prosthesis must match the dimensions of the original bone. The design calculations in the present case study will be based on the dimensions given in Figure 11.3, which represent approximate average values for an adult. Assuming that the weight of the person is

75 kg (165 lb), and taking the alternating load as three times the weight, it can be shown that the hip prosthesis will be subjected to an alternating load of 2205 N. The maximum stress occurs in the prosthesis neck, which is about 30 mm in diameter according to Figure 11.3. From these values, the alternating stress is estimated as 3.1 MPa. Assuming that the static load due to muscle contraction is about 300 N, the static stress at the neck of the prosthesis is estimated as 0.42 MPa. Using the modified Goodman relationship as given by Equation 6.12:

$$\left(\frac{n_m K_t S_m}{\text{UTS}}\right) + \left(\frac{n_a K_f S_a}{S_e}\right) = 1$$

where

n_m = Factor of safety for static load, taken as 2 in this case
n_a = Factor of safety for alternating load, taken as 3 in this case
K_t = Stress concentration factor for static load, taken as 2.2
K_f = Stress concentration factor for alternating stress, taken as 3.5
S_m = Static stress = 0.42 MPa, as calculated
S_a = Alternating stress = 3.1 MPa, as calculated
UTS = Ultimate tensile strength of the prosthesis material
S_e = Endurance limit of the prosthesis material

Taking the endurance ratio of the prosthesis material as 0.35, the value of S_e = 0.35 UTS.

Substituting these values in the mentioned Goodman relationship, UTS is found to be about 95 MPa. From the assumed value of endurance ratio of 0.35, S_e is estimated as 33.25 MPa. These values represent the minimum strengths that an implant material must have to be considered for making the pin of the hip joint prosthesis. Using materials with higher strengths will not make it possible to reduce the cross-sectional area of the prosthesis, but will have the advantage of being less likely to fail in service.

11.3.4 Wear Considerations

Wear of the prosthesis parts (cup and pin) should also be considered when designing a hip joint prosthesis. Wear debris are often found in tissue surrounding joints and could cause adverse effects due to sensitivity of the patient to the material. In addition, material wear in the total hip prosthesis, especially the enlargement of the cup (acetabular concavity) causes poor articulation of the joint. The pressure between the mating surfaces of the prosthesis can be estimated by dividing the maximum force by the projected area of the cup. In the present case, the maximum force is given by 2205 + 300 = 2505 N, and the projected area of the cup is 1385 mm^2. This gives an average pressure of 1.8 MPa. If both the pin and cup are made of metallic materials, the high friction coefficient could lead to mechanical difficulties. For this reason, the cup is usually made of a low-friction plastic such as high-density polyethylene and PTFE. The compressive strengths of most polymeric materials are higher than the calculated contact pressure, which means that selecting the material of the cup is mostly based on their biocompatibility.

11.3.5 Analysis of Implant Material Requirements

Earlier discussion has shown that the total hip joint prosthesis consists of a pin, which replaces the head of the femur bone, and a cup, which replaces the acetabular concavity. The pin and cup are fixed to the surrounding bone structure by an adhesive cement. As would be expected, the requirements for each of these components are different as they perform different functions. In the present case study, only the material requirements for pin will be discussed. Similar procedure can be applied to the cup and cement.

11.3.5.1 Tissue Tolerance

In general, biological requirements represent the major constraints in selecting implant materials. The action of the implant material on the body tissues can range from toxicity, in which case the implant material is totally rejected, to inertness. Even with inert material, their presence impedes the normal healing sequence at the implant site and leads to fibrocartilaginous membrane of low cellularity, which isolates the implant from normal tissue. This membrane is the body's response to the stimulus of an inert foreign material that is impervious to body fluids. The thickness of the membrane is proportional to the degree of implant material dissolution. Toxicity is usually signaled by a large population of inflammatory cells. This requirement is usually called tissue tolerance and is difficult to quantify. For the sake of comparison, materials are given a rating of 10 for the best material and 1 for the worst. Only materials with tissue tolerance of 7 or more will be considered in the present case study. The tissue tolerance requirements apply equally well to the pin, cup, and cement materials.

11.3.5.2 Corrosion Resistance

Another important implant material requirement is corrosion resistance. This is because body fluids are aqueous salt solutions with concentrations roughly comparable to seawater and with a pH value of 7.4. This environment is hostile and could cause corrosion of many metallic materials. Such corrosion should be avoided as it may induce deleterious effects in the surrounding tissues or even distant organs. The combination of corrosive action of the body fluids and fatigue loading could result in SCC, which imposes strict limitations on the implant material surface finish and structural homogeneity. As in the case of tissue tolerance, corrosion resistance is rated according to a scale of 10 to 1. Only materials with a corrosion resistance rating of 7 or better are considered in this case study. The corrosion resistance requirements apply equally well to the pin, cup, and cement materials.

11.3.5.3 Mechanical Behavior

The preceding design analysis shows that the strength requirements of implant materials are relatively easy to meet. Other material requirements include toughness, wear resistance, elastic compatibility, and specific gravity. Brittle materials are undesirable in hip joint prosthesis in view of the possible shock loading that may result in service. High wear resistance is necessary to avoid the accumulation of wear debris in the surrounding tissue or other organs of the body. Both toughness and wear resistance will be rated on a scale of 10 to 1, as in the case of tissue tolerance and corrosion resistance. Only materials with a wear resistance equal to or better than 7 and toughness equal to or better than 2 are considered in the present application.

11.3.5.4 Elastic Compatibility

Elastic compatibility of an implant material with the surrounding bone structure is also an important requirement. This is because large mismatches can lead to deterioration of the interface between the two materials. Unfortunately, the elastic moduli of the currently available materials for bone replacement are much higher than that of bone. Although this problem can be partially overcome by selecting the appropriate cementing agents, it is preferable for the implant material to be elastically compatible with the bone. The elastic modulus of bone (14 GPa) will be taken as a target value when candidate implant materials are evaluated.

11.3.5.5 Weight

Similarity between the specific gravity of the implant material and bone is also desirable to keep the weight of the implant material as close as possible to that of the original bone. The specific gravity of compact bone is about 2.1, which is less than that of the metallic implant materials in current use. The specific gravity of bone will be taken as a target value when candidate implant materials are evaluated.

11.3.5.6 Cost

Reasonable cost is another requirement for the hip joint prosthesis. The total cost includes the cost of the stock material and the cost of processing and finishing. As the volume of the prosthesis is independent of the material, it is more appropriate to compare materials on the basis of the cost per unit volume rather than the cost per unit mass. In view of the small numbers produced, mass production method cannot be used in making the hip joint prosthesis. This means that the costs of processing and finishing are expected to represent a large proportion of the total cost.

Based on the preceding discussion, the material requirements for the pin of the hip joint prosthesis can be summarized as shown in Table 11.4. Higher values of the first six properties are more desirable and should comply with the lower limits outlined in the discussion. According to the limits on properties method, which is discussed in Section 9.5, these properties are lower-limit properties. Elastic modulus and specific gravity are target-value properties, as materials should match the bone as closely as possible. The cost is an upper-limit property.

The relative importance of the different material requirements is given in Table 11.4. The digital logic method can be used to assist in allocating the weighting factors.

TABLE 11.4
Main Requirements and Weighting Factors for the Pin of the Hip Joint Prosthesis

Property	Weighting Factor
Tissue tolerance (lower-limit property)	0.2
Corrosion resistance (lower-limit property)	0.2
Tensile strength (lower-limit property)	0.08
Fatigue strength (lower-limit property)	0.12
Toughness (lower-limit property)	0.08
Wear resistance (lower-limit property)	0.08
Elastic modulus (target value property)	0.08
Specific gravity (target value property)	0.08
Cost (upper-limit property)	0.08

11.3.6 Classification of Materials and Manufacturing Processes for the Prosthesis Pin

A survey of the literature shows that possible materials for the pin of the hip joint prosthesis include stainless steels, cobalt- and chromium-base alloys, titanium alloys, tantalum, and FRP. The latter materials have been under study for several years and are now being considered for approval by food and drug administration (FDA) for use in the United States. Although several ceramic materials have been developed for surgical implants, their use as hip joint prosthesis is still in the experimental stage. These materials should be seriously considered when their performance is better characterized.

When the pin is made of a metallic material, it can be manufactured either by casting or by forging. In both cases, the part is finished to the required final dimensions and given a mirror finish. This means that both cast and wrought alloys can be considered for the pin of the hip joint prosthesis. In the case of FRP, a processing method should be devised to orient the fibers in the direction of maximum loading. The matrix material can then be cast to infiltrate the fibers.

Table 11.5 gives the properties of representative examples of possible candidates for the pin of the hip joint prosthesis.

11.3.7 Evaluation of Candidate Materials

The limits of properties method, discussed in Section 9.5, are used to evaluate the candidate materials listed in Table 11.5. Following the notations of Equation 9.8, the lower limits, Y_i; upper limits, Y_j; and target values, Y_k, used in the calculations were as follows:

1. Tissue tolerance, lower limit, $Y_i = 7$
2. Corrosion resistance, lower limit, $Y_i = 7$
3. Tensile strength, lower limit, $Y_i = 95$ MPa
4. Fatigue strength, lower limit, $Y_i = 33.25$ MPa

TABLE 11.5
Properties of Selected Surgical Implant Materials

Material	Tissue Tolerance	Corrosion Resistance	Tensile Strength (MPa)	Fatigue Strength (MPa)	Elastic Modulus (GPa)	Relative Toughness	Relative Wear Resistance	ρ	C
Stainless steels									
316	10	7	517	350	200	8	8.0	8.0	1.0
317	9	7	630	415	200	10	8.5	8.0	1.1
321	9	7	610	410	200	10	8.0	7.9	1.1
347	9	7	650	430	200	10	8.4	8.0	1.2
Co–Cr alloys									
Cast alloy (1)	10	9	655	425	238	2	10.0	8.3	3.7
Wrought alloy (2)	10	9	896	600	242	10	10.0	9.1	4.0
Titanium alloys									
Unalloyed titanium	8	10	550	315	110	7	8.0	4.5	1.7
Ti–6Al–4V	8	10	985	490	124	7	8.3	4.4	1.9
Composites (fabric reinforced)									
Epoxy–70% glass	7	7	680	200	22	3	7.0	2.1	3
Epoxy–63% carbon	7	7	560	170	56	3	7.5	1.6	10
Epoxy–62% aramid	7	7	430	130	29	3	7.5	1.4	5

Note: ρ is the specific gravity; C is total relative cost and includes cost of material per unit volume, processing cost, and finishing cost; cast alloy (1): 27–30 Cr, 2.5 Ni, 5–7 Mo, 0.75 Fe, 0.36 max C and Si, Rem. Co.; wrought alloy (2): 20 Cr, 10 Ni, 15 W, 0.13 Mo, 3 max Fe, 0.1 C, 2 max Mn, 0.48 Si, Rem. Co.

TABLE 11.6
Merit Parameter (m) and Ranking of Candidate Materials for the Pin of a Hip Joint Prosthesis

Material	Merit Parameter (m)	Rank
Ti–6Al–4V	0.554	1
Co–Cr wrought alloy	0.555	2
Unalloyed titanium	0.563	3
316 stainless steel	0.593	4
347 stainless steel	0.597	5
317 stainless steel	0.607	6
321 stainless steel	0.608	7
Epoxy–70% glass fabric	0.615	8
Co–Cr cast alloy	0.622	9
Epoxy–62% aramid fabric	0.649	10
Epoxy–63% carbon fabric	0.720	11

5. Toughness, lower limit, $Y_i = 2$
6. Wear resistance, lower limit, $Y_i = 7$
7. Elastic modulus, target value, $Y_k = 14$ GPa
8. Specific gravity, target value, $Y_k = 2.1$
9. Relative total cost, upper limit, $Y_j = 10$

11.3.8 Results

Equation 9.8 and the preceding limits were used to calculate the merit parameter (m) for the candidate materials, and the results are shown in Table 11.6. The table shows that the Ti–6Al–4V alloy ranks as number one followed by Co–Cr alloy. Stainless steels, although commonly used as metal plates for repairing fractures, ranked 4, 5, 6, and 7 for the present application. If less weight is given to corrosion resistance and more weight is given to the cost of the prosthesis, as in the case of temporary implants, stainless steels would occupy top ranks.

11.4 DESIGN AND MATERIALS SELECTION FOR LUBRICATED JOURNAL BEARINGS

11.4.1 Introduction

A journal bearing is a machine element designed to transmit loads or reaction forces from a rotating shaft to the bearing support. Besides carrying loads, the bearing material is subjected to the sliding movement of the shaft. The friction forces that result from the sliding motion are normally reduced by lubrication. In the case of lubricated journal bearings, a continuous oil film is formed between the shaft and the bearing as shown in Figure 11.4. When the shaft is at rest, metal-to-metal contact occurs at point (x). As the shaft rotates slowly, the point of contact moves to

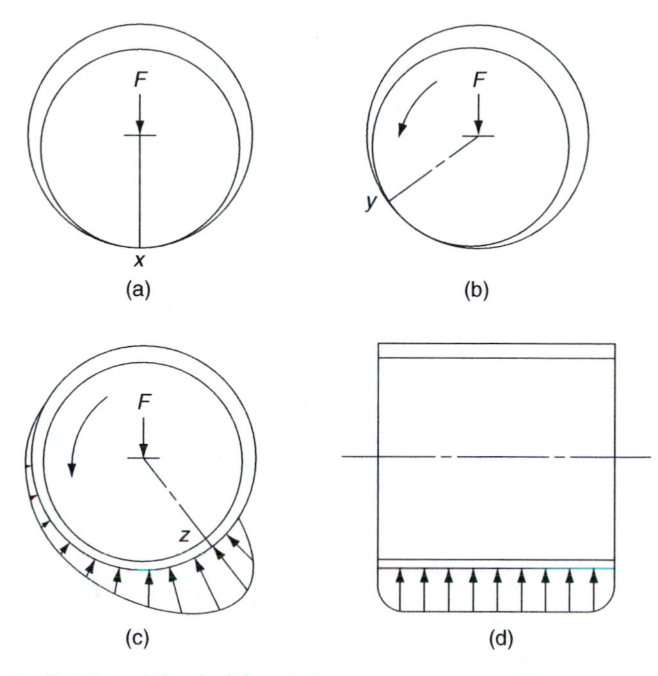

FIGURE 11.4 Position of the shaft in relation to the bearing and the pressure distribution in a lubricated journal bearing. (a) Shaft at rest; (b) shaft rotating at slow speed; (c) shaft rotating at high speed; and (d) distribution of pressure in the axial direction.

position (y). A thin adsorbed film of lubricant may partially separate the surfaces, but a continuous film will not exist because of the slow speed. With the increasing speed of rotation, a continuous lubricant film is established and the center of the shaft is moved so that the minimum film thickness is at (z). The wedge shape of the film helps to build up sufficient pressure to support the external force. The pressure distribution in both the radial and axial directions is shown in Figures 11.4c and 11.4d. The drop in pressure at the edges of the bearing in the axial direction is due to the leakage of the lubricant from the sides.

Experience shows that higher lubricant pressure builds up in the bearing as the clearance between the shaft and bearing decreases. In addition, the bearing gives better guidance to the shaft as the clearance decreases. However, allowance must be made for manufacturing tolerances in the journal and sleeve, deflection of the shaft, and space to permit foreign particles to pass through the bearing. The clearance (c) is usually related to the diameter of the shaft (d) and usually ranges between $c/d = 0.001$ and 0.0025.

The bearing length should be as long as possible to reduce the compressive stresses on the bearing material. Also a longer bearing will reduce the side leakage of the lubricant from the bearing, thus allowing higher loads to be supported without suffering metal-to-metal contact between the journal and bearing. However, space requirements, manufacturing tolerances, and shaft deflection are better met with a shorter bearing length. The length of the bearing is usually related to the diameter of the shaft and usually ranges between $L/d = 0.8$ and 2.0.

In the present case study, it is required to design and select the materials for the two journal bearings that support the rotor of a centrifugal pump in its casing. The mass of the rotor is equally distributed between the two bearings, which carry a load of 7500 N each. The diameter of the rotor shaft is 80 mm and its speed of rotation is 1000 rpm.

11.4.2 Design of the Journal Bearing

The magnitude and nature of the load are the main factors that affect the bearing design and the performance of the bearing material. Heavy loads require high compressive yield strength to avoid plastic deformation, whereas cyclic loads require high fatigue strength to avoid fracture. An approximate value of the mean compressive stress (S_c) acting on the bearing material is given by

$$S_c = \frac{F}{LD} \tag{11.13}$$

where

F = The force acting on the bearing (7500 N)
L = Axial length of the bearing
D = Diameter of the bearing ($d + c$)
d = Shaft diameter (80 mm)
c = Clearance

For the present case study, it is reasonable to assume that L/d is 1.25, that is, L = 100 mm, and c/d is 0.001. These values are within the practical limits discussed. Based on these assumptions, it can be shown that the compressive stress acting on the bearing material is about 0.93 MPa. This value is within 0.6 and 1.2, which are the limits recommended by Shigley and Mitchell (1983) for centrifugal pumps.

The coefficient of friction (f) is an important parameter in bearing design as it affects both the power loss and temperature rise due to friction. The coefficient of friction for a well-lubricated journal bearing can be expressed by the McKee empirical relation (reported by Black and Adams, 1983):

$$f = \frac{K(ZNd)}{(S_c c)} + K' \tag{11.14}$$

where

Z = Viscosity of the lubricant (cP)
N = Rotational speed of the shaft (rpm)
K = Constant whose value depends on the system of units used
K' = Constant to account for end leakage of the lubricant (0.002 for L/d in the range 0.75–2.8).

The quantity ZN/S_c is called the bearing characteristic number and is dimensionless if the quantities are expressed in a consistent system of units. Higher values of this parameter ensure the continuity of the lubricant film and avoid metal-to-metal contact between the bearing material and the journal. The value of Z depends on the grade of the oil and its temperature at the bearing–journal interface. This temperature is a function of the heat balance between the heat generated due to friction, H_1, and the heat dissipated by the bearing to the surroundings, H_2. Under the thermal equilibrium, heat will be dissipated at the same rate that it is generated in the lubricant.

The usual procedure in bearing design is to assume a bearing temperature and then estimate H_1 and H_2. If thermal equilibrium is indicated, then the assumed bearing temperature is correct. If H_1 and H_2 are not approximately equal, the designer must assume a different operating temperature, a different oil, or different values of L and d. For the present case study, it is reasonable to assume an ambient temperature, t_a, of 20°C (68°F) and a bearing surface temperature, t_b, of 70°C (158°F). Black and Adams (1983) reported that experimental work shows that

$$t_b - t_a = 0.5(t_o - t_a) \tag{11.15}$$

where t_o is the temperature of the oil film.

Using the preceding assumptions and Equation 11.15, t_o = 120°C (248°F). Selecting a lubricating oil of grade SAE No. 30, gives Z = 5.5 cP at 120°C. Under these conditions, the bearing characteristic number is

$$\frac{ZN}{S_c} = 5.5 \times \frac{1000}{0.93} = 5914$$

This value is within the range 4,300–14,300, which is recommended by Jain (1983) for centrifugal pumps.

From Equation 11.14,

$$f = 0.00395$$

The heat generated due to friction, H_1, is given by

$$H_1 = fFV \tag{11.16}$$

where

F = Force acting on the bearing = 7500 N
V = Rubbing velocity (m/s) = $\pi dN/60$ = 4.187 m/s.

From Equation 11.16,

$$H_1 = 0.00395 \times 7500 \times 4.187 = 124 \text{ W}$$

According to Black and Adams (1983), the heat dissipated by the bearing, H_2, is given by the empirical relationship:

$$H_2 = CA(t_b - t_a) = \frac{1}{2}CA(t_o - t_a) \tag{11.17}$$

where

A = Projected area of the bearing (Ld)
C = Heat-dissipation coefficient
= 3.75 ft. lb/min/in.^{2}/°F (244 W/m^2/°C) for average industrial unventilated bearings
= 5.8 ft. lb/min/in.^{2}/°F (377 W/m^2/°C) for well-ventilated bearings

Using the assumed values of temperature and Equation 11.17, it can be shown that

Under average industrial unventilated conditions $H_2 = 97.6$ W
Under well-ventilated conditions $H_2 = 150.8$ W

Comparing the calculated values of H_1 and H_2 shows that thermal balance can be achieved under moderate ventilation conditions.

Having determined the bearing loads, lubricant, and operating temperatures, it is now possible to select the optimum material for the bearing.

11.4.3 Analysis of Bearing Material Requirements

The preceding discussion shows that the compressive strength of the bearing material at the operating temperature (120°C or 248°F in the present case) must be sufficient to support the load acting on the bearing. If the material is not strong enough, it could suffer considerable plastic deformation by extrusion. Fatigue strength also becomes important under conditions of fluctuating load. Both compressive and fatigue strengths are known to increase as the thickness of the bearing material decreases. This is achieved in practice by bonding a thin layer of the bearing material (0.05–0.15 mm or 0.002–0.006 in.) to a strong backing material to form a bimetal structure. Common examples include lead and tin alloys on steel or bronze backs. An intermediate layer of copper or aluminum alloys may also be introduced between the bearing material and the steel back to produce a trimetal structure. In such cases, the bearing material can be made as thin as 0.013 mm (0.0005 in.).

Conformability of the bearing material allows it to change its shape to compensate for slight deflections, misalignments, and inaccuracies in the journal and bearing housing. Bearing materials with lower Young's modulus will undergo larger deflections under lower loads and are, therefore, more desirable.

Embeddability is the ability of the bearing material to embed grit, sand, hard metal particles, or similar foreign materials and thus prevents them from scoring and wearing the journal. Such foreign materials can be introduced with the lubricant or ventilating air. Materials with lower hardness are expected to have better embeddability.

Wear resistance of bearing materials is an important parameter in cases where the position of the journal is to be kept within narrow tolerances, or where the bearing material is a thin layer on a hard backing. Generally, the wear rate depends on the tendency of the system toward adhesive weld formation, and on the resistance of the bearing material to abrasion by asperities on the journal surface. The rate of wear can be an important factor in determining the bearing life.

Thermal conductivity becomes an important factor in selecting bearing materials under conditions of high speeds or high loads, where heat is generated at high rates (see Equation 11.16). The ability of the bearing material to conduct heat away from friction surfaces reduces the operating temperature, and thus reduces the possibility of lubricant-film failure, melting of the bearing material, and seizure of the bearing.

Corrosion resistance of the bearing material becomes an important parameter when the lubricating oil is likely to contain acidic products or to be contaminated by corrosive materials.

As many of the bearing materials contain expensive elements, the cost may become a deciding factor in selection. Using a thin layer of the expensive bearing material in bimetal or trimetal structures can reduce the total cost of the bearing.

The preceding discussion shows that the main requirements for a bearing material are

1. Compressive strength at the operating temperature. This is a lower-limit property, which means that for a candidate material to be considered, its strength should exceed a given minimum value. This value is determined from the bearing design. In the present case study, the minimum strength of the bearing material at the operating temperature of 120°C (24°F) is

$$S_m = S_c n$$

where

S_m = Minimum compressive strength of the bearing
n = Factor of safety, which can be taken as 2 in the present case

From the preceding design calculations:

$$S_m = 1.86 \text{ MPa}$$

As information on the strength of bearing materials at 120°C is not readily available, comparison between the different materials will be based on room temperature properties. For the present case study, the minimum allowable room-temperature compressive strength will be taken as 20 MPa. Most available metallic bearing alloys can meet this requirement.

2. Fatigue strength at the operating temperature. As in the case of the compressive strength, this is a lower-limit property. As fatigue is not expected to be the main selection criterion in the present case study, no special calculations are needed. It will be assumed that materials that satisfy the lower limit of the compressive strength will also satisfy the lower limit of the fatigue strength. Using a similar reasoning as in the case of compressive strength, the minimum allowable fatigue strength will be taken as 20 MPa.
3. Hardness is an upper-limit property, which means that for a candidate material to be considered, its hardness should be below a given maximum value. This maximum value depends on the hardness of the journal material. For

the present case study, it will be assumed that the maximum allowable bearing material hardness is 100 BHN. This will allow the use of most well-known bearing materials except the hardest copper-base alloys.

4. Young's modulus is also an upper-limit property. In this case, the maximum allowable Young's modulus will be taken as 100 GPa, which will allow most well-known bearing materials to be considered.
5. Wear resistance is a lower-limit property that is system-dependent. It depends on the journal materials, lubricant, surface roughness of the journal, and cleanliness of the service environment. This property is usually given as excellent (5), very good (4), good (3), fair (2), and poor (1). In the present case study, materials with a rating of poor will not be considered.
6. Corrosion resistance is a lower-limit property and is usually described by a rating system similar to that used for wear resistance. In the present case study, materials with a corrosion resistance rating of poor will not be considered.
7. Thermal conductivity is a lower-limit property and, in view of the high rotational speeds of the shaft, a relatively high minimum conductivity of 20 W/m K will be specified, and this means that all nonmetallic bearing materials are excluded.
8. Cost of the bearing material, backing material, and fabrication should be considered. In the present case study, a single value for the cost of the material on the job, which includes all the mentioned factors, will be given. For the present case study, the maximum allowable cost will be taken as that of tin-base ASTM B23 grade 5.

11.4.4 Classification of Bearing Materials

Many alloy systems have been specially developed to accommodate the conflicting requirements that have to be satisfied by bearing materials. They are used in relatively small quantities and are produced by a relatively small number of manufacturers. Although the composition and processing methods of most commercial systems are of proprietary nature, widely used bearing materials can be classified as follows:

1. White metals (babbitt alloys). These are either tin- or lead-base alloys with additions of antimony and copper. Iron, aluminum, zinc, and arsenic are also usually present in small amounts, as shown in Table 11.7. The relevant properties of a selected number of these alloys are given in Table 11.8.
2. Copper-base bearing alloys offer a wider range of strengths and hardness than white metals. Lead and tin are the main alloying elements, but silver, iron, zinc, phosphorus, and nickel are sometimes found in small quantities. Tables 11.7 and 11.8 give the composition and properties of a selected number of copper-base bearing alloys.
3. Aluminum-base bearing alloys are suitable for high-duty bearings in view of their high strengths and thermal conductivities. They can be used in single metal, bimetal, or trimetal systems. Tables 11.7 and 11.8 also give the composition and properties of selected aluminum-base bearing alloys.

TABLE 11.7
Composition of Some Bearing Alloys (%)

Alloy Grade	Sn	Sb	Pb	Cu	Fe	Zn	Al	Others
			White Metals ASTM B23 (Tin Base)					
1	91	4.5	0.35	4.5	0.08	0.005	0.005	0.08 Bi, 0.1 As
2	89	7.5	0.35	3.5	0.08	0.005	0.005	0.08 Bi, 0.1 As
3	84	8.0	0.35	8.0	0.08	0.005	0.005	0.08 Bi, 0.1 As
4	75	12	10	3.0	0.08	0.005	0.005	0.15 As
5	65	15	18	2.0	0.08	0.005	0.005	0.15 As
			White Metals ASTM 23 (Lead Base)					
6	20	15	63.5	1.5	0.08	—	—	0.15 As
7	10	15	75.0	0.5	0.1	—	—	0.6 As
8	5	15	80.0	0.5	—	—	—	0.2 As
10	2	15	83.0	0.5	—	—	—	0.2 As
11	—	15	Rem.	0.5	—	—	—	0.25 As
15	1	15	Rem.	0.5	—	—	—	1.4 As
			Copper-Base Alloys SAE (Copper–Lead)					
48	0.25	—	28	Rem.	0.35	0.1	—	1.5 Ag, 0.025 P
49	0.5	—	24	Rem.	0.35	—	—	
480	0.5	—	35	Rem.	0.35	—	—	15 Ag
			Copper-Base Alloys ASTM B22 (Bronze)					
A	19	—	0.25	Rem.	0.25	0.25	—	1 P
B	16	—	0.25	Rem.	0.25	0.25	—	1 P
C	10	—	10	Rem.	0.15	0.75	—	0.1 P, 1 Ni
			Aluminum-Base Alloys					
770	6	—	—	1	0.7	—	Rem.	1 Ni, 0.7 Si
780	6	—	—	1	0.7	—	Rem.	0.5 Ni, 1.5 Si
MB7	7	—	—	1	0.6	—	Rem.	1.7 Ni, 0.6 Si

Note: Rem. = Remainder

4. Nonmetallic bearing alloys are mostly based on polymers or polymer–matrix composites. They are widely used under conditions of light loading. The major disadvantage of this group of bearing materials is their low thermal conductivities. In view of the high speed of rotation encountered in the present case study, this group will not be considered further.

11.4.5 Selection of the Optimum Bearing Alloy

Based on the preceding analysis and design considerations, the weighting factors were estimated using the digital logic approach described in Section 9.5. Table 11.9 gives the different weighting factors for the present case study.

The table shows that the yield strength is considered as one of the most important requirements. Ensuring that extensive yielding will not take place in the bearing material will ensure the uniformity of the lubricant film and would avoid vibrations

TABLE 11.8
Properties of Some Bearing Materials

Alloy Grade	Yield Strength (MPa)	Fatigue Strength (MPa)	Hardness (BHN)	Corrosion Resistance	Wear Resistance	Thermal Conduction (W/m K)	Young's Modulus (GPa)	Relative Cost
			White Metals ASTM B23 (Tin Base)					
1	30.0	27	17	5	2	50.2	51	7.3
2	42.7	34	25	5	2	50.2	53	7.3
3	46.2	37	37	5	2	50.2	53	7.3
4	38.9	31	25	5	2	50.2	53	7.3
5	35.4	28	23	5	2	50.2	53	7.5
			White Metals ASTM 23 (Lead Base)					
6	26.6	22	21	4	3	23.8	29.4	1.3
7	24.9	28	23	4	3	23.8	29.4	1.2
8	23.8	27	20	4	3	23.9	29.4	1.1
10	23.8	27	18	4	3	23.9	29.4	1.0
11	21.4	22	15	4	3	23.9	29.4	1.0
15	28.0	30	21	4	3	23.9	29.4	1.0
			Copper-Base Alloys SAE (Copper–Lead)					
48	40	45	28	3	5	41.8	75	1.5
49	45	50	35	3	5	41.8	75	1.5
480	38	42	26	3	5	41.8	75	1.5
			Copper-Base Alloys ASTM B22 (Bronze)					
A	168	120	100	2	5	41.8	95	1.8
B	126	100	100	2	5	41.8	95	1.8
C	119	91	65	2	3	42.0	77	1.6
			Aluminum-Base Alloys					
770	173	150	70	3	2	167	73	1.5
780	158	135	68	3	2	167	73	1.5
MB7	193	170	73	3	2	167	74	1.5

TABLE 11.9
Weighting Factors of the Selection Criteria for the Bearing Material of Centrifugal Pump

Property	Weighting Factor
Yield strength	0.20
Fatigue strength	0.14
Hardness	0.08
Corrosion resistance	0.11
Wear resistance	0.11
Thermal conductivity	0.20
Young's modulus	0.08
Cost	0.08
Total	1.00

at the high operating speeds of the pump. The thermal conductivity is considered equally important to the yield strength to ensure adequate conduction of heat away from the bearing–journal interface. With the relatively high speeds in the present case, sharp temperature rise as a result of temporary failure of the lubricant film could be serious. As no mention of excessive load fluctuating was made, it was assumed that fatigue is not expected to represent a serious problem and was, therefore, given a lower weighting factor than yield strength.

Corrosion and wear resistances were given moderate weighting factors as the danger of contamination and foreign particles was not emphasized in the service conditions. Hardness was given one of the lowest weights as it is expected that the rotor shaft will be adequately hardened. Young's modulus is treated similarly as misalignment, and deflection of the rotor shaft is not expected to be excessive. As the cost of the bearing material is expected to represent a small part of the total cost of the centrifugal pump, this factor was given a low weighting factor.

The candidate bearing materials of Tables 11.7 and 11.8 were evaluated using the limits on properties method described in Section 9.5. The lower and upper limits for the different properties were discussed above and can be summarized as follows:

Lower limit of yield strength = 20 MPa
Lower limit of fatigue strength = 20 MPa
Lower limit on thermal conductivity = 20 W/m K
Lower limit on corrosion resistance = 2
Lower limit on wear resistance = 2
Upper limit on hardness = 100 BHN
Upper limit on Young's modulus = 100 GPa
Upper limit on relative cost = 7.5

The mentioned lower and upper limits were used to calculate the merit parameters (m) of the different materials using Equation 9.8. The results of evaluation are given in Table 11.10.

11.4.6 Conclusion

The results in Table 11.10 show that aluminum-base alloys are most suitable for the present application and were given suitability ratings of 1 and 2. The high conductivity, high compressive and fatigue strengths, and moderate cost are their main attractions. Copper-base alloys would also be adequate and were given suitability ratings of 3 and 4.

11.5 Analysis of the Requirements and Substitution of Materials for Tennis Rackets

11.5.1 Introduction

Leading sports and recreational industries are now using sophisticated materials and high-technology production methods to manufacture their products. In addition, biomechanics is also being used to gain better understanding of the human body to

TABLE 11.10
Merit Parameter and Suitability of Bearing Materials

Material	Merit Parameter (m)	Suitability
	White Metals ASTM B23 (Tin Base)	
1	0.59	8
2	0.54	5
3	0.54	5
4	0.56	6
5	0.56	7
	White Metals ASTM 23 (Lead Base)	
6	0.63	11
7	0.61	9
8	0.62	10
10	0.61	9
11	0.66	12
15	0.58	6
	Copper-Base Alloys SAE (Copper–Lead)	
48	0.47	3
49	0.47	3
480	0.49	4
	Copper-Base Alloys ASTM B22 (Bronze)	
B	0.49	4
C	0.49	4
	Aluminum-Base Alloys	
770	0.37	1
780	0.38	2
MB7	0.37	1

enhance player comfort and to optimize equipment performance. As a result, the shape and the materials used in making many sports equipment and leisure products have undergone considerable change. For example, tennis rackets are now available in many shapes and sizes, as shown in Figure 11.5, although they all comply with the international tennis federation (ITF) rules, which limit the total length to a maximum of 32 in. (81.3 cm) and the strung surface, called the head, to a maximum of 15.5 in. (39.4 cm) in length and 11.5 in. (29.2 cm) in width. Although there are no limitations on the racket weight, it usually ranges between 13 and 15 oz (368.5 and 425 g).

11.5.2 Analysis of the Functional Requirements of the Tennis Racket

From an engineering point of view, a tennis racket can be considered as an implement for transmitting power from the arm of the player to the ball. This should be done as efficiently as possible to allow the player to deliver the fastest balls with the least effort. Tennis players usually call this characteristic the power of the racket. In addition to power, players evaluate rackets in terms of playability, which is a

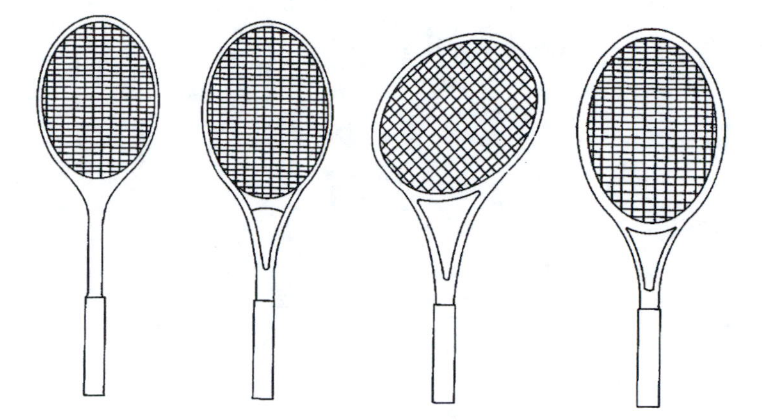

FIGURE 11.5 Examples of different tennis racket shapes.

subjective evaluation of the overall performance of the racket. For a more objective evaluation, playability may be considered as a function of control and vibrations.

Control is the ability to give the ball the desired speed and spin and to place it in the desired area of the court. Control can be considered a function of weight, balance, stability, and area of the sweet spot. These parameters are mainly affected by the material, shape, and design of the racket. The weight of the racket is mainly a function of the cross-sectional area of the racket, the shape, and the density of the material. Balance is a function of the position of the center of gravity of the racket in relation to the player's hand. Stability can be defined as the ability of the racket to resist twisting due to off-center hits. It depends primarily on the weight distribution in the racket head. In some cases, balancing weights are added to the sides of the racket head to improve stability. The sweet spot is defined as the area of maximum ball rebound velocity. The size of this area depends on the size and shape of the racket head.

Vibrations take place in the strings as a result of hitting the ball, and are then transmitted to the player's arm through the racket frame. If the racket material does not sufficiently dampen the vibrations, the player may develop tennis elbow.

In addition to the shape and the frame material, the performance of the racket is influenced by the material of the strings and the tension in the strings. With higher tension, the strings will absorb less power in deflection and deliver more power to the ball. The limiting value of the tension in the strings is decided by their own strength and will not be taken as a factor in selecting the material of the racket frame. The present case study will only consider the effect of the racket design and material of the frame on the performance of the racket.

11.5.3 Design Considerations

From the stress analysis point of view, the tennis racket can be modeled as a cantilever with the handle as the fixed end, as shown in Figure 11.6. The stiffness of the

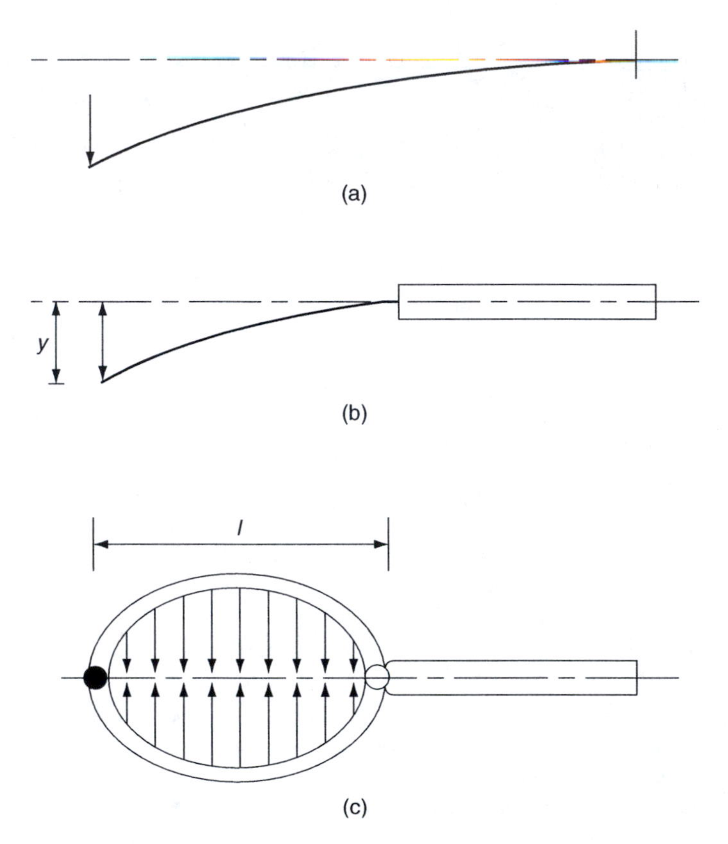

FIGURE 11.6 Simple modeling of tennis racket as a cantilever. (a) Deflection of a cantilever beam of uniform cross section under a concentrated load acting on its end; (b) deflection of tennis racket as a result of hitting the ball with its tip; and (c) forces acting on the racket head as a result of tension in the strings.

racket, that is, deflection as a result of hitting the ball, will determine the power of the racket. A stiffer racket will absorb less energy in deflection and will deliver more power to the ball. The maximum deflection will take place when the ball is hit with the outermost tip of the head. As the cross-sectional area and dimensions of the racket head are normally much smaller than those of the handle, it can be assumed that most of the deflection will take place in the head, as shown in Figure 11.6. The maximum deflection (y) is given by

$$y = \frac{(Fl^3)}{(3EI)} \tag{11.18}$$

where

F = Force acting on one side of the racket head as a result of hitting the ball, assumed to be constant for all racket designs and materials
l = Length of the racket head
E = Young's modulus of the racket material
I = Moment of inertia of the cross section of the racket head in the direction of the applied force

The cross-sectional area of the racket frame can be complex, especially in the case of hollow sections. For simplicity of analysis, it will be considered as a rectangle with outer dimensions of H and B and inner dimensions of h and b for all materials. From Figure 4.3,

$$I = \frac{(BH^3 - bh^3)}{12}$$

Equation 11.18 shows that rackets with larger head lengths will suffer larger deflection, that is, will be less powerful, unless materials with higher E are used in their manufacture. Larger heads mean larger sweet spot and better control.

As shown, balance is a function of the position of the center of gravity. Changing the size of the racket head is expected to change the position of the center of gravity. This means that the density of the materials has to be taken into account when designing the shape of the racket. The density also affects the balance of the racket. Using lighter materials will allow the use of balancing weights at the appropriate points of the head without increasing the total weight of the racket.

Combining the stiffness and low weight requirements, it can be concluded that materials with higher specific stiffness will allow the design of rackets with higher power and better control. From the vibration damping point of view, materials with lower elastic modulus provide better damping.

Cost is an important consideration in selecting a tennis racket, especially when catering for beginners and amateur players. In this case, the cost can be considered as a function of the cost of materials and processing.

11.5.4 Classification of Racket Materials

Tennis rackets can be made of several widely different materials including various types of wood, aluminum alloys, steels, and fiber-reinforced composite materials. With their much better performance at reasonable price, CFRP currently represent the favorable material for the great majority of tennis rackets.

11.5.5 Material Substitution

To continue to improve the performance of their sports equipment, manufacturers are currently examining carbon nanotube reinforced plastics (CNTRP) as possible substitutes for CFRP. CNTRP provide better properties but are more expensive. This case study, which is based on a paper by Esawi and Farag (2007), uses the cost–benefit analysis to evaluate CNTRP as a possible substitute for CFRP in tennis rackets. In this case, the cost is taken as cost of the material per unit mass, whereas the benefit is considered to consist of two elements, power and damping. As shown earlier, materials with higher specific modulus provide higher power, whereas materials with lower elastic modulus provide better damping.

11.5.6 Ranking of Alternative Substitutes

Table 11.11 gives the properties of the traditional epoxy + 65%CF as well as some experimental CNTRP composites, which are being considered as possible substitutes. For the present analysis, the AHP is used to assess the cost–benefit analysis of

TABLE 11.11
Properties of Candidates for Making a Tennis Racket

	E_c (GPa)	Density ρ_c (g/cc)	Specific Modulus (E_c/ρ_c) (GPa)/(g/cc)	Cost C_c[a] ($/kg)
Epoxy + 5% CNT	130.4	1.843	70.75421	2152.357
Epoxy + 20% CNT	425.6	1.852	229.8056	8579.429
Epoxy + 30% CNT	622.4	1.858	334.9839	12864.14
Epoxy + 50% CF	136.0	1.870	72.72727	92.5
Epoxy + 55% CF	146.4	1.873	78.16337	100.75
Epoxy + 60% CF	156.8	1.876	83.58209	109
Epoxy + 65% CF	167.2	1.879	88.9835	117.25
Epoxy + 1% CNT + 64% CF	184.8	1.879	98.35019	544.0714
Epoxy + 3% CNT + 62% CF	220.0	1.879	117.0836	1397.714
Epoxy + 5% CNT + 60% CF	255.2	1.879	135.8169	2251.357
Epoxy + 10% CNT + 55% CF	343.2	1.879	182.6503	4385.464
Epoxy + 15% CNT + 50% CF	431.2	1.879	229.4838	6519.571

Source: Based on a paper by Esawi, A.M.K. and Farag, M.M., Carbon nanotube reinforced composites: potential and current challenges, *Mater. Design*, 28(9), 2394–2401, 2007.

[a] Cost C_c is calculated using the cost of individual components and applying the rule of mixtures.

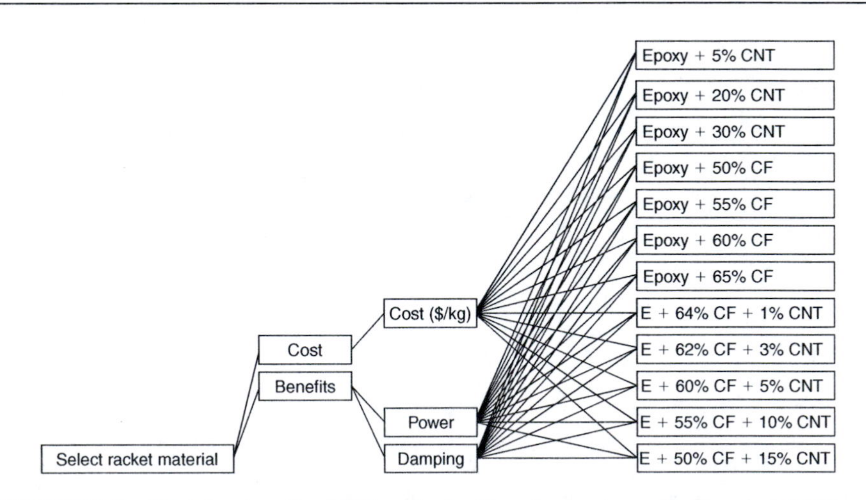

FIGURE 11.7 Decision tree for AHP analysis. (Based on a paper by Esawi, A.M.K. and Farag, M.M., Carbon nanotube reinforced composites: potential and current challenges, *Mater. Design*, 28(9), 2394–2401, 2007.)

the materials in Table 11.11 in making tennis rackets. As discussed in Section 9.5, AHP is an approach to solving multicriteria decision-making problems that depend on pairwise comparison of alternatives with respect to the selection criteria. The Criterium Decision Plus version 3.04 by InfoHarvest was used to build a decision tree as shown in Figure 11.7. The cost is taken to be directly related to the price of the material in dollars per kilogram, as shown in Table 11.11, since the cost of processing is expected to be similar for the different materials. The benefits side consists of improved power, which is taken to linearly increase with the specific modulus; and improved damping, which is categorized as high, medium, and low for elastic modulus values less than 200, 200–300, and above 300 GPa, respectively. Various scenarios for material substitution are developed by allocating different weights to the cost, power, and damping. The resulting rankings are shown in Table 11.12.

11.5.7 Conclusion

Table 11.12 shows that the material ranking is sensitive to the importance and weight, allocated to each of the main variables: cost, power, and damping. Higher weight for the cost and less emphasis on power tend to favor the traditional epoxy–CF composites. Epoxy-CF-carbon nanotubes (CNT) hybrid composites become more viable as the weight allocated to cost is reduced and the emphasis on power is increased. The epoxy–CNT composites become viable alternatives only at the two lowest weights for cost, and the consequent highest emphasis on benefits, with higher emphasis on power.

11.6 MATERIALS SUBSTITUTION IN AUTOMOTIVE INDUSTRY

11.6.1 Introduction

The materials used in making a motorcar cover almost all classes of engineering materials including metals and alloys, polymers and composites, elastomers, and ceramics. The relative amounts of these materials have changed over the years and Kandelaars and van Dam (1998) have shown that the percentage of the lighter materials, aluminum and plastics, in the total car weight has steadily increased compared with the heavier material, steel, during the period 1960–1992. For example, in 1960, aluminum and plastics constituted 2% and 1% of the total weight, respectively, whereas in 1986, these fractions increased to 4% and 7%.

The major driving forces behind material substitution in the automotive industry are cost reduction, better fuel economy, improved esthetics and comfort, and compliance to new legislation such as the End-of-Life Vehicle Directive, European Directive 2000/53/EC. An important factor in improving fuel efficiency, which is defined as the distance driven divided by energy used, is weight reduction by using higher-performance materials or substituting lightweight materials for the traditional ferrous materials used in the body, chassis, and power train components. Such substitution must be made cost-effectively while conforming to the increasingly severe safety and quality standards, and without unduly restricting the freedom of the stylist.

This case study, which is based on a paper by Farag (2007), gives an analysis of the different factors involved in materials substitution in automotive components. An interior panel is used for illustration, but the procedure can be applied to other parts of the motorcar. PVC is assumed to be the currently used material for the panel in this case study.

TABLE 11.12
Ranking of Materials According to AHP

	Weights		Top-Ranking Materials			
Alternative	Cost/ Benefit	Power/ Damping	Highest	Second	Third	Fourth
1	50%/50%	50%/50%	Epoxy + 65% CF	Epoxy + 60% CF	Epoxy + 1% CNT + 64% CF Epoxy + 55% CF	Epoxy + 50% CF
2	30%/70%	60%/40%	Epoxy + 1% CNT + 64% CF	Epoxy + 65% CF	Epoxy + 60% CF	Epoxy + 55% CF
3		70%/30%	Epoxy + 1% CNT + 64% CF	Epoxy + 65% CF	Epoxy + 60% CF	Epoxy + 55% CF
4	25%/75%	60%/40%	Epoxy + 1% CNT + 64% CF	Epoxy + 65% CF	Epoxy + 60% CF	Epoxy + 55% CF
5		70%/30%	Epoxy + 1% CNT+ 64% CF	Epoxy + 65% CF	Epoxy + 60% CF	Epoxy + 55% CF
6	20%/80%	60%/40%	Epoxy + 1% CNT + 64% CF	Epoxy + 65% CF	Epoxy + 60% CF	Epoxy + 55% CF
7		70%/30%	Epoxy + 1% CNT + 64% CF	Epoxy + 65% CF	Epoxy + 60% CF	Epoxy + 55% CF Epoxy + 30% CNT
8	15%/85%	70%/30%	Epoxy + 30% CNT	Epoxy + 1% CNT + 64% CF	Epoxy + 65% CF	Epoxy + 60% CF
9	10%/90%	70%/30%	Epoxy + 30% CNT	Epoxy + 1% CNT + 64% CF	Epoxy + 65% CF	Epoxy + 60% CF

Source: Based on a paper by Esawi, A.M.K. and Farag, M.M., Carbon nanotube reinforced composites: potential and current challenges, *Mater. Design*, 28(9), 2394–2401, 2007.

11.6.2 Materials and Manufacturing Processes for Interior Panels

PVC, which is assumed to be the currently used material for the interior panel, is one of the cheapest and most versatile polymers. This explains its wide use in a wide variety of applications ranging from pipes and fittings; flexible and rigid packaging; artificial leather and car upholstery; to cladding panels for flooring, doors, and windows. PVC can be easily shaped by injection or compression molding. However, there are environmental concerns about the chloride monomer.

FRP are being increasingly used in automotive industries as a result of their superior strength/weight and stiffness/weight ratios. The fibers in the composite can either be long in the form of continuous roving, woven fabrics, or preimpregnated tapes and sheets, or short in the form of chopped mats. The matrix can either be thermosetting plastic or thermoplastic. Both the matrix and fibers have strong influence on the properties of the composite and its manufacturing processes. Although thermosetting epoxy resins are normally used as the matrix material in advanced composites, there is increasing interest in thermoplastics such as polypropylene (PP), polyesterether ether ketone (PEEK), polyphenylene sulfone (PPS), and polyether imide (PEI). The main advantages of thermoplastics are better damage and environmental tolerance rather than improved mechanical performance. In addition, thermoplastics are easier to form and recycle. Generally, however, FRP containing carbon or glass fibers are not recyclable and their production is energy-intensive and polluting.

Recently, natural fibers such as flax, hemp, and jute have been considered for reinforcing plastics as they need much less energy to grow, are renewable, and are biodegradable after use. Natural fiber-reinforced plastics (NFRP) have the potential of vehicle weight reduction while satisfying the increasingly stringent environmental criteria, and some auto manufacturers have already started using them in some of their models, as in the case of Mercedes Benz A-Class and Ford Model U hybrid-electric car.

Filament winding and hand or machine lay up of continuous fibers are used for advanced composites with relatively short production runs. Resin transfer molding (RTM) is used for intermediate production volumes of about 30,000 parts annually. Larger production volumes can be achieved using sheet-molding compound (SMC), which is a mixture of chopped fiber roving and matrix material. SMC can be shaped using compression molding and is increasingly used for making panels in the automotive industry. Reinforced reaction injection molding (RRIM) is also suitable for large-scale production of motorcar body panels.

Wood and cork, being natural materials, are renewable sources that require little energy for their production and are easy to recycle. In addition to its good stiffness to weight ratio, wood has excellent tactile and texture qualities and is perceived to be warmer and softer than many other materials (Ashby and Johnson, 2002). Cork has similar esthetic qualities to wood in addition to having excellent sound-damping qualities. Currently, wood is used for panels and veneer of several interior components in luxury motorcars such as the Rolls-Royce Phantom and the Jaguar xj6. Granulated cork and laminated wood can be pressed into sheets and panels, but they both need to be polished and sealed if they are to be used for motorcar interior applications.

11.6.3 Performance Indices of Interior Panels

Performance indices for a motorcar interior panel can be divided into four groups as follows: (a) technical characteristics, (b) cost considerations, (c) esthetics and comfort issues, and (d) environmental considerations. Following is an analysis of the performance indices under each group.

11.6.3.1 Technical Characteristics

Technical characteristics of interior panels include rigidity and resistance to buckling, light weight, resistance to thermal distortion, and resistance to weather conditions and direct sunlight.

For simplicity of analysis in the present case study, the body panel is considered to be relatively flat and rectangular with the following dimensions: length, $l = 100$ cm (39.4 in.) and width, $b = 50$ cm (19.7 in.). The thickness of the currently used PVC panel is 3.7 mm.

The material performance index (m) for a stiff light structural member can be represented as

$$m = E^{1/3}/\rho \qquad (11.19)$$

where

E = Elastic modulus
ρ = Density

The thickness of another panel of equal stiffness and resistance to buckling is given as

$$t_n = t_o\left(\frac{E_o}{E_n}\right)^{1/3} \qquad (11.20)$$

where

t_n and t_o = Thickness of new and currently used panels, respectively
E_n and E_o = Elastic constants of new and currently used materials, respectively

The values of the elastic modulus, density, and the weight of the panels, based on the thickness calculated using Equation 11.20, are given in Table 11.13 for the different candidate materials.

The mass (M) of the panel is

$$M = \rho tbl \qquad (11.21)$$

where

ρ = Density of the panel material
l, b, and t = Length, width, and thickness of the panel, respectively

For the internal panel, Matos and Simplicio (2006) gave the thermal distortion index $M2$ as a function of thermal conductivity divided by thermal expansion coefficient. Materials with higher $M2$ are preferred as they are expected to suffer less thermal distortion. The values of thermal conductivity and thermal expansion coefficient are

TABLE 11.13
Technical, Cost, and Environmental Considerations

Material	*E* (GPa)	Density (g/cc)	Weight of Panel (kg)	Material Cost (USD)	Manufacturing Cost (USD)	Running Cost (USD)	*M*2 (W/m)
PVC	2.00	1.30	2.4	3.3	2.0	12.2	250,000
PP + glass fibers (40%)	7.75	1.67	3.3	6.0	2.0	21.8	60,000
Epoxy + carbon fibers (60%)	69	1.60	1.46	27.2	5.0	9.6	700,000
Cork	0.02	0.20	1.74	14.0	17.0	11.5	400
Wood (ash/willow)	10	0.85	0.927	1.18	17.0	6.12	60,000
PP + flax fibers (40%)	4.65	1.19	1.67	1.9	2.0	11.0	50,000
PP + hemp fibers (40%)	6.00	1.236	1.60	1.8	2.0	10.6	50,000
PP + jute fibers (40%)	3.96	1.174	1.76	1.9	2.0	11.6	50,000

Note: Values based on Ashby, M. and Johnson, C., *Materials and Design: The Art and Science of Material Selection in Product Design*, Butterworth-Heinemann, Oxford, 2002; Ermolaeva et al., 2004; Matos, M.J. and Simplicio, M.H., *Mater. Design*, 27, 74–78, 2006.

taken from Matos and Simplicio (2006), Ashby and Johnson (2002), or calculated using the rule of mixtures in the case of composite materials. The calculated values of *M*2 for the different candidate materials are given in Table 11.13.

The candidate materials under consideration are expected to satisfy the resistance to weather conditions and direct sunlight, and will not be ranked according to this requirement.

11.6.3.2 Cost Considerations for Interior Panel

The total cost (C_t) of a panel is considered to consist of four elements:

$$C_t = C_1 + C_2 + C_3 + C_4 \tag{11.22}$$

where

C_1 = Cost of material
C_2 = Cost of manufacturing and finishing
C_3 = Cost over the entire life of the component
C_4 = Cost of disposal and recycling

The cost of the material in the panel is based on its weight and the price of material per unit weight. The manufacturing cost is roughly estimated assuming that compression molding is used for PVC, GFRP, and NFRP; hand or machine lay up for CFRP; and cutting, shaping, sealing, and polishing for wood and cork.

The weight of a motorcar plays a major role in determining its fuel efficiency as measured by the distance traveled per unit volume of fuel (km/L). Reducing the weight of a given component or a subsystem (primary weight reduction) is expected to result in a secondary weight reduction in other supporting components or subsystems. According to Das (2005), a ratio of 2:1 can be assumed for primary to secondary weight savings. Das (2005) also found that reducing the vehicle curb weight by 10% results in a 6.6% increase in fuel efficiency, measured in kilometers per liter. Taking the life of the car to be 5 years, the total distance traveled as 200,000 km, the cost of fuel as $3/gal, and using the figures provided by Das (2005) of 8.62 km/L for a 1782 kg vehicle, the saving in fuel cost over the entire life of the vehicle can be estimated as about $6.6/kg reduction in curb weight of the vehicle. This amount can also be taken as the share in the running cost of a component weighing 1 kg over the entire life of the vehicle. For simplicity, the cost of disposal and recycling will not be considered in the present case study.

The different cost items for the candidate materials are shown in Table 11.13.

11.6.3.3 Esthetics and Comfort

This group of indices evaluates materials according to their esthetic and tactile characteristics as well as comfort considerations, which include heating and cooling insulation, and vibration and sound damping. For the present case study, esthetics are considered as a combined function of the tactile feel and warmth of the material, as defined by Ashby and Johnson (2002). Wood, being the material of choice for the interior panels of the Rolls-Royce Phantom, is considered to have the optimum combination and is given a value of 100 units. Other materials are ranked according to their distance from it on a tactile warmth–tactile softness chart provided by Ashby and Johnson (2002) and the values are given in Table 11.14.

TABLE 11.14
Esthetic and Comfort Characteristics of Candidate Materials

Indices	Esthetics	*M1* (W $s^{1/2}/m^2$ K) (Normalized)	Sound Damping/Loss Coefficient (Normalized)
PVC	80	780 (19.2)	0.08 (40)
PP + glass fibers (40%)	75	1460 (10.3)	0.03 (15)
Epoxy + carbon fibers (60%)	70	1460 (10.3)	0.03 (15)
Cork	85	150 (100)	0.20 (100)
Wood (ash/willow)	100	730 (20.5)	0.05 (25)
PP + flax fibers (40%)	80	1460 (10.3)	0.08 (40)
PP + hemp fibers (40%)	80	1460 (10.3)	0.08 (40)
PP + jute fibers (40%)	80	1460 (10.3)	0.08 (40)

Note: Values based on Ashby, M. and Johnson, C., *Materials and Design: The Art and Science of Material Selection in Product Design*, Butterworth-Heinemann, Oxford, 2002; Ermolaeva et al., 2004; Matos, M.J. and Simplicio, M.H., *Mater. Design*, 27, 74–78, 2006.

Thermal insulation of a material is a function of its thermal conductivity and thermal diffusivity. According to Matos and Simplicio (2006), the thermal insulation index *M*1 is given by the thermal conductivity divided by the square root of thermal diffusivity. Materials with lower values of *M*1 are considered to provide better comfort for the motorcar passenger when used for making the interior panels. The values of *M*1 for the different candidate materials are shown in Table 11.14 and have been calculated based on the values in a thermal conductivity–thermal expansion chart provided by Ashby and Johnson (2002), the values given by Matos and Simplicio (2006), or estimated using the rule of mixtures in the case of composite materials.

Sound and vibration damping for a candidate material is represented by its ability to absorb energy from a falling object on its surface, loss coefficient in an elastic modulus–loss coefficient chart provided by Ashby and Johnson (2002). Materials with higher values of loss coefficient are considered better for the interior panel application. In the case of composite materials, the rule of mixtures was used to estimate this property. The values of sound and vibration damping for the candidate materials are given in Table 11.14.

11.6.3.4 Environmental Considerations

Environmental considerations are becoming increasingly a more influential factor in materials selection and substitution as a result of the increasing awareness of the public and the legislation on environmental impact, as in the case of "EI 99," for example. According to ISO 14001, LCA, which evaluates the environmental impact of a given product over its entire life cycle, can be divided into three main phases:

a. Production phase including energy requirements for primary and secondary materials used and all the processes involved in manufacturing them into a finished product
b. Use or operation phase including the energy, fuel, and emissions over the entire lifetime of the product
c. End-of-life phase including the energy used in disposal of the discarded product and whatever energy is gained from its recycling

Ermolaeva et al. (2004), calculated the environmental impact of the three phases for the case of motorcars, LCA; and the results show that the energy consumed during the vehicle operation phase is about 85–92% of the total depending on the environmental properties of the materials and the manufacturing process used in making the motorcar. This is also in agreement with Ashby and Johnson (2002), who showed that of the total energy consumed over the life cycle of a motorcar, 94% is consumed during the use phase, 4% in the production of the materials, 1% in manufacturing, and 1% in disposal. These figures indicate that reducing the weight of the motorcar is the most important factor in reducing the environmental impact, as it has a direct relation to increased fuel efficiency (distance traveled per unit of fuel) and reduction of the total energy consumption over its life cycle. To simplify the analysis in this case study, the environmental impact of a given panel will be taken as proportional to its weight.

TABLE 11.15
Weight and Cost of Panels and *M*2 of Materials

Material	Weight of Panel (kg) (Normalized)	Total Cost of Panel (USD) (Normalized)	*M*2 (W/m) (Normalized)
PVC	2.4 (38.6)	17.5 (82.3)	250,000 (35.7)
PP + glass fibers (40%)	3.3 (28.2)	29.8 (48.3)	60,000 (8.6)
Epoxy + carbon fibers (60%)	1.46 (63.5)	41.8 (34.4)	700,000 (100)
Cork	1.74 (53.3)	42.5 (33.9)	400 (0.057)
Wood (ash/willow)	0.927 (100)	24.3 (59.3)	60,000 (8.6)
PP + flax fibers (40%)	1.67 (55.5)	14.9 (96.6)	50,000 (7.1)
PP + hemp fibers (40%)	1.6 (57.9)	14.4 (100)	50,000 (7.1)
PP + jute fibers (40%)	1.76 (52.7)	15.5 (92.9)	50,000 (7.1)

11.6.4 Comparison of Candidate Materials

Having estimated the values of the different performance requirements of the candidate materials, the next step is to normalize them. The lowest value for weight, total cost, and thermal insulation index (*M*1) are given 100 and other values are given proportionate quantities. In the case of esthetics, sound and vibration damping, and thermal distortion index (*M*2), the highest value is given 100 and others are given proportionate quantities. Tables 11.14 and 11.15 give the normalized property values in brackets.

Both the performance/cost and the compound objective function methods (see Section 10.5) are used to rank the candidate substitute materials in Table 11.14. The digital logic method (see Section 9.5) is used to assign the weighting factors.

11.6.5 Performance/Cost Method of Substitution

The performance of a material in this method is taken as the weighted sum of the normalized values of its esthetics, thermal insulation index (*M*1), sound and vibration damping, weight, and thermal distortion (*M*2), as shown in Table 11.16. The performance values of the candidate materials are then compared with the currently used material, PVC, and categorized as lower, similar, or higher performance. The results in Table 11.16 show that PP + glass fibers (40%) gives lower performance at a higher cost and is, therefore, rejected.

PP + flax fibers (40%), PP + hemp fibers (40%), and PP + jute fibers (40%) give similar performance to PVC but at a lower cost. These candidates would be preferable if the main objective of substitution is cost reduction. Of the three candidates, PP + hemp fibers (40%) is given the top ranking as it has the highest performance/cost.

Epoxy + carbon fibers (60%), cork, and wood give higher performance at a higher cost than PVC. They would be preferable if the object of substitution is to raise performance. Of the three candidates, wood is the best as it has the highest performance/cost.

TABLE 11.16
Results of the Performance/Cost Method of Substitution

Index	Esthetics	*M*1	Sound Damping	Weight of Panel	*M*2				
Weighting Factor	**0.2**	**0.2**	**0.1**	**0.4**	**0.1**	**Performance**	**Cost (USD)**	**Perform/ Cost**	**Evaluation and Rank**
PVC	16	3.8	4.0	15.5	3.6	42.9	17.5	2.5	Current material
PP + glass fibers (40%)	15	2.1	1.5	11.3	0.9	30.8	29.8	1.03	Lower performance and higher cost (reject)
Epoxy + carbon fibers (60%)	14	2.1	1.5	25.4	10	53.0	41.8	1.3	Better performance rank 3
Cork	17	20	10	21.3	0.01	68.3	42.5	1.6	Better performance rank 2
Wood (ash/willow)	20	4.1	2.5	40.0	0.9	67.5	24.3	2.8	Better performance rank 1
PP + flax fibers (40%)	16	2.1	4.0	22.2	0.7	45.0	14.9	3.0	Lower cost rank 2
PP + hemp fibers (40%)	16	2.1	4.0	23.2	0.7	46.0	14.4	3.2	Lower cost rank 1
PP + jute fibers (40%)	16	2.1	4.0	21.1	0.7	43.9	15.5	2.8	Lower cost rank 3

11.6.6 The Compound Objective Function Method

In applying this method, two substitution scenarios were created by changing the weighting factors as follows:

a. More emphasis on cost with less emphasis on esthetics and comfort, economy model, where technical and economic aspects represent 75% of the weight, whereas esthetic and comfort aspects represent 25%. The weights are allocated as follows: weight of panel (0.25), total cost (0.4), *M*2 (0.1), esthetics (0.1), *M*1 (0.08), and sound damping (0.07).
b. Less emphasis on cost and more emphasis on esthetics and comfort, luxury model, where technical and economic aspects represent 50% of the weight, whereas esthetic and comfort aspects represent 50%. The weights are allocated as follows: weight of panel (0.2), total cost (0.2), *M*2 (0.1), esthetics (0.25), *M*1 (0.15), and sound damping (0.1).

The compound objective functions for each of the materials for the two scenarios were calculated using the values in Tables 11.14 and 11.15, and the results are given in Table 11.17.

The results of Table 11.17 show that for the economy model, PP + hemp fibers (40%) and PP + flax fibers (40%) receive first and second ranks, respectively; their biggest asset being their low cost. This is in agreement with Marsh (2003), who reported that PP + flax fiber composites replaced GFRP in underbody components in vehicles such as the Mercedes Benz A-Class and the Ford Model U hybrid-electric car. For the luxury model, the results show that cork and wood have close compound objective function (COF) values and, therefore, share the top rank. Their biggest asset is excellent esthetic qualities and heat insulation. As discussed earlier, wood is used for body panels and veneers in the Rolls-Royce Phantom.

TABLE 11.17
Results of the Compound Objective Function Method of Substitution

	Scenario I (Economy Model)		Scenario II (Luxury Model)	
Material	**COF**	**Rank**	**COF**	**Rank**
PVC	58.5	5	54.7	6
PP + glass fibers (40%)	36.6	8	38.1	8
Epoxy + carbon fibers (60%)	48.6	7	50.5	7
Cork	50.3	6	63.8	1
Wood (ash/willow)	63.0	3	63.4	1
PP + flax fibers (40%)	64.8	2	56.7	4
PP + hemp fibers (40%)	66.8	1	57.9	3
PP + jute fibers (40%)	62.7	4	55.4	5

11.6.7 Conclusion

The two proposed methods are used to examine the case of material substitution for interior motorcar panels and they yielded consistent results. In both methods, PP + hemp fibers (40%) and PP + flax fibers (40%) rank highest for the economy models, where cost is important, whereas wood and cork rank highest for the luxury models, where esthetics and comfort are important. These results are consistent with the current trends in industry.

Based on the results of the case study presented here, it is expected that NFRP would be increasingly used as automotive materials. More research needs to be done on how to improve the performance and economics of these emerging materials, as they also have excellent potential for wider industrial applications.

BIBLIOGRAPHY AND FURTHER READING

Ashby, M.F., Performance indices, in *ASM Handbook Vol. 20: Materials Selection and Design*, Dieter, G.E., Editor. ASM International, Materials Park, OH, 1997, pp. 281–290.

Ashby, M. and Johnson, C., *Materials and Design: The Art and Science of Material Selection in Product Design*, Butterworth-Heinemann, Oxford, 2002.

Black, P.H. and Adams, O.E., *Machine Design*, 3rd Ed., McGraw-Hill, London, 1968.

Black, P.H. and Adams, O.E., *Machine Design*, 3rd Ed., McGraw-Hill, London, 1983.

Boyer, H.E. and Gall, T.L., *Metals Handbook*, Desk Edition, ASM, Materials Park, OH, 1985.

Das, S., Life cycle energy impacts of automotive liftgate inner, *Resour. Conservat. Recycl.*, **43**, 375–390, 2005.

De Gee, A.W., Selection of materials for lubricated journal bearings, *Wear*, **36**, 33–61, 1976.

Ermolaeva, N.S., Castro, M.B.G., and Kandachar, P.V., Materials selection for an automotive structure by integrating structural optimization with environmental impact assessment, *Mater. Design*, **25**, 689–698, 2004.

Ermolaeva, N.S., Kaveline, K.G., and Spoormaker, J.L., Materials selection combined with optimal structural design: concept and some results, *Mater. Design*, **23**, 459–470, 2002.

Esawi, A.M.K. and Farag, M.M., Carbon nanotube reinforced composites: potential and current challenges, *Mater. Design*, **28**(9), 2394–2401, 2007.

European Directive 2000/53/EC—End-of-Life Vehicle, September 2000.

Farag, M.M., *Selection of Materials and Manufacturing Processes for Engineering Design*, Prentice-Hall, London, New York, 1989.

Farag, M.M., *Materials Selection for Engineering Design*, Prentice-Hall, London, 1997.

Farag, M.M., Quantitative methods of materials selection, in *Mechanical Engineers Handbook: Materials and Mechanical Design*, Kutz, M., Editor. 3rd Ed., Wiley, New York, 2006, pp. 466–488.

Farag, M.M., Quantitative methods of materials substitution: application to automotive components, *Mater. Design*, 2007. Available at www.sciencedirect.com.

Farag, S., Sidky, B., Arafa, K., Nosseir, K., Moghrabi, M., and Idris, S., *Tennis Racket Project*, American University in Cairo, 1986.

Forrester, P.G., Selection of plain bearing materials, in *Engineering Materials*, Sharp, H.J., Editor. Heywood, London, 1964, pp. 255–270.

Giudice, F., La Rosa, G., and Risitano, A., Materials selection in the life-cycle design process: a method to integrate mechanical and environmental performances in optimal choice, *Mater. Design*, **26**, 9–20, 2005.

Graham, J.W., Biomedical materials emerge through teamwork, *Adv. Mater. Process*, January, p. 41, 1988.

InfoHarvest, Inc., www.infoharvest.com, 2005.

ISO 14001: 1998, Environmental management—life cycle assessment—goal and scope definition and inventory analysis, International Organization for Standardization, 2000.

ISO 5839, Surgical implants, ISO Bulletin, January, p. 6, 1986.

ISO 5839, Orthopedic joint prostheses, *ISO Bulletin*, September, p. 5, 1985.

Jain, R.K., *Machine Design*, 3rd Ed., Khanna Pub., Delhi, 1983.

Jones, C., *How to Play Tennis*, The Hamlyn Pub. Group Ltd., London, 1981.

Kandelaars, P.A.H. and van Dam, J.D., An analysis of variables influencing the material composition of automobiles, *Resour. Conservat. Recycl.*, **24**, 323–333, 1998.

Lenel, U.R., *Materials Selection in Practice*, The Institute of Metals Handbook, London, 1986, pp. 165–178.

Lockwood, P.A., Composites for industry, *ASTM Standardization News*, December, 28–31, 1983.

Marsh, G., Next step for automotive materials, *Mater. Today*, April, 36–43, 2003.

Matos, M.J. and Simplicio, M.H., Innovation and sustainability in mechanical design through materials selection, *Mater. Design*, **27**, 74–78, 2006.

Mohamedein, A.A., Mehenny, D.S., and Abdel-Dayem, H.W., *Turnbuckle Project*, American University in Cairo, 1988.

Parmley, R.O., *Standard Handbook of Fastening and Joining*, McGraw-Hill, New York, 1977.

Shigley, J.E. and Mitchell, L.D., *Mechanical Engineering Design*, 4th Ed., McGraw-Hill, London, New York, 1983.

Vaccari, J.A., Scoring with materials innovations, *Des. Eng.*, July, **51**, 31–38, 1980.

Part IV

Appendices

INTRODUCTION

The appendices in this section are intended to provide reference information that should be helpful to the reader in

- identifying the composition and properties of some commonly used engineering materials (Appendix A),
- converting one system of units to another and one hardness scale to another (Appendix B), and
- finding the meaning of technical terms in the area of materials and manufacturing as mentioned in the glossary (Appendix C).

Many of the ferrous materials in Appendix A are classified according to the system used by the AISI and the SAE, which consists of four or five, digits designating the composition of the alloy. The alloy system is indicated by the first two digits, whereas the nominal carbon content is given in hundredths of a percent by the last two, or three, digits. Aluminum alloys are classified according to the designation system employed by the Aluminum Association (AA), which uses four digits to identify wrought aluminum and its alloys and three digits to identify cast alloys. Unalloyed aluminum, 99% and greater, is designated by the first digit of 1, the second digit indicates a modification of impurity limits, and the last two digits indicate purity. For aluminum alloys, the first digit is between 2 and 9 and indicates the group, the second digit indicates a modification of the original alloy, and the last two digits indicate the specific alloy.

The UNS has been developed by the ASTM and SAE and several other technical societies, trade associations, and the U.S. government agencies. The UNS number is a designation of the chemical composition and consists of a letter and five numerals. The letter indicates the broad class of the alloy and the numerals define specific alloys within that class. Existing systems, such as the AISI–SAE system for steels and the AA for aluminum, have been incorporated into UNS designations. The following table gives a sample of the UNS numbering system:

Name	Number	Example	
Ferrous Materials		AISI-SAE	UNS
Plain carbon steel	10xx	1015	G10150
Free machining steel	11xx	1118	G11180
Manganese steel	13xx	1335	G13350
Molybdenum steel	40xx	4027	G40270
Chromium–molybdenum steel	41xx	4118	G41180
Nickel–chromium steel	43xx	4320	G43200
Chromium steel	50xx	5015	G50150
Chromium–vanadium steel	61xx	6118	G61180
Wrought Auminum Alloys		AA	UNS
Aluminum 99% and greater	1xxx	1060	A91060
Two-phase Al–Cu alloys	2xxx	2014	A92014
One-phase Al–Mn alloys	3xxx	3003	A93003
Two-phase Al–Si alloys	4xxx	4032	A94032
One-phase Al–Mg alloys	5xxx	5052	A95052
Two-phase Al–Mg–Si alloys	6xxx	6061	A96061
Two-phase Al–Zn alloys	7xxx	7075	A07075
Cast Aluminum Alloys		AA	UNS
Al–Cu alloys	2xx.x	108.0	A02080
Al–Si–Cu alloys	3xx.x	333.0	A03330
Al–Si alloys	4xx.x	B443.0	A24430
Al–Mg alloys	5xx.x	520	A05200
Al–Zn alloys	7xx.x	A712.0	A17120
Al–Sn alloys	8xx.x	850.0	A08500

Appendix A: Composition and Properties of Selected Engineering Materials

TABLE A.1
Physical and Chemical Data for Selected Elements

Element	Symbol	Atomic Mass (amu)	Density of Solid (at 20°C) (Mg/m^3 = g/cm^3)	Crystal Structure (at 20°C)	Melting Point (°C)	Atomic Number
Hydrogen	H	1.008			−259.34 (TP)	1
Helium	He	4.003			−271.69	2
Lithium	Li	6.941	0.533	bcc	180.6	3
Beryllium	Be	9.012	1.85	hcp	1289	4
Boron	B	10.81	2.47		2092	5
Carbon	C	12.01	2.27	Hex.	3826 (SP)	6
Nitrogen	N	14.01			−210.0042 (TP)	7
Oxygen	O	16.00			−218.789 (TP)	8
Fluorine	F	19.00			−219.67 (TP)	9
Neon	Ne	20.18			−248.587 (TP)	10
Sodium	Na	22.99	0.966	bcc	97.8	11
Magnesium	Mg	24.31	1.74	hcp	650	12
Aluminum	Al	26.98	2.70	fcc	660.452	13
Silicon	Si	28.09	2.33	Dia.cub.	1414	14
Phosphorus	P	30.97	1.82 (white)	Ortho.	44.14 (white)	15
Sulfur	S	32.06	2.09	Ortho.	115.22	16
Chlorine	Cl	35.45			−100.97 (TP)	17
Argon	Ar	39.95			−189.352 (TP)	18
Potassium	K	39.10	0.862	bcc	63.71	19
Calcium	Ca	40.08	1.53	fcc	842	20
Scandium	Sc	44.96	2.99	fcc	1541	21
Titanium	Ti	47.90	4.51	hcp	1670	22
Vanadium	V	50.94	6.09	bcc	1910	23
Chromium	Cr	52.00	7.19	bcc	1863	24
Manganese	Mn	54.94	7.47	Cubic	1246	25
Iron	Fe	55.85	7.87	bcc	1538	26

(Continued)

TABLE A.1 (Continued)
Physical and Chemical Data for Selected Elements

Element	Symbol	Atomic Mass (amu)	Density of Solid (at 20°C) (Mg/m^3= g/cm^3)	Crystal Structure (at 20°C)	Melting Point (°C)	Atomic Number
Cobalt	Co	58.93	8.8	hcp	1495	27
Nickel	Ni	58.71	8.91	fcc	1455	28
Copper	Cu	63.55	8.93	fcc	1084.87	29
Zinc	Zn	65.38	7.13	hcp	419.58	30
Gallium	Ga	69.72	5.91	Ortho.	29.7741 (TP)	31
Germanium	Ge	72.59	5.32	Dia cub.	938.3	32
Arsenic	As	74.92	5.78	Rhomb.	603 (SP)	33
Selenium	Sc	78.96	4.81	He2x.	221	34
Bromine	Br	79.90			−7.25 (TP)	35
Krypton	Kr	83.80			−157.385	36
Rubidium	Rb	85.47	1.53	bcc	39.48	37
Strontium	Sr	87.62	2.58	fcc	769	38
Yttrium	Y	88.91	4.48	hcp	1522	39
Zirconium	Zr	91.22	6.51	hcp	1855	40
Niobium	Nb	92.91	8.58	bcc	2469	41
Molybdenum	Mo	95.94	10.22	bcc	2623	42
Technetium	Tc	98.91	11.50	hcp	2204	43
Ruthenium	Ru	101.07	12.36	hcp	2334	44
Rhodium	Rh	102.91	12.42	fcc	1963	45
Palladium	Pd	106.4	12.00	fcc	1555	46
Silver	Ag	107.87	10.50	fcc	961.93	47
Cadmium	Cd	112.4	8.65	hcp	321.108	48
Indium	In	114.82	7.29	fct	156.634	49
Tin	Sn	118.69	7.29	bct	231.9681	50
Antimony	Sb	121.75	6.69	Rhomb.	630.755	51
Tellurium	Te	127.60	6.25	Hex.	449.57	52
Iodine	I	126.90	4.95	Ortho.	113.6 (TP)	53
Xenon	Xe	131.30			−111.7582 (TP)	54
Cesium	Cs	132.91	1.91 (−10°)	bcc	28.39	55
Barium	Ba	137.33	3.59	bcc	729	56
Lanthanum	La	138.91	6.17	Hex.	918	57
Cerium	Ce	140.12	6.77	fcc	798	58
Praseodymium	Pr	140.91	6.78	Hex.	931	59
Neodymium	Nd	144.24	7.00	Hex.	1021	60
Promethium	Pm	(145)		Hex.	1042	61
Samarium	Sm	150.4	7.54	Rhomb.	1074	62
Europium	Eu	151.96	5.25	bcc	822	63
Gadolinium	Gd	157.25	7.87	hcp	1313	64
Terbium	Tb	158.93	8.27	hcp	1356	65
Dysprosium	Dy	162.50	8.53	hcp	1412	66
Holmium	Ho	164.93	8.80	hcp	1474	67

TABLE A.1 (Continued)
Physical and Chemical Data for Selected Elements

Element	Symbol	Atomic Mass (amu)	Density of Solid (at 20°C) ($Mg/m^3 = g/cm^3$)	Crystal Structure (at 20°C)	Melting Point (°C)	Atomic Number
Erbium	Er	167.26	9.04	hcp	1529	68
Thulium	Tm	168.93	9.33	hcp	1545	69
Ytterbium	Yb	173.04	6.97	fcc	819	70
Lutetium	Lu	174.97	9.84	hcp	1663	71
Hafnium	Hf	178.49	13.28	hcp	2231	72
Tantalum	Ta	180.95	16.67	bcc	3020	73
Tungsten	W	183.85	19.25	bcc	3422	74
Rhenium	Re	186.2	21.02	hcp	3186	75
Osmium	Os	190.2	22.58	hcp	3033	76
Iridium	Ir	192.22	22.55	fcc	2447	77
Platinum	Pt	195.09	21.44	fcc	1769.0	78
Gold	Au	196.97	19.28	fcc	1064.43	79
Mercury	Hg	200.59			−38.836	80
Thallium	Tl	204.37	11.87	hcp	304	81
Lead	Pb	207.2	11.34	fcc	327.502	82
Bismuth	Bi	208.98	9.80	Rhomb.	327.502	83
Uranium	U	238.03	19.05	Ortho.	1135	92

Source: Shackelford, J.F. *Introduction to Materials Science for Engineers,* Macmillan, New York, 1992.

TABLE A.2
Relation between Modulus of Elasticity and Melting Point for Some Metallic Elements

Material	Modulus of Elasticity (GPa)	Melting Point (°C)
Lead (Pb)	14	327
Magnesium (Mg)	45	649
Aluminum (Al)	70	660
Copper (Cu)	130	1084
Iron (Fe)	210	1538
Molybdenum (Mo)	300	2610
Tungsten (W)	403	3410

TABLE A.3
Comparison of Thermal Conductivity and Linear Thermal Expansion Coefficient for Selected Materials

Material	Thermal Conductivity (W/m°C)	Linear Expansion (10^6/°C)
Aluminum	230	22.5
Copper	400	17
Steel 1020	50	11.7
Al_2O_3	29	9
SiC	12	4.68
Fireclay	0.8	4.5
Soda-lime glass	0.96	9.2
LDPE	0.34	198
PVC	0.12	144

TABLE A.4
Mechanical Properties of Some Carbon Steels

Specification and Grade		Tensile Strength		Yield Strength		Elongation (%)
		MPa	ksi	MPa	ksi	
Hot-rolled sheet and strip structural-quality low-carbon steel						
ASTM A570	A	310	45	170	25	23–27
	B	340	49	205	30	21–25
	C	360	52	230	33	18–23
	D	380	55	275	40	15–21
	E	400	58	290	42	13–19
Special-quality hot-rolled steel bars						
ASTM A675	45	310–380	45–55	155	22.5	33
	50	345–415	50–60	170	25	30
	60	415–495	60–72	205	30	22
	70	485–585	70–85	240	35	18
	80	550 min	80 min	275	40	17
Steel castings						
ASTM A27-77	60–30	415	60	205	30	24
	70–36	485	70	250	37	22
A148-73	80–40	552	80	276	40	18
	90–60	621	90	414	60	20
	120–95	827	120	655	95	14
	175–145	1207	175	1000	145	6

TABLE A.5
Mechanical Properties of Selected HSLA Steels

ASTM No. and Type	UNS Designation	Tensile Strength		Yield Strength		Elongation (%)
		MPa	ksi	MPa	ksi	
A242 type 1	K11510	435–480	63–70	290–345	42–50	21
A572 grade 50	—	450	65	345	50	21
A607 grade 60	—	520	75	415	60	16–18
grade 70	—	590	85	485	70	14
A618 grade 1	K02601	483	70	345	50	22
A717 grade 60	—	485	70	415	60	20–22
grade 70	—	550	80	485	70	18–20
grade 80	—	620	90	550	80	16–18

TABLE A.6
Mechanical Properties of Some Ultra High-Strength Steels

Designation or Grade	Tempering Temp.		Tensile Strength		Yield Strength		Elongation (%)
	°C	°F	MPa	ksi	MPa	ksi	
Medium-carbon low-alloy water quenched and tempered							
4130	205	400	1765	256	1520	220	10
	315	600	1570	228	1340	195	13
	425	800	1380	200	1170	170	16.5
4340	205	400	1980	287	1860	270	11
	315	600	1760	255	1620	235	12
	425	800	1500	217	1365	198	14
Medium-alloy air-hardening steel							
H13	527	980	1960	284	1570	228	13
	593	1100	1580	229	1365	198	14.4
18Ni maraging steels (solution treated 1 h at 820°C (1500°F), then aged 3 h at 480°C (900°F))							
18Ni(200)			1500	218	1400	203	10
18Ni(250)			1800	260	1700	247	8
18Ni(300)			2050	297	2000	290	7

TABLE A.7
Composition and Properties of Selected Stainless Steels

AISI	Nominal Composition (%)							Tensile Strength		Yield Strength		Elongation (%)	Hardness (BHN)
	C	Cr	Ni	Mn	Si	Others[a]		MPa	ksi	MPa	ksi		
Austenitic													
201	0.15	17	4.5	6.50	1.0			805	117	385	56	55	185
301	0.15	17	7	2.0	1.0			770	112	280	41	60	162
302	0.15	18	9	2.0	1.0			630	91	280	41	50	162
304	0.08	19	9.5	2.0	1.0			588	85	294	43	55	150
316	0.08	17	12	2.0	1.0	2.5	Mo	588	85	294	43	50	145
330	0.08	18.5	35.5	2.0	1.0			630	91	266	39	45	150
Ferritic													
405	0.08	13	—	1.0	1.0	0.2	Al	490	71	280	41	30	150
430	0.12	17	—	1.0	1.0			525	76	315	46	30	155
442	0.2	21.5	—	1.0	1.0			560	81	315	46	20	185
Martensitic													
403	0.15	12.5	—	1.0	0.5			525	76	280	41	35	153
416	0.15	13	—	1.25	1.0	0.6	Mo	525	76	280	41	30	153
431	0.2	16	2.0	1.0	1.0			875	127	665	97	20	260
502	0.1	5	—	1.0	1.0	0.55	Mo	455	66	175	25	30	150
Precipitation hardening													
17–7 PH[b]	0.07	17	7			1.0	Al	1484	215	1346	195	9	465
PH 13–8Mo[b]	0.05	14	8.5			2.5	Mo	1552	225	1414	205	12	465
						1.0	Al						
AM 350[c]	1.0	16.5	4.3			2.75	Mo	1518	220	1311	190	13	450

[a] Most steels contain 0.035 and 0.04–0.06 P except for type 416, which contains 0,15 S.
[b] Aged at 510°C.
[c] Aged at 450°C.

TABLE A.8
Composition and Mechanical Properties of Some Tool Steels

	Composition (%)						Hardness (RC)		Toughness	
Grade	C	Cr	V	W	Mo	Others	Room Temperature	(560°C) (1040°F)	J	ft lb
Water hardening										
Wl	0.6–1.4						63	10	68	50.2
W2	0.6–1.4		0.25				63	10	68	50.2
Shock resisting										
S1	0.50	1.5		2.5			60	20	95	70.1
S2	0.50				0.50	1.0 Si	63	20	95	70.1
Oil hardening										
Ol	0.9	0.5		0.5		1.0 Mn	63	20	54	39.9
Air hardening										
A2	1.0	5.0			1.0		63	30	48	35.4
Tungsten high speed										
T1	0.70	4.0	1.0	18.0			66	52	61	45
T2	0.85	4.0	2.0	18.0			65	52	61	45
Molybdenum high speed										
M2	0.85	4.0	2.0	6.25	5.00		65	52	68	50.2
M3	1.00	4.0	2.4	6.00	5.00		67	52	48	35.4
M4	1.3	4.0	4.0	5.50	4.50		67	52	48	35.4
M10	0.85	4.0	2.0		8.00		65	52	68	50.2

TABLE A.9
Composition and Properties of Selected Cast Irons

Specification	Class or Grade	UNS No.	Composition (CE = %C + 0.3[%Si + P])	Tensile Strength		Yield Strength		Average (BHN)	Elongation (%)
				MPa	ksi	MPa	ksi		
Gray iron									
ASTM A48–74	20		CE = 4.34	140	20.3	140	20.3	170	—
	25		CE = 4.08	175	25.38	175	25.38	190	—
	35		CE = 3.77	245	35.53	245	35.53	220	—
	40		CE = 3.65	280	40.6	280	40.6	225	—
	50		CE = 3.45	350	50.76	350	50.76	250	—
	60		CE = 3.37	420	60.9	420	60.91	270	—
Nodular (ductile) iron									
ASTM A536	60–40–18	F32800	Chemical composition is subordinate to	420	60.9	280	40.6	170	18
	65–45–12	F33100	mechanical properties. However, the content	455	65.98	315	45.68		12
	80–55–06	F33800	of any chemical element may be specified by	560	81.2	385	55.8	215	6
	100–70–03	F34800	mutual agreement	700	101.5	490	71.06		3
	120–90–02	F36200		840	121.8	630	91.36	270	2
ASTM A395	60–40–18	F32800	CE = 3.77	420	60.9	280	40.6	165	18
AS1M A476	80–60–03	F34100	CE = 3.8–4.5	560	81.21	420	60.9	201 min	3
Malleable iron									
ASTM	32510		TC, 2.5%; Si, 1.3%; S, 0.11%, P, 0.18% max	350	50.76	230	33.35	110–145	10
A47 (ferritic)	35018		TC, 2.3%; Si, 1.2%; S, 0.11%; P, 0.18% max	370	53.66	245	35.5		18
ASTM A220 (pearlitic)	40010		TC, 2.3%; Si, 1.3%; S, 0.11%; P, 0.18% max	420	60.9	280	40.6	180–240	10

TABLE A.10
Some Physical Properties of Selected Nonferrous Metals and Alloys

Metal/Alloy	Density (kg/m³)	Melting Point (°C)	Cost[a] (Relative to Mild Steel)
Light metals and alloys			
Mg	1700	650	3–4
Mg alloys	1750–1800	610–660	
Al	2700	660	3–4
Al alloys	2600–2800	475–660	
Ti	4500	1670	20–30
Ti alloys	4400–4700	1550–1650	
Copper and alloys			
Cu	8930	1085	5–6
Cu alloys	7400–8950	880–1260	
High-temperature metals and alloys			
Ni	8910	1453	20–30
Ni alloys	7750–8900	1100–1450	
Co	8832	1495	35–40
Mo	1022	2610	150–200
Nb	8570	2468	100–150
Ta	16,600	2996	
W	19,250	3410	50
Low-melting metals and alloys			
Zn	7113	420	3–4
Zn alloys	6640–7200	385–525	
Pb	11,350	327	2
Pb alloys	8850–11,350	180–327	
Sn	5765	232	20–30
Cd	8642	321	5–6
Bi	9808	271	10–20
In	7286	157	100–150
Precious metals and alloys			
Ag	10,490	962	500–800
Au	19,302	1064	$5–8 \times 10^4$
Pt	21,450	1769	$8–10 \times 10^4$

[a] Costs vary significantly depending on supply and demand, quantity purchased, size and shape, delivery time, and several other factors.

TABLE A.11
Composition and Properties of Selected Aluminum Alloys

Alloy	Temper	Nominal Composition (%)	Tensile Strength		Yield Strength		Elongation (%)	Hardness (BHN)
			MPa	ksi	MPa	ksi		
Wrought alloys								
1060	0	99.6 + A1	70	10	28	4	43	19
	H18		133	19	126	18	6	35
2014	0	4.4 Cu, 0.8 Si, 0.8 Mn, 0.4 Mg	189	27	98	14	18	45
	T6		490	71	420	61	13	135
3003	0	1.2 Mn	112	16	42	6	40	28
	H18		203	29	189	27	10	55
4032	T6	12.5 Si, 1.0 Mg, 0.9 Cu, 0.9 Ni	385	56	322	47	9	120
5052	0	2.5 Mg, 0.25 Cr	196	28	91	13	30	47
	H38		294	43	259	38	8	77
6061	0	1.0 Mg, 0.6 Si, 0.25 Cu, 0.25 Cr	126	18	56	8	30	30
	T6		315	46	280	41	17	95
7075	0	5.5 Zn, 2.5 Mg, 1.5 Cu, 0.3 Cr	231	34	105	15	16	60
	T6		581	84	511	74	11	150
Casting alloys								
208.0	Sand cast	4 Cu, 3 Si	147	21	98	14	2.5	55
356.0	T51	7 Si, 0.3 Mg	175	25	140	20	2	60
	T6		231	34	168	24	3.5	70
B443.0	Sand cast	5 Si	133	19	56	8	8	40
	Die cast		231	34	112	16	9	50
520.0	T4	10 Mg	336	49	182	26	16	75
850.0	T5	6.5 Sn, 1 Cu, 1 Ni	161	23	77	11	10	45

TABLE A.12
Composition and Properties of Selected Magnesium Alloys

Alloy	Temper	Nominal Composition (%)	Tensile Strength		Yield Strength		Elongation (%)	Hardness (BHN)
			MPa	ksi	MPa	ksi		
Wrought alloys								
AZ31B	0	3.0 Al, 0.2 Mn, 1.0 Zn	224	33	115	17	11	56
	H24		255	37	165	24	7	73
AZ61A	F	6.5 Al, 0.15 Mn, 1.0 Zn	266	39	140	20	8	55
HK31A	H24	0.7 Zr, 3.2 Th	235	34	175	25	4	57
HM21A	T8	0.8 Mn, 2.0 Th	224	33	140	20	6	55
ZK40A	T5	4.0 Zn, 0.45 Zr	280	41	250	36	4	60
Casting alloys								
AM 00A	F	10.0 Al, 0.1 Mn	140	20	70	10	6	53
	T4		238	35	70	10	6	52
	T6		238	35	105	15	2	52
AZ63A	F	6.0 Al. 0.15 Mn, 3.0 Zn	182	26	77	11	4	50
	T6		238	35	112	16	3	73
EZ33A	T5	2.6 Zn, 0.7 Zr, 3.2 Re	140	20	98	14	2	50
HK31A	T6	0.7 Zr, 3.2 Th	189	27	91	13	4	55
HZ32A	T5	2.1 Zn, 0.7 Zr, 3.2 Th	189	27	91	13	4	55
ZK61A	T6	6.0 Zn, 0.8 Zr	280	41	182	26	5	70

TABLE A.13
Properties and Applications of Selected Wrought Titanium Alloys

Alloy and Composition	Tensile Strength		Yield Strength		Elongation (%)	Hardness (RC)	Application
	MPa	ksi	MPa	ksi			
Unalloyed							
ASTM Grade 1	240	35	170	25	—	—	Excellent corrosion resistance
ASTM Grade 4	550	80	480	70	—	—	
Alpha alloys							
5 Al, 2.5 Sn	875	127	819	119	16	36	Weldable, aircraft engine compressor blades and ducts, steam turbine blades
8 Al, 1 Mo, 1 V	1029	149	945	137	16	—	
Alpha + beta alloys							
3 Al, 2.5 V	700	101	595	86	20	—	Aircraft hydraulic tubes
6 Al, 4 V	1008	146	939	136	14	36	Rocket motor cases, blades, and discs for turbines
7 Al, 4 Mo	1120	162	1050	152	16	38	Airframes and jet engine parts
6 Al, 2 Sn, 4 Zr, 6 Mo	1288	187	1190	173	10	42	Components for advanced jet engines
10 V, 2 Fe, 3 Al	1295	188	1218	177	10	—	Airframe structures requiring toughness and strength
Beta alloys							
13 V, 11 Cr, 3 Al	1239	180	1190	173	8	—	High-strength fasteners
8 Mo, 8 V, 2 Fe, 3 Al	1330	193	1260	183	8	40	Aerospace components
3 Al, 8 V, 6 Cr, 4 Mo, 4 Zr	1470	213	1400	203	7	42	High-strength fasteners
11.5 Mo, 6 Zr, 4.5 Sn	1407	204	1337	194	11	—	High-strength sheets for aircraft

TABLE A.14
Composition and Properties of Selected Wrought Copper Alloys

Alloy	Nominal Composition (%)	Treatment	Tensile Strength		Yield Strength		Elongation (%)	Rockwell Hardness
			MN/m²	psi	MN/m²	psi		
Pure copper								
C10200	99.95 Cu	—	221–455	33–66	69–365	10–53	55	—
Dilute copper alloys								
Beryllium copper	97.9 Cu, 1.9 Be	Annealed	490	71	—	—	35	RB 60
	0.2 Ni or Co	HT (hardened)	1400	203	1050	152	2	RC 42
Brass								
Gilding, 95%	95 Cu, 5 Zn	Annealed	245	36	77	11	45	RF 52
		hard	392	57	350	51	5	RB 64
Red brass, 85%	85 Cu, 15 Zn	Annealed	280	41	91	13	47	RF 64
		hard	434	63	406	59	5	RB 73
Cartridge brass,70%	70 Cu, 30 Zn	Annealed	357	52	133	19	55	RF 72
		hard	532	77	441	64	8	RB 82
Muntz metal	60 Cu, 40 Zn	Annealed	378	55	119	17	45	RF 80
		half-hard	490	71	350	51	15	RB 75
High-lead brass	65 Cu, 33 Zn, 2 Pb	Annealed	350	51	119	17	52	RF 68
		hard	318	46	420	61	7	RB 89

(Continued)

TABLE A.14 (Continued)
Composition and Properties of Selected Wrought Copper Alloys

Alloy	Nominal Composition (%)	Treatment	Tensile Strength		Yield Strength		Elongation (%)	Rockwell Hardness
			MN/m^2	psi	MN/m^2	psi		
Phosphor bronze, 5%	95 Cu, 5 Sn	Annealed	350	51	175	25	55	RB 40
		hard	588	85	581	84	9	RB 90
Phosphor bronze, 10%	90 Cu, 10 Sn	Annealed	483	70	250	36	63	RB 62
		hard	707	103	658	95	16	RB 96
Aluminum bronze	95 Cu, 5 Al	Annealed	420	61	175	25	66	RB 49
		cold rolled	700	102	441	64	8	RB 94
Aluminum bronze (2)	81.5 Cu, 9.5 Al,	Soft	630	91	—	—	12	—
	5 Ni, 2.5 Fe, 1 Mn	hard	735	107	420	61	12	RB 105
High-silicon bronze	96 Cu, 3 Si	Annealed	441	64	210	31	55	RB 66
		hard	658	95	406	59	8	RB 93
Copper nickel								
Cupro nickel, 30%	70 Cu, 30 Ni	Annealed	385	56	126	18	36	RB 40
		cold rolled	588	85	553	80	3	RB 86
Nickel silver								
Nickel silver	65 Cu, 23 Zn, 12 Ni	Annealed	427	62	196	28	35	RB 55
(German silver)		hard	595	86	525	76	4	RB 89

TABLE A.15
Composition and Properties of Some Zinc Die-Casting Alloys

Material	Zamak 3 ASTM AG40A(XXIII)	Zamak 5 ASTM AG41A(XXV)
Composition (%)		
Copper	0.25	0.75–1.25
Aluminum	3.50–4.30	3.50–4.30
Magnesium	0.02–0.05	0.03–0.08
Iron (max)	0.10	0.10
Zinc	Rem	Rem
Properties		
Tensile strength MPa (ksi)	287 (42)	329 (48)
Elongation (%)	10.00	7.00
Charpy impact (J) (ft lb)	58.3 (43)	66 (49)
Hardness (BHN)	82	91
Specific gravity	6.6	6.7

Note: Rem = Remainder.

TABLE A.16
Chemical Composition of Selected Ni-Base and Co-Base Alloys

	Nominal Composition (wt %)										
Material	C	Mn	Si	Cr	Ni	Co	W	Nb	Zr	Fe	Others
Fe–Ni-base alloys											
Incoloy 80	0.05	0.8	0.5	21	32.5					45.7	0.38 Al, 0.38 Ti
Inconel 718	0.08			19	52.5			5.1		18.5	3 Mo, 0.9 Ti, 0.5 Al, 0.15 Cu
Nickel-base alloys											
DS-nickel					bal						2.1 ThO_2
Hastelloy X	0.1	1.0	1.0	21.8	bal	1.5	0.6			18.5	
Inconel 600	0.08	0.5	0.2	15.5	76					8.0	
Inconel 617	0.1	0.5	0.5	22.0	52	12.5				1.5	9 Mo, 0.3 Ti, 1.2 Al, 0.2 Cu
MAR-M200, c	0.15			9.0	bal	10	12.5	1.8	0.05		2Ti, 5 Al, 0.015 B
TRW VI A, c	0.13			6.0	bal	7.5	5.8	0.5	0.13		2 Mo, 1 Ti, 0.02 B, 5.4 Al, 9 Ta, 0.5 Re, 0.4 Hf
Cobalt-base alloys											
AiRcsist 13, c	0.45	0.5		21	1	bal	11	2		2.5	3.5 Al, 0.1 Y
X-40, c	0.5	0.5	0.5	25	10	bal	7.5			1.5	
MAR-M302, c	0.85	0.1	0.2	21.5		bal	10		0.15		9 Ta, 0.005 B
MAR-M918	0.05	0.2	0.2	20	20	bal			0.1	0.5	7.5 Ta

Note: bal = balance; c = cast alloy.

TABLE A.17
Rupture Strength of Selected Nickel-Base and Cobalt-Base Alloys

	Rupture Strength (MPa) (ksi)					
	650°C (1200°F)		815°C (1500°F)		1093°C (2000°F)	
Material	**100 h**	**1000 h**	**100 h**	**1000 h**	**100 h**	**1000 h**
Nickel-base alloys						
DS-Nickel	162	155	131	120.6	61.3	51
	(23.5)	(22.5)	(19)	(17.5)	(8.9)	(7.4)
Hastelloy X	330	234	96	69	8.2	4.1
	(48)	(34)	(14)	(10)	(1.2)	(0.6)
Inconel 600			55.2	38.6	9.6	6.2
			(8)	(5.6)	(1.4)	(0.9)
Inconel 617	414	324	145	96	18.6	10.3
	(60)	(47)	(21)	(14)	(2.7)	(1.5)
MAR-M200, c			524	414	75.8	44.8
			(76)	(60)	(11)	(6.5)
TRW VI A, c	1000	896	552	420.6	82.7	
	(145)	(130)	(80)	(61)	(12)	
Cobalt-base alloys						
AiResist 13, c			172.4	117.2	30.3	
			(25)	(17)	(4.4)	
X-40, c	390	339	179.3	137.9	27.6	
	(57)	(49)	(26)	(20)	(4)	
MAR-M302, c			276	206.9	41.4	27.6
			(40)	(30)	(6)	(4)
MAR-M918	462		206.9	137.9	17.2	
	(67)		(30)	(20)	(2.5)	

Note: c = cast alloy.

TABLE A.18
Composition and Properties of Some Refractory Metals and Alloys

		Test Temperature		UTS at Test Temperature		10 h Rupture Stress	
Alloy	Nominal Additions (%)	°C	°F	MPa	ksi	MPa	ksi
Niobium and alloys							
Unalloyed Nb		1366	2491	70	10	38	5.5
SCB291	10 Ta, 10 W	1366	2491	224	33	63	9.1
C129Y	10 W, 10 Hf, 0.1 Y	1590	2894	182	26	105	15.2
FS85	28 Ta, 11 W, 0.8 Zr	1590	2894	161	23	84	12.2
Molybdenum and alloys							
Unalloyed Mo		1366	2491	182	26	102	14.8
TZM	0.5 Ti, 0.08 Zr, 0.015 C	1590	2894	371	54	154	22.3
WZM	25 W, 0.1 Zr, 0.03 C	1590	2894	504	73	105	15.2
Tantalum and alloys							
Unalloyed		1590	2894	60	8.7	17.5	2.5
Ta-10 W	10 W	1590	2894	350	50.8	140	20.3
T-222	9.6 W, 2.4 Hf, 0.01 C	1590	2894	280	40.6	266	38.6
Tungsten and alloys							
Unalloyed		1922	3492	175	25.4	48	7.0
W-2 ThO_2	2 ThO_2	1922	3492	210	30.5	126	18.3
W-15 Mo	15 Mo	1922	3492	252	36.6	84	12.2
GE 218	Doped	1922	3492	406	58.9	—	—

TABLE A.19
Some Mechanical Properties of Selected Plastics

Material	Tensile Strength		Tensile Modulus		Elongation (%)	Hardness	Izod Impact	
	MN/m²	ksi	MN/m²	ksi			J	ft lb
I Thermoplastics								
Polyethylene								
Low density	7–21	1.0–3.0	140	20.0	50–800	Rr 10	27	19.93
Medium density	14–21	2.0–3.0	280	40.0	50–800	Rr 14	2.7–22	1.99–16.24
High density	21–35	3.0–5.0	700–1400	100–200	1030	Rr 65	1.36–6.8	1–5.02
Polypropylene								
Homopolymers	35	5.0	1200	171.4	150	Rr 90	0.54–2.0	0.4–1.48
High-impact copolymers	27	3.8–6	—	—	400	Rr 65	5.4–8.1	4–5.98
Talc filled, 40%	35	5.0	3600	514.2	5.0	Rr 95	0.54	0.4
Polystyrene								
General purpose	35–56	5.0–8.0	3300	471.4	1.5–4.0	Rm 74	0.27–0.54	0.2–0.4
High impact	23–32	3.3–4.6	2200	314.2	25–60	Rm 60	1.36–3.8	1–2.8
30% glass fiber	77–100	11.0–14.3	8800	1257.1	—	Rm 90	3.8	2.8
Polyvinyl chloride								
General purpose	7–28	1.0–4.0	20	2.8	400	Sa 75	—	—
Flexible PVC	14–21	2.0–3.0	15	2.1	250	Sa 85	—	—
Rigid PVC	35–63	5.0–9.0	200–4200	285.7–600	100	Rr 115	1.4–2.7	1.03–1.99

Acrylic (MMA)								
Cast, general purpose	42–84	6.0–12.0	2800	400	—	Rm 91	0.68	0.5
Molding grade	63–77	9.0–11.0	2800	400	—	Rm 95	0.5	0.37
High impact	42–63	6.0–9.0	2100	300	25–40	Rm 45	1.4–5.4	1.0–3.99
Nylons								
Nylon 6/6	84	12	3300	471.4	60–300	Rr 118	1.36–2.7	1–1.99
Nylon 6/12	62	8.8	21000	3000	150–340	Rr 114	1.36–2.7	1–1.99
Acetal								
Homopolymer	70	10.0	3700	528.6	25–75	Rm 94	1.9–3.1	1.4–2.29
Homopolymer 22% TFE fiber	53	7.5	2800	400.0	12–21	Rm 78	0.95–2.3	0.7–1.7
Copolymer	62	8.8	2900	414.2	60–75	Rm 80	1.63–3.5	1.2–2.58
Polycarbonate								
Unfilled	63	9.0	2400	342.8	110	Rm 70	16–22	11.81–16.24
ABS								
Medium impact	46	6.6	2500	357.1	6–14	Rr 111	5.4	3.99
High impact	42	6.0	2300	328.5	10–35	Rr 103	8.8	6.49
Very high impact	33.6	4.8	1750	250.0	15–50	Rr 88	10.8	7.97
Heat resistant	50.4	7.2	2450	350.0	5–20	Rr 111	3.12	2.3
Fluoroplastics								
PCTFE	28–42	4.0–6.0	1400	200.0	160	Sd 76	4	2.95
PTFE	14–49	2.0–7.0	700	100.0	100–450	Sd 58	6.1	4.5
High-temperature plastics								
Polyamide unfilled	91	13.0	3150	450.0	7–9	Rm 97	1.36	1
Polyamide 40% graphite	53	7.6	5300	757	2–3	Rm 73	—	—
Polysulfone	71.4	10.2	2500	357	50–100	Rr 120	1.63	1.2

(*Continued*)

TABLE A.19 (Continued)
Some Mechanical Properties of Selected Plastics

Material	Tensile Strength		Tensile Modulus		Elongation (%)	Hardness	Izod Impact	
	MN/m^2	ksi	MN/m^2	ksi			J	ft lb
II Thermosetting plastics								
Phenolic								
General purpose	35–63	5.0–9.0	800	114.2	—	Re 95	0.44	0.32
Shock and heat	28–63	4.0–9.0	14,000	2000	—	Re 85	2.17	1.6
Heat (mineral)	35–49	5.0–7.0	11,000	1571.4	—	Re 85	0.54	0.4
Electrical (mineral)	28–56	4.0–8.0	6000	857.1	—	Re 87	0.41	0.59
Epoxy								
Cast rigid	63–105	9.0–15.0	3200	457.1	—	Rm 106	0.5	0.37
Molded	56–140	8.0–20.0	14,000	2000	—	B78	1.3–5.5	0.96–4.06
High-strength laminate	350–490	50.0–70.0	32,000	4571.4	—	B71	1.36–41	1–30.26
Polyester								
Unfilled	56	8.0	2400	342.8	200–300	Rm 117	1.63	1.2
30% glass fiber	123	17.6	7700	1100	3	Rm 90	2.3	1.7
Alkyd								
Granular (mineral)	42–63	6.0–9.0	16,000	2285.7	—	Re 85	0.42	0.31
Silicone								
30% glass fiber	42	6	17,500	2500	—	Rm 90	4–20.3	2.95–14.98
Silica reinforced	28	4	11,000	1571.4	—	Rm 82	0.41	0.3

TABLE A.20
Some Physical Properties and Uses of Selected Plastics

Material	Expansion Coefficient		Heat Deflection Temperature		Specific Gravity	Relative Cost
	$\times 10^5$m/m/°C	$\times 10^5$in./in./°F	°C	°F		
I Thermoplastics						
Polyethylene						
Low density	19.8	11	36	96.8	0.92	1.0
Medium density	16.2	9	44.4	111.92	0.93	1.0
High density	13.5	7.5	49	120.2	0.96	1.0
Polypropylene						
Homopolymers	8.6	4.78	57	134.6	0.91	0.8
High-impact copolymers	—	—	50	122	0.9	1.1
Talc filled, 40%	—	—	81	177.8	1.23	1.0
Polystyrene						
General purpose	7.4	4.11	67–100	152.6–212	1.05	1.0
High impact	7.0	3.89	67–97	152.6–206.6	1.04	1.0
30% glass fiber	3.2	1.78	97	206.6	1.29	1.75
Polycarbonate						
Unfilled	6.75	3.75	132	269-6	1.2	3.6
ABS						
Medium impact	8.46	4.7	94	201.2	1.05	1.4
High impact	9.5	5.28	99	210.2	1.04	1.6
Very high impact	11.0	6.11	96	204.8	1.02	1.7
Heat resistant	6.7	3.72	114	237.2	1.05	1.8
Fluoroplastics						
PCTFE	4.5	2.5	—	—	2.1	20
PTFE	9.9	5.5	—	—	2.16	10

(*Continued*)

TABLE A.20 (Continued)
Some Physical Properties and Uses of Selected Plastics

Material	Expansion Coefficient		Heat Deflection Temperature		Specific Gravity	Relative Cost
	x10^{5}m/m/°C	x10^{5}in./in./°F	°C	°F		
High-temperature plastics						
Polyamide unfilled	5.0	2.78	360	680	1.43	—
Polyamide 40% graphite	3.2	1.78	360	680	1.65	—
Polysulfone	5.6	3.11	174	345.2	1.24	6.3
Polyvinyl chloride						
General purpose	14.4	8	—	—	1.40	1.6
Flexible PVC	14.4	8	—	—	1.35	1.6
Rigid PVC	9–11	5–6.11	72	161.6	1.40	1.3
Acrylic (MMA)						
Cast, general purpose	5.4–7.2	3–4	99	210.2	1.19	1.4
Molding grade	5.4–7.2	3–4	88	190.4	1.18	1.7
High impact	7.2–11	4–6.11	77	170.6	1.11	2.2
Nylons						
Nylon 6/6	8.1	4.5	104	219.2	1.14	3.2
Nylon 6/12	9	5	82	179.6	1.07	5.8
Acetal						
Homopolymer	9–14.4	5–8	124	255.2	1.42	2.5
Homopolymer 22% TFE fiber	9–14.4	5–8	100	212	1.52	15
Copolymer	8.46	4.7	110	230	1.41	1.9
II Thermosetting plastics						
Phenolic						
General purpose	3.8	2.11	174	345.2	1.38	—
Shock and heat	2.3	1.28	154	309.2	1.83	—
Heat (mineral)	1.8	1	177	350.6	1.53	—
Electrical (mineral)	3.96	2.2	154–204	309.2–399.2	1.52–1.67	—

Epoxy						
Cast rigid	5.6	3.11	166	330.8	1.20	1.3
Molded	3.6	2	191	375.8	1.91	1.3
High-strength laminate	3.6	2	—	—	1.84	1.3
Polyester						
Unfilled	11.3	6.28	54	129.2	1.31	2.7
30% glass fiber	2.3–9.7	1.28–5.39	213	415.4	1.54	2.9
Alkyd						
Granular (mineral)	2.9	1.61	146–191	294.8–375.8	2.2	1.4
Silicone						
30% glass fiber	5.76	3.2	482	899.6	1.88	—
Silica reinforced	4–8	2.22–4.44	482	899.6	1.93	—

TABLE A.21
Some Properties and Applications of Selected Rubbers

Material	Tensile Strength		Elongation (%)	Hardness (Shore A)	Specific Gravity
	MN/m^2	ksi			
Natural rubber	28	40	700	30–90	0.92
Styrene butadiene	24.5	35	600	40–90	0.94
Neoprene	28	40	600	30–90	1.24
Butyl	21	30	800	40–80	0.92
Silicone	8.4	12	700	30–85	0.98
Fluorocarbon	17.5	25	300	60–90	1.85
Hypalon	21	30	500	50–90	1.18

TABLE A.22
Properties and Preparation of Some Common Adhesives

	Curing Temperature		Service Temperature		Lap-Shear Strength[a]		Peel Strength at RT	
Adhesive	**°C**	**°F**	**°C**	**°F**	**MPa at °C**	**ksi at °F**	**N/cm**	**lb/in.**
Acrylics	RT	RT	Up to 120	Up to 250	17.2–37.9	2.5–5.5	17.5–105	10–60
Anaerobics	RT	RT	Up to 166	Up to 330	15.2–27.6	2.2–4.0	17.5	10
Butyral-phenolic	135–177	275–350	−51 to 79	−60 to 175	17.2 6.9(80)	2.5 1.0(175)	17.5	10
Cyanoacrylates	RT	RT	Up to 166	Up to 330	15.2–27.6	2.2–4.0	17.5	10
Epoxy (RT cure)	RT	RT	−51 to 82	−60 to 180	17.2 10.3(80)	2.5 1.5(175)	7	4
Epoxy (HT cure)	90–175	195–350	−51 to 175	−60 to 350	17.2 10.3(175)	2.5 1.5(350)	8.8	5
Epoxy–nylon	120–175	250–350	−250 to 82	−420 to 180	41 13.8(80)	6.0 2.0(180)	123	70
Epoxy–phenolic	120–175	250–350	−250 to 260	−420 to 500	21–28 6.9(80)	3.0–4.0 1.0(175)	8.8	5
Polyurethanes	149	300	Up to 66	Up to 150	24	3.5	123	70
Silicones	149	300	Up to 260	Up to 500	0.3	0.04	43.8	25
Hot melts (general)	—	—	Up to 120	Up to 250	1.4–4.8	0.2–0.7	35	20
(polyamides)	260–370	500–700	Up to 315	Up to 600	13.8	2.0	17.5	10

Note: RT = room temperature; HT = high temperature.

[a] RT if temperature is not specified.

TABLE A.23
Properties of Some Commercial Polystyrene Foams

Polymer	Density		Compressive Strength		Compressive Modulus	
	kg/m³	lb/ft.³	kN/m²	psi	kN/m²	psi
Polystyrene (molded)	16.0	1.0	138	20	1725	250
	32.0	2.0	242	35	5175	750
	64.0	4.0	483	70	12,075	1750
Polystyrene (extruded)	30.4	1.9	242	35	6900	1000
	46.4	2.9	449	65	20,700	3000
	70.4	4.4	897	130	34,845	5050

TABLE A.24
Some Properties of Oxide and Carbide Ceramics

Material	Specific Gravity	Thermal Expansion Coefficient		Tensile Strength		Compressive Strength		Young's Modulus	
		10^{-6}/°C	10^{-6}/°F	MPa	ksi	MPa	ksi	GPa	lb/in.$^2 \times 10^6$
I Oxides									
Alumina (85%)	3.39	—	—	155	22.5	1930	280	221	32
(90%)	3.60	—	—	221	32	2482	360	276	40
(99.5%)	3.89	8.8	4.9	262	38	2620	380	372	54
Beryllia (98%)	2.90	9.0	5.0	85	12.3	1980	287	230	33
(99.5%)	2.88	9.0	5.0	98	14.2	2100	304	245	35.5
Magnesia	3.60	13.5	7.5	140	20.3	840	122	280	40.6
Zirconia	5.8	9.8	5.4	147	21.3	2100	304	210	30.4
Thoria	9.8	9.2	5.1	52	7.5	1400	203	140	20.3
II Carbides									
SiC (silicate bond)	2.57	4.68	2.6	—	—	105	15.2	21	3.0
SiC (SiN bond)	2.62	4.68	2.6	—	—	150	21.7	44	6.4
SiC (self-bonded)	3.10	3.78	2.1	—	—	1400	203	175	25.4
Boron carbide	2.51	4.5	2.5	—	—	2900	420	308	44.6
Tungsten carbide	16.0	5.95	3.3	—	—	—	—	—	—

TABLE A.25
Typical Compositions and Properties of Commercial Glasses

	Fused Silica	96% Silica (Vycor)	Soda-Lime–Silica	Boro-Silica (Pyrex)	Alumino-Silicate	Lead-Alkali
Chemical composition (%)						
SiO_2	99.5+	96	72.6	80.2	57	56.5
Al_2O_3			1.6	2.6	20.6	1.4
K_2O			0.9	0.3		8.25
Na_2O			13.1	4.5	1.0	4.25
CaO			3.7	0.1	5.4	
MgO			8.0		12.0	
B_2O_3		4		12.3	4	
PbO						29.6
Fe_2O_3			0.1			
Properties						
Young's modulus GPa (lb/in.^{2} × 10^6)	73.5 (10.7)	67.2 (9.8)	70 (10.2)	66.5 (9.6)	89.6 (13)	63 (9.1)
Poisson's ratio	0.17	0.18	0.24	0.20	0.26	—
Specific gravity	2.20	2.18	2.46	2.23	2.53	3.04
Linear expansion coefficient, × (07)/°C(/°F)	5.6 (3.1)	8 (4.4)	92 (51)	32.5 (18)	42 (23)	90 (50)
Maximum service temperature (°C) (°F)	1197 (2187)	1097 (2007)	460 (860)	490 (914)	637 (1179)	380 (716)
Volume resistivity (Ω cm) at 250°C (482°F)	1.6×10^{12}	5×10^9	2.5×10^6	1.3×10^8	3.2×10^{13}	8×10^8
Dielectric constant	3.8	3.8	7.2	4.6	6.3	6.7
Refractive index	1.458	—	1.510	1.474	1.634	1.583

TABLE A.26
Some Properties of Selected Composite Materials

Matrix Material	Fiber Material	ff (%)	Specific Gravity	Tensile Strength		Elastic Modulus	
				MPa	ksi	GPa	lb/in.$^2\times10^6$
Epoxy	S glass[a,b]	70	2.11	2100	304.4	62.3	9.0
	S glass[c]	70	2.11	680	98.6	22	3.2
	S glass[a,b]	14	1.38	500	72.5	—	—
	E glass[a,b]	73	2.17	1642	238	55.9	8.1
	E glass[a,b]	56	1.97	1028	149	42.8	6.2
	E glass[a,d]	56	1.97	34.5	5	10.4	1.5
	Carbon[a,b]	63	1.61	1725	250	158.7	23
	Carbon[a,d]	63	1.61	41.4	6	11.0	1.6
	Aramid[a,b]	62	1.38	1311	190	82.8	12
	Aramid[a,d]	62	1.38	39.3	5.7	5.5	0.8
Polyester	E glass[c]	65	1.8	340	49.3	19.6	2.8
	Glass[e]	40	1.55	140	20.3	8.9	1.3
Polystyrene	Glass[e]	30	1.28	97	14	8.2	1.2
Polycarbonate	Glass[e]	20	1.31	107	15.5	6.2	0.9
	Glass[e]	40	1.44	131	19	10.4	1.5
Nylon 66	Glass[e]	20	1.31	152	22	8.3	1.2
	Glass[e]	40	1.41	200	29	11.0	1.6
	Glass[e]	70	—	207	30	21.4	3.1
Aluminum	Carbon[a,b]	40		1242	180		
	SiO_2[a,b]	48		870	126		
	B[a,b]	10		297	43		
Nickel	B[a,b]	8		2650	384		
	W[a,b]	40		1100	159		
Copper	Carbon[a,b]	65		794	115		
	W[a,b]	77		1760	255		

[a] Continuous fibers.
[b] Fibers aligned in loading direction.
[c] Fabric.
[d] Fibers at 90° to loading direction.
[e] Discontinuous fibers.

TABLE A.27
Some Properties of Selected Fiber Materials

Fiber Material	Specific Gravity	Tensile Strength GPa	Tensile Strength ksi	Elastic Modulus GPa	Elastic Modulus $lb/in.^2 \times 10^6$	Relative Cost[a]
E glass	2.54	3.5	507	73.5	10.7	1
S glsss	2.49	4.6	667	85.5	12.4	3–4
Carbon (HS)	1.9	2.5	263	240	34.8	20–40
Carbon (HM)	1.9	2.1	305	390	56.6	20–40
Aramid	1.5	2.8–3.4	406–493	66–130	9.6–18.8	4–12
Steel	7.8	4.2	609	207	30	
W	19.4	4.1	594	413	59.9	
Rene 41	8.26	2.0	290	168	24.3	
Mo	10.2	2.2	319	364	52.8	
Boron	2.6	3.5	507	420	60.9	
SiC	4	2.1	305	490	71	
$A1_2O_3$	3.15	2.1	305	175	25.4	
SiO_2	2.19	6	870	73.5	10.7	

[a] Cost of E glass is taken as unity.

Appendix B: Conversion of Units and Hardness Values

TABLE B.1
Conversions to SI units

Quantity	Multiply Number of	By	To Obtain Number of
Length	Inches	25.4	mm
	Feet	0.3048	Meters (m)
	Yards	0.9144	m
Area	Square inches	645.16	mm^2
	Square feet	0.092903	m^2
	Square yards	0.836130	m^2
Volume	Cubic inches	16387.1	mm^3
	Cubic feet	0.0283168	m^3
	Cubic yards	0.764555	m^3
Mass	Ounces	0.0283495	Kilograms (kg)
	Pounds (lb)	0.45359237	kg
	Short tons	907.185	kg
	Long tons	1016.05	kg
Density	lb/in.^{3}	27679.9	kg/m^3
	lb/ft.^{3}	16.0185	Kg/m^3
Force	Pounds force (lbf)	4.44822	Newtons (N)
	Tons force (long)	9964.02	N
	Dynes	10^{-5}	N
	kgf	9.80665	N
Stress	lbf/in.^{2}	6894.76	N/m^2
	tonf/in.^{2}	15.4443×10^6	N/m^2
	kgf/cm^2	98.0665×10^3	N/m^2
Work	ft. lbf	1.35582	Joules (J)
	hp/h	2.68452×10^6	J
	BTU	1.05506×10^3	J
	kw/h	3.6×10^6	J
	kcal	4.1868×10^3	J
	kgf/m	9.80665	J
Power	ft./lbf s	1.35582 watts (W)	
	Horsepower (hp)	745.7	W
	Metric hp (CV)	735.499	W
	BTU/h	0.293071	W
Thermal conductivity	BTU/h ft. °F	1.73073	W/m^3
	BTU in./h ft.^{2} °F	0.144228	W (m K)
	Kcal/m h °C	1.163	

Note: Temperature °C $= \frac{5}{9}$(°F − 32).
°F $= \frac{9}{5}$°C + 32.

TABLE B.2
Hardness Conversions: Soft Steel, Gray and Malleable Cast Iron, and Most Nonferrous Metals

Rockwell Scale			BHN 500 kg (10 mm Ball)	VHN 10 kg and BHN 3000 kg	Tensile Stength (MN/m^2)
B	A	30T			
100	61.5	82.0	201	240	800
98	60.0	81.0	189	228	752
96	59	80	179	216	710
94	57.5	78.5	171	205	676
92	56.5	77.5	163	195	641
90	55.5	76.0	157	185	614
88	54	75	151	176	586
86	53	74	145	169	559
84	52	73	140	162	538
82	50.5	71.5	135	156	517
80	49.5	70	130	150	497
78	48.5	69	126	144	476
74	46	66	118	135	448
70	44	63.5	110	125	420
66	42	60.5	104	117	392
62	40.5	58	98	110	—
58	38.5	55	92	104	—
54	37	52.5	87	98	—
50	35	49.5	83	93	—
46	33.5	47	79.5	88	—
42	31.5	44	76	86	—
38	30	41.5	73	—	—
34	28	38.5	70	—	—
30	26.5	36	67	—	—
20	22	29	61.5	—	—
10	—	22	57	—	—

TABLE B.3
Hardness Conversions: Hardened Steel and Hard Alloys

Rockwell Scale					
C	A	30T	VHN 10 kg	BHN 3000 kg	Tensile Stength (MN/m²)
80	92	92	1865	—	—
75	89.5	89	1478	—	—
70	86.5	86	1076	—	—
65	84	82	820	—	—
64	83.5	81	789	—	—
62	82.5	79	739	—	—
60	81	77.5	695	614	2310
58	80	75.5	655	587	2205
56	79	74	617	560	2065
54	78	72	580	534	2006
52	77	70.5	545	509	1889
50	76	68.5	513	484	1758
48	74.5	66.5	485	460	1634
46	73.5	65	458	437	1524
44	72.5	63	435	415	1427
42	71.5	61.5	413	393	1338
40	70.5	59.5	393	372	1255
38	69.5	57.5	373	352	1179
36	68.5	56	353	332	1117
34	67.5	54	334	313	1054
32	66.5	52	317	297	992.9
30	65.5	50.5	301	283	937.7
28	64.5	48.5	285	270	889.4
26	63.5	47	271	260	848
24	62.5	45	257	250	807
22	61.5	43	246	240	772
20	60.5	41.5	236	230	745

Appendix C: Glossary

Abrasive wear: Occurs in a soft surface when the asperities of a hard surface rub against it.

Activation energy: The energy required to cause a reaction to occur.

Additive: Material added to a polymer to modify its characteristics.

Adhesive wear: Loss of material from a surface as a result of tearing off of its asperities when they form a bond with a stronger material.

Age hardening (precipitation hardening): Increase of hardness as a result of the precipitation of a hard phase from a supersaturated solid solution.

Allotropy (polymorphism): Change of lattice structure with temperature, for example, iron changes from bcc to fcc at 910°C.

Alloy: A metal containing one or more additional metallic or nonmetallic elements.

Amorphous (noncrystalline): Atoms or molecules of the material are not arranged according to a repetitive pattern or exhibit long-range order.

Anion: Negative ion, which is formed when an atom gains one or more electrons.

Anisotropic: Exhibiting different properties in different directions.

Annealing (steels): Heating to the austenite range, then cooling slowly enough to form ferrite and pearlite.

Annealing (strain-hardened metal): Heating a cold-worked metal to the recrystallization temperature to soften the material. In the case of steels, this is called process anneal.

Anode: The electrode that supplies electrons to the external circuit in an electrochemical cell. It is the electrode that undergoes corrosion or the negative electrode.

Anodizing: Electrochemically coating the surface with an oxide layer, by making the part an anode in an electrolytic bath.

Artificial ageing: Heating a solution-treated and quenched precipitation-hardenable alloy to speed up the precipitation process.

Atomic mass unit (amu): One-twelfth of the mass of C_{12}, also equal to 1.66×10^{-24} g.

Atomic number: The number of electrons possessed by an uncharged atom; also the number of protons per atom.

Atomic radius: One-half of the interatomic distance between like atoms.

Atomic weight: Atomic mass expressed in atomic mass units, or in gram per mole.

Austempering: Quenching steel from the austenite range to a temperature just above the martensitic transformation range and holding it long enough to form bainite, which is a dispersion of carbide in a ferrite matrix.

Austenite: Fcc iron (γ phase) or an iron-rich, fcc solid solution.

Austenitic stainless steel: Corrosion-resistant alloy steel containing at least 11% chromium, which is mainly in γ phase (austenite).

Austenization: Heating steel to the austenite range to dissolve carbon into fcc iron, thereby forming austenite.
Avogadro's number: The number of atoms in a gram mole. It is equal to 6.02×10^{23} atoms/g mol.
Bainite: Microstructure of steel consisting of fine needles of Fe_3C in α iron. It is formed as a result of austempering treatment.
Binary alloy: Alloy composed of two elements.
Block copolymer: Mixture of polymers that form blocks along a single molecule chain.
Blow molding: Processing polymers or glass by expanding a parison into a mold by air pressure. Usually used to make hollow containers and bottles.
Body-centered cubic (bcc): Arrangement of atoms in a cell such that one atom is at the center of a cube in addition to eight atoms at the cube corners.
Bond energy: Energy required to separate two bonded atoms.
Branching: Addition of molecules as branches to the sides of the main polymer molecular chain.
Brass: An alloy of copper containing up to about 40 wt% zinc.
Brinell hardness: Measurement of hardness obtained by indenting the surface with a 10-mm ball under a load of 3000 kg.
Brittle: Lacking in ductility. Brittle materials break without undergoing plastic deformation.
Bronze: An alloy of copper and tin, unless otherwise specified; as, for example, in the case of aluminum bronze, which is an alloy of copper and aluminum.
Bulk modulus (K): Hydrostatic pressure per unit volume strain.
Carbide: Compound of metal and carbon, for example, iron carbide is Fe_3C.
Carbon fiber-reinforced plastic (CFRP): Composite material consisting of a polymer matrix reinforced with carbon, or graphite, fibers.
Carburization: Increasing the carbon content of a steel surface by diffusion. The purpose is usually to harden the surface.
Cast iron: Ferrous alloy containing more than 2 wt% carbon.
Casting: Shaping by pouring a liquid material in a mold and allowing it to solidify, thus taking the shape of the mold.
Cathode: The electrode in an electrochemical cell that receives electrons from the external circuit. The electrode on which electroplating is deposited.
Cation: Positively charged ion.
Cementite: The iron carbide, Fe_3C, hard brittle phase in steel.
Ceramic: Inorganic nonmetallic insulating material usually based on compounds of metals with nonmetals. They are generally characterized by their resistance to high temperatures and poor ductility.
Charpy impact test: Test to measure the toughness of materials.
Cold working: Plastic deformation below the recrystallization temperature. Measured by the percent reduction in area or thickness.
Composite: Synthetic material made by adding particulate or fibrous phases to a matrix material to modify its properties.

Compression molding: Shaping of thermosetting plastics by applying pressure and heat to allow cross-linking and setting to take place.

Coordination number (CN): Number of closest atomic or ionic neighbors.

Copolymer: Mixture of polymers containing more than one type of monomer. Block copolymer results from clustering of like mers along the chain. Graft copolymer results from attaching branches of a second type of polymer.

Coring: Segregation during solidification resulting from relatively fast cooling.

Corrosion: Surface deterioration as a result of electrochemical reaction.

Covalent bond: Attraction as a result of sharing of electrons between adjacent atoms.

Creep: Time-dependent plastic strain that occurs as a result of mechanical stresses at relatively high temperatures.

Creep rupture: Fracture as a result of creep.

Crevice corrosion: Localized corrosion in a corrosion cell near a restricted area, or a crevice.

Critical stress intensity factor (K_{IC}): The stress at the root of a crack that is sufficient to cause crack propagation.

Cross-linking: Linking of adjacent polymer molecules by chemical bond.

Crystal: A solid with a long-range repetitive pattern of atoms.

Debonding in composite materials: Uncoupling of reinforcement phase from the matrix.

Decarburization: Loss of carbon from the surface of steel.

Deformation (elastic): Temporary deformation that is eliminated when the stress is removed. It occurs as a result of atomic or molecular movement without permanent displacements.

Deformation (plastic): Permanent deformation that persists after the stress is removed. It arises from the displacement of atoms or molecules to new positions.

Deformation (viscoelastic): Combined viscous flow and elastic deformation.

Degree of polymerization: Average number of mers in a polymer molecule.

Devitrification: Crystallization of glass. Process for producing "glass ceramics."

Diffusion: The movement of atoms or molecules in a material.

Dislocation: Linear defect in a crystalline solid. Edge dislocation is at the edge of an extra crystal plane. The slip vector, which defines the direction and magnitude of deformation associated with the movement of dislocation, is perpendicular to the defect line. In screw dislocations, the slip vector is parallel to the defect line.

Dispersion strengthening: Increase in strength as a result of introducing fine particles in the material.

Drawing: Forming of wires or sheet metals by applying tension through a die as in wire drawing or sheet drawing.

Ductile iron: Type of cast iron in which the graphite phase is spheroidal rather than flakes as in GCI.

Ductile–brittle transition temperature: Temperature that separates the mode of brittle fracture from the higher temperature range of ductile fracture.

Ductility: Ability to undergo plastic deformation without fracture. Measured as elongation in length or as reduction of area.

Edge dislocation: Linear defect in a crystalline solid at the edge of an extra crystal plane. The slip vector, which defines the direction and magnitude of deformation associated with the movement of dislocation, is perpendicular to the defect line.

Elastic modulus (*E*): Stress per unit of elastic strain. Measured by the slope of the elastic part of the stress–strain diagram.

Elastic strain: Strain that is recoverable when the load is removed.

Elastomer: Polymer with a large, more than 100%, elastic strain.

Electrochemical cell: System providing electrical connection of anode, cathode, and electrolyte.

Electrode potential: Voltage developed at an electrode with reference to a standard electrode.

Electrolyte: Conductive ionic liquid or solid solution.

Electron: Negatively charged subatomic particle moving in orbits around positively charged nucleus.

Element: Fundamental chemical species as given in the periodic table of elements.

Elongation: Change in length resulting from the application of external forces.

Elongation percent: Total plastic strain at fracture. A gage length must be stated.

Enamel: Ceramic, usually vitreous, coating on a metal.

End-quench test (Jominy bar): Standardized test performed by quenching from one end only, for determining hardenability.

Endurance limit: The maximum stress allowable for unlimited stress cycling.

Endurance ratio: The endurance limit divided by the ultimate tensile strength of the material. It is about 0.5 for many of the steels.

Engineering strain: Elongation divided by the original length of sample.

Engineering stress: Load divided by the original cross-sectional area of sample.

Equilibrium: The state at which there is no net reaction because the minimum free energy has been reached.

Erosion: Wear caused by the movement of hard particles relative to the surface.

Eutectic alloy: Alloy with the lowest melting point compared with its neighboring alloys.

Eutectic reaction: Transformation of a liquid alloy to more than one solid phase simultaneously during solidification.

Eutectic temperature: Temperature of the eutectic reaction.

Eutectoid composition: Composition of the solid solution phase that possesses a minimum decomposition temperature.

Eutectoid reaction: Transformation of a solid solution to more than one solid phase simultaneously on cooling.

Eutectoid temperature: Temperature of the eutectoid reaction.

Extrusion: Shaping by pushing the material through a die. Used for metals and plastics.

Face-centered cubic (fcc): Arrangement of atoms in a cell such that one atom is at the center of each of the faces in addition to eight atoms at the cube corners.

Fatigue curve (S–N curve): Plot of the alternating stress (S) versus number of cycles to failure (N).

Fatigue strength (endurance limit): The maximum stress allowable for unlimited stress cycling.

Ferrite (α): Bcc iron or an iron-rich, bcc solid solution.

Ferritic stainless steel: Corrosion-resistant alloy steel containing at least 11% chromium, which is mainly in α phase.

Ferrous alloy: Iron-base alloy.

Fiber-reinforced plastic (FRP): Composite material consisting of a polymer matrix reinforced with fibers of glass, carbon, aramid, or a combination.

Fiberglass: Composite material consisting of a polymer matrix reinforced with glass fibers.

Filler: Particulate or fibrous additive to polymers for reinforcement, dimensional stability, or dilution.

Firing: Heating a ceramic material to create a bond between its constituents and particles.

Fracture: Breaking of materials under stress. Brittle fracture involves negligible plastic deformation and minimum energy absorption. Ductile fracture is accompanied by plastic deformation and, therefore, by energy absorption.

Fracture toughness: Critical value of the stress intensity factor, K_{IC}, for fracture propagation.

Free electrons: Electrons that are responsible for electrical conductivity in metals.

Galvanic cell: Electrochemical cell consisting of anode, cathode, and electrolyte.

Galvanic corrosion: Corrosion among dissimilar metals in electrical contact in the presence of an electrolyte.

Galvanic protection: Protection of a structure against corrosion by making it the cathode in a galvanic cell.

Galvanic series: Arrangement of metallic materials in sequence (cathodic to anodic) of corrosion susceptibility in aqueous environment such as seawater.

Galvanized steel: Steel coated with a layer of zinc. The zinc provides galvanic protection by serving as a sacrificial anode.

Glass: An amorphous (noncrystalline) solid below its transition temperature. A glass lacks long-range crystalline order, but normally has short-range order.

Glass ceramic: Crystalline ceramic material produced by controlled devitrification of glass.

Glass fiber-reinforced plastic (GFRP or fiberglass): Composite material consisting of a polymer matrix reinforced with glass fibers.

Glass transition temperature: The temperature at which a supercooled liquid becomes a rigid glassy solid.

Glaze: Glass layer on a ceramic component.

Grain: Individual crystal in a polycrystalline microstructure.

Grain boundary: The zone of crystalline mismatch between adjacent grains.

Grain growth: Increase in the average size of the grains, usually as a result of prolonged heating.

Grain size: A measure of the average size of grains in a polycrystalline microstructure.

Gray cast iron (GCI): Type of cast iron in which the graphite phase is in the form of flakes in a matrix of ferrite, pearlite, or a mixture of the two.

Hardenability: The ease with which steel can be transformed to hard martensite by quenching.

Hardness: Resistance to indentation or scratching by a hard indenter. Common hardness tests include Brinell, Rockwell, and Vickers.

Heat treatment: Controlled heating and cooling of the material to control its microstructure and properties.

Hexagonal close-packed (hcp): Arrangement of atoms in a cell with a hexagonal top and bottom, each containing six atoms in the corners and one atom in the center in addition to three atoms in a plane between the top and bottom.

High-alloy steel: Steel containing a total of more than 5 wt% alloying elements.

High-strength low-alloy steels (HSLA): Steels containing a total of less than 5 wt% alloy additions, but exhibiting relatively high strength.

Homogenization (soaking): Heat treatment to equalize composition by diffusion.

Hot isostatic pressing (HIP): Compacting powders under high temperatures to allow compaction and sintering to take place simultaneously.

Hot shortness: Melting of some parts of the alloy although the temperature is below the equilibrium melting temperature. This is usually a result of segregation.

Hot working: Deformation of the material above the recrystallization temperature so that working and annealing occur concurrently.

Hydrogen embrittlement: Loss of ductility as a result of hydrogen diffusion in the material.

Hypereutectic: Composition greater than the eutectic composition.

Hypoeutectic: Composition less than the eutectic composition.

Hypoeutectoid: Composition less than the eutectoid composition.

Immiscibility: Mutually insoluble phases.

Imperfection in crystals: Defects in crystals including point defects such as vacant atom sites and extra atoms (interstitials), line defects such as dislocations, or surface defects such as grain boundaries.

Impact strength (toughness): Resistance to fracture by impact loading and it is measured by the Izod or Charpy tests, which give the amount of energy required to fracture a standard sample.

Impressed current: Direct current applied to protect a structure by making the metal cathodic during service.

Inhibitor: An additive to an electrolyte that decreases the rate of corrosion and promotes passivation.

Injection molding: Molding of polymers under pressure in a closed die.

Interrupted quench: Two-stage quenching of steel from austenitic phase, initial quenching to a temperature above the start of martensite formation, followed by a second (slower) cooling to room temperature.

Interstitial solid solution: An alloy in which the atoms of one of the constituents are small enough to fit in the spaces between the solvent atoms.

Ion: An atom that possesses a charge because it has added or removed electrons.

Ionic bond: A primary atomic bond involving transfer of electrons between unlike atoms.

Isotropic: Exhibiting properties that do not change with the direction of measurement.

Jominy end-quenching test: Test to measure the hardenability of steels using a standard specimen and test conditions.

Laminate: Composite material in which the phases are arranged in layers.

Larson–Miller parameter: A relationship between the stress, temperature, and the rupture time in creep.

Lattice: Arrangement of atoms in a crystalline solid.

Lattice constants (lattice parameters): Dimensions of the edges of a unit cell.

Lever rule: Equation to determine the quantity of phases in an alloy under equilibrium conditions.

Long-range order: Repetition of the pattern of atomic arrangement in a crystalline solid over many atomic distances.

Low-alloy steel: Steel containing a total of less than 5 wt% alloying elements.

Macromolecules: Molecules made up of hundreds to thousands of atoms.

Malleable iron: Type of cast iron with some ductility. It is obtained by heat treating white cast iron to change its iron carbide to nodular graphite.

Martempering: Quenching steel from the austenite range to a temperature just above the martensitic transformation range, followed by a slow cool through martensitic transformation range to reduce the stresses associated with that transformation.

Martensite: A phase arising from quenching of steels from the austenite temperature range as a result of a diffusionless, shearlike phase transformation. It is a hard and brittle phase. Martensitic transformations also occur in some nonferrous alloys.

Martensitic stainless steels: Corrosion-resistant alloy steel containing at least 11% chromium and is mainly in martensitic phase.

Materials: Engineering materials are substances used for manufacturing products and include metals, ceramics, polymers, composites, semiconductors, glasses, cement, and natural substances such as wood and stone.

Matrix: The continuous phase that envelops the reinforcing phase in a composite material.

Mer: The smallest repetitive unit in a polymer.

Metal: Material characterized by its high electrical and thermal conductivities as a result of the presence of the free electrons of the metallic bond.

Metal–matrix composites (MMC): Composites in which the reinforcing phase is enveloped in a metallic material.

Metallic bond: Interatomic bonds in metals characterized by the presence of electrons that are free to move and conduct electrical current.

Metastable: A state of material that does not change with time although it does not represent true equilibrium.

Microstructure: Arrangement and relationship between the grains and the phases in a material. Generally requires magnification for observation.

Modulus of elasticity (elastic modulus, Young's modulus): Stress per unit of elastic strain. Measured by the slope of the elastic part of the stress–strain diagram.

Modulus of rigidity (shear modulus): Shear stress per unit shear strain.

Mole: Mass equal to the molecular weight of a material.

Molecular weight: Mass of one molecule expressed in atomic mass units.

Molecule: Group of atoms bonded by strong attractive forces, primary bonds.

Monomer: A molecule with a single mer. Monomers combine with similar molecules to form a polymeric molecule.

Natural ageing: Allowing the solution-treated and quenched precipitation-hardenable alloy to stay long enough at room temperature for precipitation to take place.

Neutron: Subatomic particle located in the nucleus and has a neutral charge.

Noncrystalline (amorphous): Atoms or molecules of the material are not arranged according to a repetitive pattern or exhibit long-range order.

Nondestructive testing (NDT): Inspection of materials and components without impairing their integrity.

Nonferrous alloy: Metallic alloy with a base metal other than iron. Examples include aluminum- and copper-base alloys.

Normalizing: Heating of steel into the austenite range and cooling at a rate that is faster than that used for annealing, but slower than that required for hardening. This treatment is used to produce a uniform, fine-grained microstructure.

Nucleation: The start of a new phase in phase transformation or the beginning of solidification from a liquid. Heterogeneous nucleation takes place on a preexisting surface or a "seed," whereas homogeneous nucleation occurs without the aid of a preexisting phase.

Nucleus: Central core of an atom around which electrons orbit. Also a small solid particle that forms from the liquid at the beginning of solidification.

Overageing: Continuing with the age-hardening treatment beyond the peak hardness, thus causing the hardness to decrease as a result of precipitate coarsening.

Oxidation: Reaction of a metal with oxygen.

Particulate composite: Composite material consisting of particles embedded in a matrix.

Passivation: Impeding the rate of corrosion due to the presence of an adsorbed protective film on the surface.

Pearlite: A lamellar mixture of ferrite and iron carbide formed by decomposing austenite of eutectoid composition.

Peritectic reaction: Reaction of a liquid phase with a solid phase to form a second solid phase.

Phase: A physically or chemically homogeneous part of the structure of a material.

Phase boundary: Compositional or structural discontinuity between two phases.

Phase diagram: Graphical representation of the phases present in equilibrium in an alloy system at different compositions and temperatures.

Pitting corrosion: Corrosion attack that is localized in narrow areas of the surface.

Plastic: Engineering material composed primarily of a polymer with additives.

Plastic strain: Permanent deformation that persists after the stress is removed. It arises from the displacement of atoms or molecules to new positions.

Plasticizer: An additive of small-molecular-weight molecules to a polymeric mix to reduce the rigidity.

Point defect: Crystal imperfection or disorders involving one or a small number of atoms.

Poisson's ratio: Ratio between lateral strain (contraction) and axial strain (extension) under tensile load.

Polycrystalline: Material with multiple grains and associated boundaries.

Polymer: Nonmetallic organic material consisting of macromolecules composed of many repeating units, which are called mers.

Polymer–matrix composite: Composite material in which the reinforcing phase is held together by a polymer matrix.

Polymorphism (allotropy): Change of lattice structure with temperature, for example, iron changes from bcc to fcc at 910°C.

Powder metallurgy technique: Processing powders by compaction and sintering to produce a solid product.

Precipitation: Separation of a second phase from a supersaturated solution.

Preferred orientation: Alignment of grains or inclusions in a particular direction, thus leading to anisotropy.

Precipitation hardening (age hardening): Increase in hardness as a result of the precipitation of a hard phase from a supersaturated solid solution.

Primary bond: Strong interatomic bond. Examples are covalent, ionic, or metallic bonds.

Proeutectic: A phase that separates from a liquid before the latter undergoes eutectic transformation.

Proeutectoid: A phase that separates from a solid solution before the latter undergoes eutectoid transformation.

Proportional limit: The end of the range within which the strain increases linearly with stress level.

Prosthesis: Part used to replace a body part.

Protective coating: Layer on the surface to act as a barrier or protect the material from the surrounding environment.

Proton: Positively charged subatomic particle in the nucleus of the atom. The number of protons is equal to the number of electrons in a neutral atom.

Quenching: Fast cooling, usually in water or oil to produce a nonequilibrium structure.

Radiation damage: Creation of structural defects as a result of exposure to radiation.

Recovery: A low-temperature anneal, which involves heating the cold-worked material to partially reduce structural defects leading to a slight softening.

Recrystallization: The formation of new annealed soft grains in place of previously strain-hardened ones.

Recrystallization temperature: Temperature above which recrystallization is spontaneous. It is about 0.4 T_m, where T_m is the melting point of the material expressed in Kelvin or Rankin.

Reduction: Removal of oxygen from an oxide.

Reduction of area percent: Total plastic strain at fracture expressed as percent decrease in cross-sectional area at the point of fracture.

Refractory: A material that can resist high temperatures.

Refractory metal: Metal with a melting point higher than 1700°C.

Reinforcement: The component that provides a composite material with high elastic modulus and high strength.

Relaxation time: Time required for the stress resulting from a fixed value of strain to decrease to 0.37 ($=1/e$) of the initial value.

Residual stresses: Stresses stored in the material as a result of processing.

Rockwell hardness: A measure of hardness obtained by indenting the surface by applying a force to a hard ball or a diamond cone, and measuring the depth of indention.

Rolling: Mechanical working through the use of two rolls that are rotating in opposite direction. The rolls may be cylindrical, as in the case of sheet rolling, or shaped, as in the case of section rolling.

Rupture time: The time required for a sample to fail by creep at a given temperature and stress.

Sacrificial anode: Expendable metal that is used to protect the more noble metal of a component or structure.

Scale: Surface layer of oxidized metal.

Screw dislocation: Linear defect in a crystalline solid where the slip vector, which defines the direction and magnitude of deformation associated with the movement of dislocation, is parallel to the defect line (see dislocation).

Secondary bonds: Weak interatomic bonds arising from dipoles within the atoms or molecules, for example, van der Waal's bond.

Segregation: Nonuniform distribution of alloying elements as a result of non-equilibrium conditions.

Shear modulus: Shear stress per unit shear strain.

Sheet molding: Thermal forming of sheets of FRP.

Short-range order: Specific arrangement of atoms relative to their close neighbors, but random long-range arrangements.

Sintering: Heating of compacted powders to induce bonding as a result of diffusion.

Slip casting: Shaping of ceramic parts by pouring powder–water mixture, slip into a porous mold that allows the water to escape leaving a dense mass behind, which is then extracted from the mold and dried.

Slip: Deformation of a material by the movement of dislocations through the lattice.

Slip direction: Crystal direction of the displacement vector on which slip takes place.

Slip plane: Crystal plane along which slip occurs.

Slip system: Combination of slip directions and slip planes corresponding to dislocation movement.

Slip vector: Defines the direction and magnitude of deformation associated with the movement of a dislocation. It is parallel to a screw dislocation and perpendicular to an edge dislocation.

***S–N* curve:** Plot of the alternating stress versus number of cycles to fatigue failure.

Solder: Alloys whose melting points are below 425°C and are used for joining. The Pb–Sn eutectic alloy, with a melting point of 183°C is among the commonly used soldering alloys.

Solid solution: A homogeneous crystalline phase composed of more than one element. Substitutional solid solutions are obtained when the elements involved have similar atomic size. Interstitial solid solutions are obtained when the atomic size of one of the constituent elements is so small that it can fit in the spaces between the larger atoms of the parent metal.

Solubility limit: Maximum solute addition without supersaturation.

Solute: The component that is dissolved in a solvent to form a solid solution.

Solution hardening: Increase in strength associated with the addition of alloying elements to form a solid solution.

Solution treatment: Heating a multiphase alloy to become a single phase.

Solvent: The main component of a solid solution in which the solute is dissolved.

Spalling: Cracking or flaking off of a material as a result of thermal stresses.

Stainless steel: Corrosion-resistant alloy steel containing at least 11% chromium and may also contain nickel. Depending on the composition and treatment, stainless steels can be ferritic, austenitic, or martensitic.

Steel: Basically an alloy of iron plus up to 2 wt.% carbon and may contain additional elements. Plain carbon steels are Fe–C alloys with minimal alloy content. Low-alloy steels contain up to 5% alloying elements other than carbon.

Strain: Deformation per unit length as a result of stress. Elastic strain is recoverable when the load is removed. Plastic strain results in a permanent deformation.

Strain hardening: Increase in strength as a result of plastic deformation below the recrystallization temperature.

Strength: Resistance to stress. Yield strength is the stress to initiate plastic deformation. Ultimate tensile strength is the maximum stress that can be borne by the material, calculated on the basis of the original area.

Stress: Force or load per unit area. Engineering stress is based on the original cross-sectional area. True stress is based on the actual area.

Stress corrosion: Accelerated corrosion due to the presence of stress.

Stress intensity factor (K_I): Stress intensity at the root of a crack.

Stress relaxation: Decrease with time of the stress at a fixed value of strain (see relaxation time).

Stress relief: Removal of internal stresses in a material by heat treatment.

Stress rupture: Time-dependent fracture resulting from constant load under creep conditions.

Stress–strain diagram: Plot of stress as a function of strain.

Superalloys: A group of alloys that retain their strength at high temperatures. Commonly used superalloys include iron-, nickel-, and cobalt-base alloys.

Superplasticity: The ability of a material to undergo very large amount of plastic deformation.

Tempered glass: Glass that has been heat treated to generate residual compressive stresses in the surface layers.

Tempered martensite: A microstructure composed of fine dispersion of carbide in ferrite and is obtained by heating martensite.

Tempering: Heating of martensite to increase its toughness by producing a microstructure composed of fine dispersion of carbide in ferrite.

Thermal shock: Failure of the material as a result of sudden change in temperature.

Thermoplastic: A polymer that softens with increased temperature, thus becoming moldable, and rehardens on cooling. Such polymer is usually soft and ductile at normal temperatures.

Thermosetting plastic: A polymer that hardens on heating as a result of cross-linking, that is, the formation of three-dimensional network of strongly bonded molecules. Such polymer is usually rigid and brittle.

Toughness: Resistance to fracture by impact loading and is measured by the Izod or Charpy tests, which give the amount of energy required to fracture a standard sample. It can also be represented by the total area under the stress–strain diagram.

True strain: Elongation divided by the actual length of sample.

True stress: Load divided by the actual cross-sectional area of sample.

Ultimate strength: Maximum stress based on the original area.

Unit cell: The smallest repetitive group of atoms that represents the arrangement of atoms in a crystal lattice. Common unit cells include bcc, fcc, and hcp.

Vacancy: A point defect in a crystalline structure, which is associated with an unoccupied lattice site.

Valence electrons: Electrons in the outer shell(s) of an atom that take part in atomic bonding.

Van der Waal's bond: Weak interatomic secondary bonds arising from dipoles within the atoms or molecules.

Viscoelastic deformation: Deformation involving both fluidlike viscous flow and solidlike elastic strain.

Viscosity: Ratio of shear stress to the velocity gradient of flow.

Viscous flow: Time-dependent flow in polymers and glasses above their glass transition temperature.

Vulcanization: Treatment of rubber with sulfur to cross-link elastomer chains, that is, form a three-dimensional network structure.

Wear: Loss of surface material as a result of mechanical action.

Welding: Joining of parts by local melting at the joint.

Whisker: Very small nearly perfect fiber.

White cast iron: Type of cast iron in which the carbon is present as iron carbide. It is very hard and brittle.

Wrought alloy: Alloy that can be shaped by forming processes such as forging, rolling, extrusion, and drawing.

Yield strength: Stress that causes the material to reach its elastic limit, that is, resistance to the onset of plastic deformation.

Young's modulus (modulus of elasticity, elastic modulus): Stress per unit of elastic strain. Measured by the slope of the elastic part of the stress–strain diagram.

Index

A

B

C

D

E

F

G

H

I

J

K

L

M

N

O

P

Q

R

S

T

U

V

W

X

Y

Z